Emerging technologies and international security

This book offers a multidisciplinary analysis of emerging technologies and their impact on the new international security environment across three levels of analysis.

While recent technological developments, such as artificial intelligence (AI), robotics, and automation, have the potential to transform international relations in positive ways, they also pose challenges to peace and security and raise new ethical, legal, and political questions about the use of power and the role of humans in war and conflict. This book makes a contribution to these debates by considering emerging technologies across three levels of analysis: (1) The international system (systemic level) including the balance of power; (2) the state and its role in international affairs and how these technologies are redefining and challenging the state's traditional roles; and (3) the relationship between the state and society, including how these technologies affect individuals and non-state actors. This provides specific insights at each of these levels and generates a better understanding of the connections between the international and the local when it comes to technological advance across time and space.

The chapters examine the implications of these technologies for the balance of power, examining the strategies of the US, Russia, and China to harness AI, robotics, and automation (and how their militaries and private corporations are responding); how smaller and less powerful states and non-state actors are adjusting; the political, ethical, and legal implications of AI and automation; what these technologies mean for how war and power is understood and utilized in the 21st century; and how these technologies diffuse power away from the state to society, individuals, and non-state actors.

This volume will be of much interest to students of international security, science and technology studies, law, philosophy, and international relations.

Reuben Steff is a Senior Lecturer in International Relations and International Security at the University of Waikato, New Zealand. He is the author of *Security at a Price: The International Politics of US Ballistic Missile Defense* (Rowman & Littlefield, 2017) and *Strategic Thinking, Deterrence and the US Ballistic Missile Defense Project: from Truman to Obama* (Routledge, 2014). He has a number of journal articles published in the *Journal of Strategic Studies, Pacific*

Review, *Contemporary Security Policy*, *Defense and Security Analysis*, *National Security Journal*, *New Zealand International Review* and the *Australian Journal of International Affairs*. His research addresses emerging technologies and international security, nuclear deterrence and ballistic missiles defense, great power competition, US foreign policy, and the role small states. His forthcoming book is *US Foreign Policy in the Age of Trump: Drivers, Strategy and Tactics* (Routledge, October 2020).

Joe Burton is a Senior Lecturer in International Security at the New Zealand Institute for Security and Crime Science, University of Waikato, New Zealand. He holds a Doctorate in International Security and a Master's degree in International Studies from the University of Otago and an undergraduate degree in International Relations from the University of Wales, Aberystwyth. Joe is the recipient of the US Department of State SUSI Fellowship, the Taiwan Fellowship, and has been a visiting researcher at the NATO Cooperative Cyber Defence Centre of Excellence (CCDCOE) in Tallinn, Estonia. He is the author of *NATO's Durability in a Post-Cold War World* (SUNY Press, 2018) and his work has been published in *Asian Security*, *Defence Studies*, *Political Science* and with a variety of other leading academic publishers. Joe is currently a Marie Curie fellow (MSCA-IF) at Université libre de Bruxelles (ULB) completing the two-year European Commission-funded project *Strategic Cultures of Cyber Warfare* (CYBERCULT).

Simona R. Soare is Senior Associate Analyst at the European Union Institute of Security Studies (EUISS). Her research focuses on transatlantic and European security and defence, EU-NATO cooperation and defence innovation. Prior to joining EUISS, Simona served as advisor to the Vice-President of the European Parliament (2015–2019), working on European defence initiatives (EDF, military mobility, EU-NATO cooperation), CSDP and transatlantic relations, and as an analyst with the Romanian Ministry of Defence. She has lectured in international relations at the National School for Political and Administrative Studies in Romania and she is a regular contributor to CSDP courses with the European Security and Defence College (ESDC). Since 2016, Simona has been an associate fellow with the Institut d'études européennes (IEE) at Université Saint-Louis Bruxelles where she works on defence innovation and emerging technologies. Simona holds a PhD (2011) in Political Science and she is a US Department of State fellow. She has published extensively on American and European security and defence, including defence capability development, emerging technologies and defence innovation, arms transfers, export controls and regional defence.

Routledge Studies in Conflict, Security and Technology
Series Editors: Mark Lacy, *Lancaster University*; Dan Prince, *Lancaster University*; and Sean Lawson, *University of Utah*

The *Routledge Studies in Conflict, Security and Technology* series aims to publish challenging studies that map the terrain of technology and security from a range of disciplinary perspectives, offering critical perspectives on the issues that concern public, business, and policy makers in a time of rapid and disruptive technological change.

National cyber emergencies
The return to civil defence
Edited by Greg Austin

Information warfare in the age of cyber conflict
Edited by Christopher Whyte, A. Trevor Thrall, and Brian M. Mazanec

Emerging security technologies and EU governance
Actors, practices and processes
Edited by Antonio Calcara, Raluca Csernatoni and Chantal Lavallée

Cyber-security education
Principles and policies
Edited by Greg Austin

Emerging technologies and international security
Machines, the state, and war
Edited by Reuben Steff, Joe Burton, and Simona R. Soare

For more information about this series, please visit: https://www.routledge.com/Routledge-Studies-in-Conflict-Security-and-Technology/book-series/CST

Emerging technologies and international security

Machines, the state, and war

Edited by Reuben Steff, Joe Burton, and Simona R. Soare

Routledge
Taylor & Francis Group
LONDON AND NEW YORK

First published 2021
by Routledge
2 Park Square, Milton Park, Abingdon, Oxon OX14 4RN

and by Routledge
52 Vanderbilt Avenue, New York, NY 10017

Routledge is an imprint of the Taylor & Francis Group, an informa business

British Library Cataloguing-in-Publication Data
A catalogue record for this book is available from the British Library

Library of Congress Cataloging-in-Publication Data
A catalog record has been requested for this book

ISBN: 978-0-367-40739-1 (hbk)
ISBN: 978-0-367-80884-6 (ebk)

Typeset in Times New Roman
by Deanta Global Publishing Services, Chennai, India

Contents

Figures

Tables

Contributors

Khusrow Akkas Abbasi is a PhD student at the Department of Political Science and Public Policy, the University of Waikato. He has served as a research fellow at the Centre for International Strategic Studies (CISS), Islamabad, Pakistan. He earned an M. Phil in Strategic Studies from the Quaid-i-Azam University (QAU), Islamabad. His areas of interest include international security, emerging nuclear China, conflict and cooperation in the Asia Pacific, and emerging technologies.

Sean Ainsworth is a PhD candidate at Victoria University of Wellington. His research interests include emerging technologies, international security, and conflict, with a specific focus on cyber and information warfare as emergent means of interstate competition.

Curtis Barnes is a founder and director at the Brainbox Institute, where he conducts and coordinates research into subjects at the intersection of emerging technologies and law. He has legal expertise in artificial intelligence applications, including strategic uses of synthetic media technologies and online disinformation. He has further research interests in computational law and access to justice. His research has generated significant policy impacts and has been covered in domestic and international media.

Tom Barraclough is a founder and director at the Brainbox Institute. He is a legal researcher specializing in law, policy, and emerging technologies. His current research interests relate to computational law, disinformation, and synthetic media, as well as assorted health policy issues relating to access to justice for people with disabilities. His research has generated significant policy impacts and been covered in domestic and international media.

Andrew M. Colarik is a consultant, teacher, researcher, author, and inventor of information security technologies. He has published multiple security books and research publications in the areas of cyber terrorism, cyber warfare, and cyber security. For more information on Dr. Colarik, visit his website at www.AndrewColarik.com.

Peter Cook currently works for the New Zealand Defence Force (NZDF) as a Space Analyst responsible for analysing space capabilities within New Zealand. He has worked for NZDF for nine years in various engineering and analytical roles having moved to New Zealand from the UK. He has degrees in engineering and international security from the UK and New Zealand and has a keen interest in emerging and disruptive technology. Having significant experience in 3D Printing/Additive Manufacturing, he has been instrumental in introducing and developing the technology within NZDF.

Nathan Cooper is an academic lawyer working on questions around the compatibility of `rights claims' with ecologically sustainable governance at the University of Waikato, New Zealand. I am interested in the way that formal law shapes norms and interacts with other norms, and in the role of `vernacular law' and law-like emanations, in pursuing (ecosystems) solidarity, and in achieving (ecologicial) justice and (ecocentric) welfare. My current focus is on the governance of socio-economic necessities - in particular on water governance - through international human rights law, domestic constitutions, development goals, and grass-roots organisation.

Alek Hillas is a researcher in the School of Global, Urban and Social Studies at RMIT University, where he graduated with a first-class honors degree in International Studies. His research interests are in global security and international humanitarian law, including artificial intelligence and lethal robotics and Australian foreign policy.

William Hoverd is a Senior Lecturer in People Environment and Planning, College of Humanities, and Social Sciences, Massey University. He is a social scientist with a specific interest in critical research into New Zealand security issues and religious diversity. He has published a variety of qualitative and quantitative research publications in security, sociology, religious studies and psychology. His most recent books include *New Zealand National Security (2017)* and *The Critical Analysis of Religious Diversity (2018)*. Between 2015 and 2017, he taught into the New Zealand Defence Force Command and Staff College. In 2012 he was a successful co-recipient of the NZ $231,000 Danish Research Council funded Critical Analysis of Religious Diversity Network. In 2011/2012, he was a DFAIT Government of Canada Post-Doctoral Fellow, at the Religion and Diversity Project at Ottawa University, Ontario.

James Johnson is an Assistant Professor in the Department of Law and Government at Dublin City University, and a non-resident fellow with the Modern War Institute at West Point. He is the author of *The US–China Military & Defense Relationship during the Obama Presidency*. His latest book project is entitled, *Artificial Intelligence & the Future of Warfare: USA, China, and Strategic Stability*. James is fluent in Mandarin.

Ryan K. I. Ko is a computer scientist specializing in cyber security and systems research. His research on homomorphic encryption, data provenance and user data control are central to many cloud security, data tracking and security information and event management (SIEM) tools across open source (e.g. OpenStack, Kali Linux) and industry tools (e.g. ArcSight) today. He is currently Chair Professor and Director of Cyber Security at the University of Queensland, Australia. Prior to his role in Australia, he created the national diploma curriculum in cyber security for New Zealand, and established the New Zealand Cyber Security Challenge – the national cyber security competition in New Zealand. Ryan advises ministers and governments (New Zealand, Australia, Singapore, Tonga), has held directorships and advisory roles across academia, industry, stock exchange-listed companies, INTERPOL, ISO and the governments. He is a Fellow of Cloud Security Alliance (CSA) and recipient of the (ISC)2 Information Security Leadership Award.

Francis Okpaleke is a PhD Candidate and Sessional Academic at the Department of Politics and Public Policy at the University of Waikato. His research interests include automated weapons systems, grand strategy, artificial intelligence, contemporary security studies, and climate security. He has written a number of articles and conference papers on these subject areas.

Aiden Warren is Associate Professor of International Relations at RMIT University, Melbourne, Australia. He is the 2018–19 Fulbright Scholar in Australia-US Alliance Studies, sponsored by the Australian Government's Department of Foreign Affairs & Trade (DFAT). Dr. Warren's teaching and research interests are in the areas of international security, US national security and foreign policy, US politics (ideas, institutions, contemporary, and historical), international relations (especially great power politics), and issues associated with weapons of mass destruction (WMD) proliferation, nonproliferation, and arms control. He has spent extensive time in Washington DC completing fellowships at the James Martin Center of Nonproliferation, the Arms Control Association (ACA), and Institute for International Science and Technology Policy (IISTP) at George Washington University. Dr. Warren is the sole author, coauthor, and editor of seven books. He is editor of *Rethinking Humanitarian Intervention in the 21st Century* and is also the series editor of the Weapons of Mass Destruction (WMD) book series with Rowman and Littlefield, New York.

Dan Weijers is a senior lecturer in the Philosophy Programme at the University of Waikato. He specializes in interdisciplinary wellbeing research, normative ethics, and the ethics of new technologies. Weijers is the managing editor of the *International Journal of Wellbeing*, International Editorial Board Member of Rowman and Littlefields's book series on Behavioral Applied Ethics, and Editorial Review Board Member for the *International Journal of Technoethics*. He has provided policy advice to the United Nations, the New Zealand Treasury, Statistics New Zealand, and the XPRIZE Foundation.

Richard Wilson has been an officer in the United States Army since 2005. A graduate of the University of Idaho and Massey University, he has practiced international security across the world through four deployments. Wilson is currently working in Schofield Barracks, Hawaii in the 25th Infantry Division to improve the military's ability to provide Multi-Domain Fire Support into Indo-Pacific Army Operations.

Acknowledgments

This book project started with *The Waikato Dialogue* – a symposium held at the University of Waikato in 2018. We gratefully acknowledge the University of Waikato, the New Zealand Institute for Security and Crime Science, the Faculty of Arts and Social Sciences, and the Political Science and Public Policy program for supporting this collaborative, multidisciplinary endeavor.

Introduction

Machines, the state, and war

Reuben Steff, Joe Burton, and Simona R. Soare

The world stands at the cusp of a new era of technological change: A range of emerging technologies, such as artificial intelligence (AI), robotics, automation, 3D printing, deepfakes, and blockchain are primed to have an impact on all aspects of society. In aggregate, these technologies have potentially transformative implications for the international balance of power, alliances and security organizations, how governments control information, how international actors compete militarily and economically, and how they wage war (Horowitz, 2018; Allen & Chan, 2017). They will challenge the political accountability for decision-making, human control of conflict escalation, the relationship between the state and its citizens and national monopolies on the legitimate use of force. Ultimately, they are likely to prove pivotal in how states define what is in their vital national interests.

Incentivized by the great promise held by these emerging technologies for both civilian and military fields, and perhaps fearful of falling behind others, many nations and corporations are racing to invest in them. These efforts have gained greater urgency in light of the global threat of the COVID-19 pandemic, with nations rushing to develop apps to track cases, fears of excessive governmental digital surveillance, and the emergence of new conspiracy theories relating to the rollout of 5G technologies and their impact on the spread of the virus. The paradox of the need for new technological tools to deal with evolving threats to security and the simultaneous risks posed by those technologies are at the forefront of political debates once again. While emerging technologies offer immense promise, their proliferation poses challenges to international peace and security, and they raise new ethical and political questions about the use of power and the role of humans in conflicts and war and the way states are interacting with their citizens. They could benefit the most disadvantaged countries in the world (Cummings et al, 2018) or accentuate existing inequalities if first-movers acquire an unassailable lead (Lee, 2017). The speed of technological development also seems to have overtaken the ability of governments and societies to keep pace in adapting existing legislation, developing ethical and safety standards, or creating international norms that regulate their development.

Artificial intelligence: Disruptor and enabler

AI, in particular, has been singled out by prominent tech companies, philosophers, and political leaders as the most transformative of these emerging technologies and in need of intense and sustained investigation (Future of Life Institute, 2020). AI is a broadly applicable *technological enabler*: Analogous to electricity or the combustion engine, it is useful in virtually any digital device or system, and, like steam, oil, gas, and electricity, it will prove instrumental to states' national security and economic prosperity, acting as the central component of the Fourth Industrial Revolution (Horowitz, Kania, Allen & Scharre, 2018). Indeed, AI's contribution to the global economy by 2030 is estimated to be $15.7 trillion (PWC, 2018). Meanwhile, the core fuel of AI algorithms, *data*, is expanding at an exponential rate as billions of more devices come online every year. According to *The Economist* (2017), data has now replaced oil as "the world's most valuable resource". In this context, AI is essential for state and non-state actors to be able to filter and make sense of these immense data-sets, disciplining them for practical economic and military applications (Ryan, 2018).

In a 2017 study, the US Army declared AI to be "the most disruptive technology of our time", (US Army, 2017, p. 8) while the Belfer Center suggests that AI has the "potential to be a transformative national security technology, on a par with nuclear weapons, aircraft, computers, and biotech" (Allen and Chan, 2017). Russia's President Vladimir Putin has claimed, "Whoever becomes the leader in this sphere (AI) will become the ruler of the world" (Dougherty and Jay, 2017); and Tesla's CEO Elon Musk maintains that competition for AI superiority at the national level could spark World War III (Fiegerman, 2017). Smaller states have also taken notice with the United Arab Emirates (UAE) becoming the first country to appoint a Minister of State for AI (Dutton, 2018).

In recent years, AI technology has achieved key milestones and surpassed previous projections. For example, in 2014, the designer of the "best" Go-playing program (Go is exponentially more complex than chess) estimated it would take ten years until an AI could defeat it; instead it was beaten only a year later by DeepMind (Allen and Chan, 2017). In January 2019, DeepMind's Alphastar AI defeated the world's best players in the real-time strategy game, Starcraft II (Deepmind, 2019). Additionally, AI applications have beaten poker players, showed better-than-human voice and image recognition, and defeated former US Air Force pilots in combat simulations (Allen and Chan, 2017). Recent tests have shown AI to be more prudent than its human operators in making decisions with incomplete or tainted information (Tucker, 2020). Indeed, an international competition or "arms race" to develop AI is alleged to have already broken out (Geist, 2016) and since development is largely being driven by the private sector and AI research has both military and civilian uses, it is plausible that the rapid diffusion of cheap lethal AI applications will take place. If the barriers to entry decrease markedly it could empower middle and smaller states relative to larger ones by offering them better prospects for competing in the development of AI compared to the large complex military hardware of the past. New lethal options for violent non-state actors will also be on offer, while information campaigns advanced through AI-enhanced social media, fake news, and "deepfake" technologies could become more regularly used to undermine political stability in western democracies.

Lower barriers to entry mean more actors will have a stake in the regulation of emerging technologies. In turn, some actors will feel great incentives to develop emerging technologies for fear that others will be doing so covertly. At present, strategic fissures at the international level are impeding efforts by the most influential states to regulate emerging technologies in multilateral forums and existing arms control, and non-proliferation treaties that grew out of efforts during the Cold War to regulate nuclear weapons are not designed to address a new AI arms race (Bell and Futter, 2018). Under these circumstances, AI and the broader set of technologies it enables risk feeding existing or new security dilemmas and heightening strategic competition in an increasingly unstable and contested international environment (Altmann and Sauer, 2017; Horowitz et al, 2018).

Scope and objectives

This book tackles the aforementioned issues – and more. Its scope is broad, with individual chapters drilling deeper into specific issues that fall under the rubric of "emerging technologies" – defined here as technologies with lower levels of readiness and maturity and which are set to have a disruptive impact on international affairs. The book's primary focus is on artificial intelligence (AI), robotics, and automation as key emerging technologies that are impacting the international system and the security of states and their citizens. Related developments in "big data", advanced manufacturing (3D printing), cyber security, blockchain, lethal autonomous weapons systems (LAWS), prediction markets, "sustainable" technologies, and audio-visual technologies are also considered. While these advancements threaten to disrupt established processes, they may also act as technological enablers that improve existing processes or create entirely new ones, while facilitating the achievement of core national security tasks. This includes enhanced intelligence, surveillance, and reconnaissance (ISR), predictive analytics, cyber security, warfighting, and command and control (C2) operations. Indeed, emerging technologies suggest a new revolution in military affairs (RMA) may be at hand, as conflict shifts from "informatized" operations to "intelligentized" warfare, compelling leaders to rethink their professional military education systems in order to adapt, harness, and employ AI alongside other emerging technologies (Kania, 2017).

The aim of this book, therefore, is to make a valuable and novel contribution to the fledgling international relations and security studies literature on emerging technologies. It will expand our knowledge and understanding of these critically important issues by interrogating how states are already adjusting to and politically shaping technological advancements and what this means for international peace and security; considering the implications for the everyday security of states and citizens; and exploring how technological advancements can be harnessed in a positive manner to strengthen, rather than weaken, interstate security. To achieve this, chapters examine the implications of these technologies for the balance of power, examining the strategies of the US, Russia, and China to harness AI, robotics, and automation (and how their militaries and private corporations are responding); how smaller and less powerful states and non-state actors are adjusting; the political, ethical, and legal implications of AI and automation; what these technologies mean for how war and power is understood and utilized in the 21st century; and how these technologies diffuse power away from the state to society, individuals, and non-state actors.

Machines, the state, and war: A level of analysis problem

Despite the clear importance of recent technological developments to international politics, the corresponding academic literature is embryonic. The last few years have seen a proliferation of literature mainly concentrated in the think tank community, which is focused on exploring policy issues and developing policy recommendations on specific emerging technologies – most commonly AI. The need to reflect on the deeper implications of these emerging technologies for our security and defense is apparent. In particular, attempts to analyze their combined and interrelated effects on issues pertaining to war, conflict, and political authority have been minimal in their scope and ambition. This is, perhaps, understandable – these technologies are considered *emerging* – as their full implications are yet to be discerned. Yet, this lack of attention is concerning, particularly as societal and geopolitical effects are already visible.

This book seeks to make a contribution to these debates by considering emerging technologies across three levels of analysis: (1) the *international system* (*systemic level*) including the balance of power; (2) *the state* and its role in international affairs and how these technologies are redefining and challenging the state's traditional roles, and (3) the relationship between the *state and society*, including how these technologies affect individuals and non-state actors. We hope that this will yield specific insights at each of these levels and generate a better understanding of the connections between the international and the local when it comes to technological advances across time and space. Of course, this is not the first book to take this approach. Sixty-one years ago, Kenneth Waltz published his first book, *Man, the State and War*, which became the foundation of the structural realist approach to international relations – one that is still subscribed to by many leading academics and which has provided a backbone for recent analysis of emerging technologies and their implications for international relations. The subtitle of our book, *Machines, the State, and War*, is a tribute to Professor Waltz, who passed away in 2013, and a recognition of the intellectual significance of his work in international relations (IR). It is also a recognition that, although the technological features of our societies are changing at an unprecedented pace, history offers important lessons and comparisons to guide our future. While the levels of analysis debate has evolved, with global and transnational levels increasingly at play, and a continued erosion of the authority of the state and its boundaries due to globalization, we see this is a useful approach which can yield insights and lessons for IR theory and practice.

The book is also multidisciplinary, with contributions from scholars working in computer science, philosophy, law, political science, and in the policy community. We see collaborations between academic disciplines as essential to solving modern security problems, and we hope this book will contribute to much needed conversations between scholars from different intellectual backgrounds and traditions. By offering a multidisciplinary and multilevel analysis, the book begins to close an analytical gap in existing approaches to emerging technologies. It offers a comprehensive view of emerging technologies and the issues they have generated, and looks not just at individual emerging technologies in separation but, rather, at the interaction of these emerging technologies and how they are being used and understood differently by a multitude of international actors.

Chapter outline

The first chapter of the book is written by Joe Burton and titled "Histories of technologies: Society, the state, and the emergence of postmodern warfare". Burton lays a historical and conceptual foundation for the book by placing recent trends in emerging technologies in a broader historical and theoretical context. The chapter assesses the varied conceptual and theoretical lenses that exist in academia for interpreting the relationship between technological and historical change before questioning the extent to which revolutions in military affairs (RMAs) are actually revolutions. The chapter concludes by considering whether a new form of postmodern warfare has emerged, and the implications of this for states, societies, and the international system.

Part I: The machine and the international system

The first section of the book is broadly focused on the international system as a level of analysis and the balance of power between the different actors that populate that system, including the great powers, the US, and China. Key issues that animate the analyses in this section of the book include how emerging technologies affect polarity and the global balance of power, the distribution of power in the system between small and large states, and the relative advantages that will accrue to each from emerging tech. The section also covers the systemic risks posed by emerging technologies, including conflict and crisis escalation dynamics.

In Chapter 2, "Emerging technologies and the Chinese challenge to US innovation leadership", James Johnson uses "polarity" as a lens through which to view the shifting great power dynamics in AI and related enabling technologies. The chapter describes how and why great power competition is mounting within several interrelated dual-use technological fields; why these innovations are considered by Washington to be strategically vital, and how (and to what end) the US is responding to the perceived challenge posed by China to its technological hegemony. In Chapter 3, “Artificial intelligence: Implications for small states”, Reuben Steff considers the implications of AI for small states, arguing that, thus far, the bulk of analysis and commentary on AI has focused on how large powerful states are adjusting to AI while the implications of AI for small states are largely missing. On one hand, the challenges to harnessing AI are greater for small states relative to their larger peers. At the same time, AI may “level the playing field”, offering capital-rich small states asymmetric potential if they make proactive strategic decisions to position themselves as “AI powers” and come up with innovative ways of using it. If small states are unable to harness AI, Steff argues, the prospects of a world of AI “haves” vs “have-nots” will increase – with negative consequences for small state sovereignty and independence.

In Chapter 4, “Artificial intelligence and the military balance of power: Interrogating the US–China confrontation”, Reuben Steff and Khusrow Akkas Abbasi explain that there is a very strong likelihood that AI will alter the balance of military power between the existing status quo superpower, the US, and it’s great power challenger, China. They argue that, historically, technology has been a fundamental building block in the balance of power between states, that the leading states in technological AI power will likely be best at translating it into military might and global influence, and that, at the present time, the US appears to have a distinct advantage across a number of key AI metrics. Yet, there are some areas where China is ahead of the US and others where it is rapidly catching up. The focus of Chapter 5, “Mitigating accidental war: Risk-based strategies for governing lethal autonomous weapons systems”, by Aiden Warren and Alek Hillas, is the impact of the introduction of lethal autonomous weapons systems (LAWS) into war and conflict. Warren and Hillas argue that these so-called “killer robots” are not able to understand context as well as humans and could act in unintended and problematic ways in the field. Recognizing that policymakers are unlikely to be able to develop preemptive bans on these technologies, for both technical and political reasons, they lay out a range of strategies to address or mitigate the risks posed by LAWS. The chapter includes an analysis and evaluation of the likelihood and consequences of accidental use-of-force and miscalculations leading to war.

Part II: Emerging technologies, the state, and the changing character of conflict

The chapters in the second section of the book are focused on the state as a unit of analysis, including how different political regimes approach emerging technologies through their foreign policy and political doctrines, and the implications of machines for the role of the state more generally in an era of accelerating and increasingly complex technological change. Key issues covered in this section include the role of grand strategy in shaping the use of technology by states; how states’ doctrines for the use of technology evolve over time, but are also rooted in historical, ideational, and cultural patterns; the difference between democratic and authoritarian states' approaches to emerging technologies and security in a new technological age; and the implications for state sovereignty of new technologies.

The section starts with Chapter 6, "Politics in the machine: The political context of emerging technologies, national security, and great power competition". In this chapter, Simona R. Soare examines the relationship between politics and machines, and the ways in which democratic states and authoritarian states use technologies, including in their interactions with one another. Soare argues that democratic and authoritarian regimes both pursue emerging technologies for domestic and international purposes. However, their political interests, which are influenced by different institutional and political dynamics, shape their use of AI and other emerging technologies in diverging and sometimes conflictual ways. Leveraging international networks and alliances through politically driven uses of emerging technologies also creates geopolitical gains for authoritarian and democratic states, and helps them to establish "technospheres" of influence. The chapter also addresses how efficient democracies and authoritarians are in the renewed great power competition and which side is "winning" the strategic competition over emerging technologies. In Chapter 7, "Inequitable Internet: Reclaiming digital sovereignty through the blockchain", Andrew M. Colarik and Richard Wilson highlight three issues that are at the core of the challenges states face in managing technology: (1) the consolidation of market power among a few technology corporations; (2) the opaque, one-sided nature of the data economy; and (3) the fractured and increasingly vulnerable ecosystem of digital identity management. The authors propose blockchain as a potential mitigating technology, arguing that it has the potential to reallocate control of user-generated data from the collecting corporations back to users themselves.

The next two chapters return to the specific policies of two of the world's leading states, Russia and the US. In Chapter 8, "The evolution of the Russian way of informatsionnaya voyna (information warfare)", Sean Ainsworth examines the history and evolution of Russian information operations. This chapter analyzes the evolution of Russia's information warfare strategy over a 20-year period covering the First Chechen War to the sophisticated information warfare operations employed during the ongoing Ukraine crisis. Ainsworth argues that Russia has proved adept at modernizing and adapting its long history of strategic thought and military doctrine concerning information warfare to the new strategic environment of the information revolution and cyberspace. These modernization and adaptation efforts have primarily been driven by "lessons learned" from the Chechen Wars and the dominant Russian strategic understanding of the "Color Revolutions" of the early 2000s. In Chapter 9, "US grand strategy and the use of unmanned aerial vehicles during the George W. Bush administration", Francis Okpaleke and Joe Burton argue that the use of drones served to undermine key aspects of the Bush administration's offensive–liberal strategic doctrine. The authors highlight how the effects of drone strikes worked at cross purposes with the administration's stated goal of spreading democracy, highlighting the countervailing democratic reactions engendered in the aftermath of drone strikes in targeted states, such as local protests against their use, unintended civilian death, the growth of anti-American sentiments, and militant recruitment and violence.

Part III: The state, society, and non-state actors

The third section of the book examines the emergence of new technologies and how they are challenging and shaping the relationship between states and societies. Key issues animating the analyses in this section of the book include the democratization and diffusion of new technologies including 3D printing and deepfakes to non-state actors, the level of public trust in society in emerging technologies, including in attribution of cyber-attacks, the automation of cyber defense and attack, and in game changing life-saving technologies; and the need to think about security outside of a strictly military/defense sphere, including the use of environmental/sustainable technologies to enhance human security.

The section starts with Chapter 10, "Cyber autonomy: Automating the hacker – self-healing, self-adaptive, automatic cyber defense systems and their impact on industry, society, and national security". In this chapter, Ryan Ko analyzes the impact of the increasing number of automated cyber defense tools, including deception, penetration testing, and vulnerability assessment tools. Like other industries disrupted by automation, Ko argues that these trends have several implications for national security and private industry, including changes to business models and national security and human resource planning. This chapter reviews new and emerging cyber security automation techniques and tools, the perceived cyber security skills gap/shortage, implications for human rights and ethics, and the potential demise of the manual penetration testing industry in the face of automation. In Chapter 11, "The international security implications of 3D printed firearms", Peter Cook moves on to consider the impact of advanced manufacturing techniques (such as 3D printing) as an emerging technology. Using New Zealand as a case study, Cook examines the threat 3D-printed firearms pose to national security and public safety and, if needed, how legislation can be updated to mitigate the threat. Through examination of literature, statistics, and interviews with relevant experts, Cook demonstrates that although the risk is very low, due to the rapid and potentially disruptive advances in this technology, it is important to be proactive regarding threat assessment and legislation change in order to reduce future risk to public safety.

In Chapter 12, "Deepfakes and synthetic media", Curtis Barnes and Tom Barraclough examine growing concerns around the impact of audio-visual technologies. They argue that the technology already allows semiskilled users to create highly persuasive audio-visual information with a strong likelihood of increasing sophistication and the democratization of their availability in the near future. Because of their many benign and commercially valuable uses, these technologies are proliferating throughout global society, but, if used maliciously, they are a concerning addition to the continuum of tools available for disinformation and propaganda. Chapter 14 moves over to the environmental domain. In this chapter, "Disrupting paradigms through new technologies: Assessing the potential of smart water points to improve water security for marginalized communities", Nathan Cooper highlights how new water technologies could help address the global lack of access to safe drinking water and increasing water scarcity. He argues that advancements in water technology, such as "smart pumps", offer ways to achieve reliable, sustainable, and equitable water services for users in marginalized communities, but, at the same time, they represent a disruption to established relationships vis-à-vis water management. Using a diverse mix of case studies from Latin America and Africa, the chapter considers the effects of technological interventions to help achieve local water security and provides theoretical insights into the interrelational and institutional dynamics involved.

Two later chapters of the book deal with issues of trust in emerging technologies. In Chapter 13, "Cyber threat attribution, trust and confidence, and the contestability of national security policy", William Hoverd focuses on the attribution processes surrounding cyber-attacks. Hoverd highlights that the often-classified nature of the threat results in governments not being able to provide the public with an evidence base for the threat attribution. This presents a social scientific crisis where, without substantive evidence, the public is asked to trust and have confidence in a particular technological threat attribution claim without any further assurance. This chapter draws on recent "Five Eyes" (US, UK, Canada, Australia, and New Zealand) condemnation of Russia and North Korea cyber policy as a sociological case study to illustrate where and if a technological threat attribution and trust and confidence challenge might be evident. In the final chapter of the book, "'Just wrong', 'disgusting', 'grotesque': How to deal with public rejection of new potentially life-saving technologies", Dan Weijers explains how many new technologies are criticized on moral grounds, leading some potentially life-saving technologies to be left on the shelf (in a folder marked "rejected ideas"). Weijers presents a procedural framework for policy makers to use when members of the public deem a potentially beneficial new technology morally repugnant. The framework takes into account the possibility of different and conflicting moral beliefs and indicates the appropriate response to moral repugnance about potentially beneficial new technologies. The example of Policy Analysis Market (PAM), a proposed prediction market with the potential to prevent terrorist attacks that was shut down by the US government in 2003 owing to a public backlash, is used to illustrate the framework.

The conclusion highlights the novel contribution of the analysis in this volume with respect to three key aspects. First, it highlights the under-conceptualized nature of efforts to determine (and measure) the revolutionary impact of emerging technologies on the security of international actors. Second, it challenges the technocentric view of the relationship between emerging technologies and security. Finally, it argues that the adoption of emerging technologies further blurs the lines between the traditional levels of analysis – sub-state, state, and inter-state.

References

Allen, G and Chan, T (2017) *Artificial intelligence and national security*. Belfer Center for Science and International Affairs. https://www.belfercenter.org/sites/default/files/files/publication/AI%20NatSec%20-%20final.pdf, accessed May 1, 2020.

Altmann, J and Sauer, F (2017) 'Autonomous weapon systems and strategic stability', *Survival*, 59(5), pp. 117–142.

Bell, A and Futter, A (2018) 'Reports of the death of arms control have been greatly exaggerated', *War on the Rocks*. https://warontherocks.com/2018/10/reports-of-the-death-of-arms-control-have-been-greatly-exaggerated/, accessed May 1, 2020.

Buzan, B (2007) *People, states & fear: an Agenda for international security studies in the post-cold war era*. ECPR Press.

Carr, EH and Cox, M (2016) *The twenty years' crisis, 1919–1939: reissued with a new preface from Michael Cox*. Palgrave Macmillan.

Coker, C (2015) *Future war*. Polity.

Congressional Research Service (2018) *Lethal autonomous weapon systems: issues for congress*. April 14. https://www.everycrsreport.com/reports/R44466.html, accessed July 31, 2018.

Cummings, ML, Roff, HM, Cukier, K, Parakilas, J and Bryce, H (2018) *Artificial intelligence and international affairs disruption anticipated*. Chatham House. https://www.chathamhouse.org/sites/default/files/publications/research/2018-06-14-artificial-intelligence-international-affairs-cummings-roff-cukier-parakilas-bryce.pdf, accessed May 1, 2020.

DeepMind (2019) *AlphaStar: mastering the real-time strategy game Starcraft II*. January 24. https://deepmind.com/blog/article/alphastar-mastering-real-time-strategy-game-starcraft-ii, accessed April 30, 2020.

Department of Defense (2016) *Defense science board study on autonomy*. Defense Science Board. June. https://www.hsdl.org/?view&did=794641, accessed May 1, 2020.

Dougherty, J and Jay, M (2017) 'Russia tries to get smart about artificial intelligence', *Wilson Quarterly*. https://wilsonquarterly.com/quarterly/living-with-artificial-intelligence/russia-tries-to-get-smart-about-artificial-intelligence/, accessed May 7, 2020.

Dutton, T (2018) 'An overview of national AI strategies', *Medium*. June 28. https://medium.com/politics-ai/an-overview-of-national-ai-strategies-2a70ec6edfd, accessed February 1, 2019.

The Economist (2017) *The world's most valuable resource is no longer oil but data*. https://www.economist.com/leaders/2017/05/06/the-worlds-most-valuable-resource-is-no-longer-oil-but-data, accessed May 7, 2020.

Fiegerman, S (2017) 'Elon musk predicts world war III', *CNN*. September 4. http://money.cnn.com/2017/09/04/technology/culture/elon-musk-aiworld-war/index.html, accessed December 19, 2018.

Future of Life Institute (2020) *An open letter: research priorities for robust and beneficial artificial intelligence*. https://futureoflife.org/ai-open-letter/?cn-reloaded=1, accessed May 7, 2020.

Geist, E and Andrew, JL (2018) *How might AI affect the risk of nuclear war?* RAND Corporation. https://www.rand.org/pubs/perspectives/PE296.html, accessed June 8, 2018.

Geist, EM (2016) 'It's already too late to stop the AI arms race – we must manage it instead', *Bulletin of the Atomic Scientists*, 72(5), pp. 318–321.

Gruszczak, A and Frankowski, P (eds) (2018) *Technology, ethics and the protocols of modern war*. Routledge.

Hoadley, DS and Lucas, NJ (2018) *Artificial intelligence and national security*. Congressional Research Service. April 26. http://www.crs.govr45178/, accessed July 31, 2018.

Hoffman, RR, Cullen, TM and Hawley, JK (2016) 'The myths and costs of autonomous weapon systems', *Bulletin of the Atomic Scientists*, 72(4), pp. 247–255.

Horowitz, M, Kania, EB, Allen, GC and Scharre, P (2018) 'Strategic competition in an era of artificial intelligence', *CNAS*. July 25. https://www.cnas.org/publications/reports/strategic-competition-in-an-era-of-artificial-intelligence#fn14, accessed May 7, 2020.

Horowitz, MC (2018) 'Artificial intelligence, international competition, and the balance of power', *Texas National Security Review*, 1(3), May. https://tnsr.org/2018/05/artificial-intelligence-international-competition-and-the-balance-of-power/, accessed June 8, 2018.

Kania, EB (2017) 'Battlefield singularity: artificial intelligence, military revolution, and China's future military power', *CNAS*. November. https://s3.amazonaws.com/files.cnas.org/documents/Battlefield-Singularity-November-2017.pdf?mtime=20171129235805, accessed May 7, 2020.

Kiggins, R (ed.) (2018) *The political economy of robots: prospects for prosperity and peace in the automated 21st century*. Palgrave Macmillan.

Larrey, P (2018) *Connected world: from automated work to virtual wars: the future, by those who are shaping it*. Penguin.

Lee, KF (2017) 'The real threat of artificial intelligence', *New York Times*. June 24. https://www.nytimes.com/2017/06/24/opinion/sunday/artificial-intelligence-economic-inequality.html, accessed May 7, 2020.

Meier, P (2015) *Digital humanitarians: how big data is changing the face of humanitarian response*. CRC Press.

Payne, K (2018) *Strategy, evolution, and war: from apes to artificial intelligence*. Georgetown University Press.

PWC (2018) *Nations will spar over AI*. https://www.pwc.com/us/en/services/consulting/library/artificial-intelligence-predictions/ai-arms-race.html, accessed May 7, 2020.

Ryan, M (2018) 'Intellectual preparation for future war: how artificial intelligence will change professional military education', *War on the Rocks*. https://warontherocks.com/2018/07/intellectual-preparation-for-future-war-how-artificial-intelligence-will-change-professional-military-education/, accessed May 7, 2020.

Scharre, P and Horowitz, M (2018) 'Artificial intelligence: what every policymakers needs to know', *CNAS*. June 19. https://www.cnas.org/publications/reports/artificial-intelligence-what-every-policymaker-needs-to-know, accessed July 31, 2018.

Scott, B, Heumann, S and Lorenz, P (2018) 'Artificial intelligence and foreign policy', Stiftung Neue Verantwortung. https://www.stiftung-nv.de/sites/default/files/ai_foreign_policy.pdf, accessed November 5, 2018.

Tucker, P (2020) 'Artificial intelligence outperforms human intel analysts in a key area', Defense One. https://www.defenseone.com/technology/2020/04/artificial-intelligence-outperforms-human-intel-analysts-one-key-area/165022/, accessed May 7, 2020.

U.S. Army (2017) *The operational environment and the changing character of future warfare. Training and doctrine command (TRADOC) G-2*. https://www.pdf-archive.com/2017/07/28/theoperationalenvironment/theoperationalenvironment.pdf

Welsh, S (2017) *Ethics and security automata: policy and technical challenges of the robotic use of force*. Routledge.

Waltz, KN (2001) *Man, the state, and war: a theoretical analysis*. Columbia University Press.

Završnik, A (ed.) (2016) *Drones and unmanned aerial systems: legal and social implications for security and surveillance*. Springer.

1 Histories of technologies

Society, the state, and the emergence of postmodern warfare

Joe Burton[*]

Introduction

The history of the last century is in many ways a history of technological change. Technology has played a profound role in the emergence of the modern international system, including the states that populate it and the relationship between those states and their citizens. The Cold War, for example, was shaped by the emergence of nuclear technologies. Nuclear weapons brought an end to the Second World War in the Pacific theatre, and the threat of their use in the European context ushered in a new age of atomic diplomacy. But the advent of the nuclear age also had an effect on the internal nature of states, with a marked change in culture and media, the emergence of an increasingly centralized and powerful "national security state", and a securitized relationship between citizens and their governments. The end of the Cold War was similarly defined by technology. It was precipitated by an increased awareness of the difference in living standards between west and east due to advances in information and communications technologies, and the strategic competition between the superpowers over new space-based technologies, which contributed to the Soviet demise. The emergence of air power in the 20th century had a similarly profound impact. Consider the advancement from zeppelins and biplanes in World War I, which were fairly negligible in determining the outcome of that war, to the pivotal aerial battles in the Pacific and Europe that so defined World War II, including the Battle of Britain, the Blitz, and the use of air power at Pearl Harbor, Tokyo, Hiroshima, and Nagasaki. Fast forward to the 21st century and drones capable of fully autonomous use of force have emerged as tools for assassination, counterterrorism, and counterinsurgency and represent powerful symbols of the technological prowess of the states that possess them and their ability to coerce and surveil societies.

* Research for this chapter received funding from the European Union's Horizon 2020 research and innovation programme under the Marie Skłodowska-Curie grant agreement No 844129.

This chapter seeks to lay a historical foundation for this book by placing the latest trends in emerging technologies in a broader historical and theoretical context. It does this because too many contemporary debates about emerging technologies have been ahistorical – without due regard to the interrelationship between technology and history and its role in shaping the present. The aims of the chapter are threefold: First, to determine whether historical lessons can indeed be learned and applied to emerging technological trends; second, to assess the ways in which technologies have shaped the different levels of analysis that are the focus of the book – society, state, and the international system (including the relationships between them); and third, to build a more nuanced understanding of the complex relationship between technology and history.

The chapter proceeds in three main sections. The first section outlines how history itself has become increasingly contested and highlights some of the different theoretical approaches to history and technology that now populate academia. In doing so, it builds the argument that there are multiple understandings and interpretations of histories "out there" and that attempts to derive concrete lessons have been complicated by this theoretical diversity. The second section questions the extent to which revolutions in military affairs (RMAs) have driven technological progress and reshaped war, the state, and the international system. In doing so, it highlights that many of the technologies that have emerged over the last century have been based on long term accumulations of scientific knowledge. Furthermore, because they have been "dual use" – that is to say they have had both applications in the military and civilian fields of action – they have resulted in changes in war and conflict but also in the broader political relationships between states and their citizens. The final section of the chapter reflects on whether we are indeed witnessing another profound techno-historical shift by focusing on the concept of postmodern warfare and its relevance to the emerging security environment.

Histories of technological change

In "The History Boys", a play depicting the experiences of a group of British schoolboys, a student is asked by his teacher, "What is history?". The student pauses, reflects for a moment, and then replies, "It's just one f**king thing after another". The same response could be applied to the seemingly relentless advances in technology, a similarly bewildering and perplexing process. We struggle to keep track of how technology affects our societies and our laws and regulations often lag behind technological advances. The effects and indeed unintended consequences of the adoption and diffusion of technology are often slow to be appreciated and can create a sense of exasperation and frustration. The answer to the question, what is history, continues to produce a variety of colorful and contradictory answers. History is at once a deeply personal phenomenon which relates to our own familial, social, and cognitive experiences, and a process that determines the shape of the world around us including our relationships with the states that we live in, the cohesiveness of our societies, and the level of peace or conflict in the international system.

To put it more academically, there is no single epistemological or ontological understanding of what history is. It cannot be simply defined. Historians themselves are wrought from myriad different cloths. Social historians, international historians, and critical historians see history in different ways and focus on different aspects of the past and how it feeds into the present. Within the field of international relations (IR), the role of history is equally fractured and contentious. Some historians suggest that IR as a field of study has become disconnected from history to its detriment. Geoffrey Roberts (2006, p. 708), for example, has argued that "IR theoretical concepts and postulates need to be buttressed and validated not just by example-mongering or selective empirical sampling, but by specific stories about the evolution and development of international society". Ian Clark (2002, p. 277) paints a similar critique, suggesting that "drawing on both the insights of history and political science" is necessary for a fuller and more nuanced understanding of the present. By these definitions, history constitutes a narrative that is constantly constructed and reconstructed by the actors that populate the world.

The history of technology is similarly diverse, and there is little agreement within different subfields of scholarly inquiry about the role of technology in shaping the modern world. According to the realist approaches to IR, military technology is a value-free tool to accomplish specific tasks and is developed by states to enhance their survival in an international system characterized by anarchy (the lack of an international sovereign or government to decisively regulate state behavior under conditions of self-help). Technology by this conception provides a path toward survival, and the most powerful states are the ones that possess the most powerful and advanced military and civilian technology. The technological might of a nation, counted in the number of tanks, guns, drones, and aircraft carriers, the ability to maintain these technologies and deploy them, contributes to the global balance of power and will determine the outcome of military conflict and, therefore, the historical trajectory of the world. Liberal scholars, conversely, place more emphasis on the role of technology in positive, normative political change over time, including democratization, the dissemination of norms and values, and the advancement of human rights, transparency, and accountability. Technology by this conception is a force for progress, not just a tool of power relations, and can be harnessed to enhance a state's power but also to progressively and incrementally make the world a more peaceful, just, and democratic place. New technologies can help us communicate and increase economic interdependencies that have a positive effect on the levels of peace and conflict in the international system. In this understanding, democracy itself is a natural historical trajectory, which is aided and abetted by technological change.

There are some other deeper philosophical divides when we examine the theories of history and their relationship with technology. Marxist historians, for example, see technology as a tool in class struggle and the ownership and access to technology as deeply inequitable. According to this conception, technology underpins the exploitative practices of modern capitalism, has driven historical colonial expansion, and has been a vehicle for exporting inequality. The oft-cited concern that millions of people will lose their jobs to machines makes sense in the context of the Marxist critique of capitalist economics. The contribution of post-structuralist and postmodern scholarship has also had a marked impact on the debate about technology and history. In the critical security studies sphere, for example, a host of analyses have emerged about new technological arms races and imperialism (Shaw, 2017), the militarization and securitization of technology (Cavelty, 2012), and the adverse effects of its adoption by states for war fighting (Burton & Soare, 2019). Technologies are mechanisms of societal control – facial recognition software and fingerprint technologies are used to construct and enforce borders, dehumanize, and undermine freedom and open societies. In these conceptions, security technologies are used to embed authoritarian practices in both democratic and authoritarian states.

Perhaps one of the deepest fault lines within the academy when it comes to the role of history and its relationship with technology has been the divide between technological determinism and the science and technology studies (STS) approach. The former places emphasis on technology having an independent role in shaping politics and societies, and that history is determined by technological change. The STS approach contends that technologies themselves are socially constructed and embedded and emerge out of very specific societal, political, and social contexts, which determine how they are used (Jasanoff et al., 1995). By this logic, after 9/11, the convergence of the US-led War on Terror with the growth of the global internet resulted in mass surveillance. That is to say, the political situation in the US determined how the technology was used, and not vice versa. This view is related to constructivist conceptions of technology, which assume that the way we use and develop technology is deeply cultural and stems from our historical practices, ideas, beliefs, and behaviors. One can hardly discount the relevance of these assumptions when considering the continued disposition in the academic and policy worlds to approach security in a clearly globalizing environment through the lens of the nation state, borders, walls, and boundaries.

If the field of history and technology that exists in the academy leads us to such different conclusions and illuminates different aspects of reality, then how can we possibly learn lessons from history when confronting the apparently transformative technological change we are now witnessing in societies? Is it possible to learn lessons from the history of technology and apply them in any meaningful way to help mitigate the apparent risks that the current environment engenders? One way forward here is to explore the major historical transformations that have occurred in the past, and to examine the role that technologies played in them. This form of *historical excavation* could tell us where, when, and how technology has had an influence. An examination of the Arab Spring, for example, a transformative event in the Middle East, might establish how social media technologies allowed dissent to emerge and how images of protests, violence, and state brutality diffused so quickly. It might help explain the organizational effectiveness of activists seeking to mobilize support and overthrow their governments. However, that reading belies the other major causes, including the broader social and economic inequalities, the nature of the authoritarian regimes themselves, and how technologies were a double-edged sword used to crush dissent. A similar analysis of the end of the Cold War might reveal certain technological influence in the demise of the Soviet Union. President Reagan's pursuit of the SDI ("Star Wars") initiative as a means to place increased pressure on the Soviet Union to compete in what had become an intensely technological competition might be highlighted. The broader role of communications technologies and television media in highlighting the divides that exist between east and west might also be brought into view. But clearly there were also deep structural weaknesses in the Soviet system and ideological holes in the fabric of communism itself, including the continued identification of people in the eastern bloc with their national identities rather than any sense of communist or class-based identity. To use these two big historical turning points, we can see technology played a role, but was certainly not the formative factor. By this form of analysis, technology is only ever likely to be one variable among many others in shaping major historical transformations, and one that will be difficult to weigh or quantify.

(R)evolutions in military affairs?

Perhaps a better way forward is to examine the emergence of particular technologies and their impact in reshaping the character of war and politics throughout the ages. Instead of focusing on the role of technology in major historical transitions, this form of *technological excavation* would look at the nature of the technological change itself and the extent to which it reshaped how nations compete and come into conflict and even the ways in which they cooperate. Examining history in this way requires considering the question of whether the technology is revolutionary itself, rather than the role technology played in revolutions. Military historians and strategists have certainly been occupied with mapping the impact of profound and sudden technological changes that have altered the strategic interactions of nation states. Nevertheless, these efforts have been complicated by determining exactly what counts as a technological revolution. There have certainly been large technological advances which have influenced modern political history; the most commonly cited examples are the crossbow, gunpowder, iron/steel clad ships, the combustion engine, the tank, the fighter aircraft and bomber, aircraft carriers, nuclear weapons, intercontinental ballistic missiles, precision-guided munitions, and, in the more modern era, ICT and cyber weapons. But a close examination of these military-technological advances raises questions about how revolutionary they actually are.

The situation is further complicated by the fact that technologies do not develop in a linear fashion. Artificial intelligence (AI) is a good example. Since the 1950s, we have had intense periods of interest in AI technologies which have ended up in so-called AI winters for reasons related to computing power and available data. Whatever progress is made in the next decade with AI will be the culmination of nearly a century of leaps forward and incremental efforts. This is not the only example – high-capacity laser technology has been a US military program since the 1970s and has been abandoned (mainly for budgetary constraints and lack of progress) and resurrected dozens of times because progress in related areas promised to facilitate breakthroughs. We also need to consider what it is about the technology that is revolutionary. In this context, the nuclear attacks against Japan followed the conventional (incendiary) bombing of Tokyo, which caused 100,000 deaths. Is the technology revolutionary, therefore, because of its destructive power or because of the changes in behavior it engenders in societies, states, and in the international system?

Nuclear revolution?

Perhaps the most cited case of a revolutionary technology that fundamentally changed the world we live in is the nuclear revolution. Oppenheimer's famous words "I have become death, the destroyer of worlds" certainly indicated a profound change, but they also proved inaccurate. Oppenheimer himself did not create the technology. It was the result of an intensified scientific effort that was again driven by the context of the time and the urgency surrounding the $2 billion Manhattan Project, which took place in the very real context of the Second World War, itself an existential struggle for survival. The agency of individual scientists has of course been hugely influential in history. Einstein's theory of relativity, Newton's work on gravity, or Stephen Hawking's theories of interstellar travel and black holes, are all associated with the scientist in question. However, nuclear weapons emerged out of a much broader and grander structural fight for survival and on the back of generations of nuclear science. Efforts to split the atom, the foundational requirement of a nuclear bomb, were long in the making, stretching back at least to the mid-1800s. In that sense, the culmination of the Manhattan Project may have been a revolutionary moment, but it was based on a scientific evolution spanning many decades. When did the nuclear revolution happen?, we might ask. In the late 1930s to early 1940s when the Nazis started working on the technology? Alternatively, in the mid-1940s when the US acquired the capability and built and used a nuclear bomb? Similarly, nuclear weapons became much more dangerous as a result of the development of associated missile technologies. Arguably, it was the ability to deliver nuclear devices accurately across continents that was the real strategic game changer.

Oppenheimer's worst fears did not come to pass either. Nuclear weapons have not destroyed worlds. They have only ever been used in combat twice, in Hiroshima and Nagasaki, and the diffusion of nuclear technologies had been widespread and, for the most part, peaceful. This is not to minimize the destructiveness and inherent dangers of nuclear weapons and materials. Certainly, it is true to say that in 1945 the human race gained the ability to destroy our own civilization for the first time in history. The accidents at Chernobyl in 1986 and the destruction of the Fukushima nuclear facility in 2011, which followed numerous other nuclear-related crises and near misses, should stand as stark warnings of the danger associated with nuclear energy. But having recognized these dangers in the early-mid Cold War period, societal and social dynamics kicked in, both within states and in the international arena. They included mechanisms of control, inspection, and verification of capabilities, sophisticated encryption processes and technologies to guard against misuse of nuclear technology, and a degree of restraint, cooperation, and the mutual recognition of a harmony of interests in managing a game changing technology.

Scientific complexity also comes into the debate here. There are still only nine states that possess nuclear weapons. Some have foregone becoming nuclear powers because of moral or normative concerns, some because of coercive measures, such as sanctions and threats, but equally important is the complex scientific knowledge and expertise needed and the high costs associated with developing the technologies themselves. The assumption may have been in 1945 that nuclear weapons would be used again and that dozens of other nations would seek to acquire them, but both these eventualities have been limited by scientific complexity and management processes. Nuclear energy, similarly, has not emerged as a replacement for fossil fuels, which still constitute the majority of the world's market share. The diffusion of technology, including dual-use technologies, is limited by high barriers of entry and tight control regimes. This remains the case today where there are multiple high barriers to entry. For example, countries with (civilian or military) nuclear capabilities must rely on other states (such as Russia) for storage and recycling facilities for nuclear waste and nuclear rods.

While it is difficult to argue that nuclear weapons revolutionized warfare itself if they have only ever been used twice, the knock-on effects of the presence of such a destructive technology at both the societal level and the level of the international system was perhaps a marked and sudden change. At the level of the international system, the conventional wisdom is that nuclear weapons led to restraint on the part of the two superpowers, who never entered into a hot war with one another but instead fought a series of proxy conflicts in places like Korea, Vietnam, and Afghanistan. Great power war has not been a feature of the post–World War II international security environment, but intra-state wars involving the great powers have. Nuclear weapons could be seen as one variable in this change. Nuclear weapons also had an impact on strategy. In the early Cold War period, US strategy was based on brinkmanship – where the US would bring conflicts to the brink of a nuclear war in the hope that the adversary backed down. This dangerous and inherently crisis-prone strategic doctrine was modified as the Cold War environment evolved, with Kennedy's "flexible response" filling the gap, which involved options short of nuclear war, including the use of tactical (smaller yield) nuclear weapons. What then was revolutionary about the technology? Its use for mass retaliation or as a warfighting weapon, or rather non-use, including for strategic deterrence, coercion, and as part of the emergent mutually assured destruction (MAD) doctrine?

At the state and societal level too, the impact of nuclear technologies was pronounced. The national security state in America emerged not least because of the presence of nuclear weapons and the need to manage them, and this model was followed by other powers. Nuclear weapons fed into the atmosphere of fear and paranoia, which characterized the Cold War environment and shaped states' surveillance of their own societies. The growth of intelligence and surveillance societies can be directly attributed to this period. This suggests nuclear weapons may not have revolutionized war itself but had broader societal impacts that do not ever get the same level of attention.

Cyber revolution?

Has there been a cyber revolution? Certainly, some scepticism is merited here, too. Cyber-attacks and nuclear attacks cannot be compared in terms of their destructive power. No single cyber-attack has directly caused a loss of life. Despite excessive fears about cyber Pearl Harbors, weapons of mass disruption, and digital 9/11s, cyber-attacks have not crippled nations, brought down skyscrapers, or caused wars. There is a variety of reasons for this. First, great powers have exercised a degree of restraint in deploying the most destructive variations of their malware. Perhaps the most sophisticated cyber-attack documented to date – Stuxnet – was used in a very targeted operation, and there has been little evidence that such capabilities are being deployed against critical infrastructure since that time. Russian attacks against Ukraine's power grid certainly signalled a worrying trend, but power was restored quickly, with no permanent damage done, at least in the physical realm. There may also be norms that are already governing the relationship between the great powers in cyberspace; that despite their not being any new international laws outlawing attacks on critical infrastructure,[1] including health, water, transport, these types of attacks are already off the table. Economic entanglement has been another explanation for the lack of more cyber disruption (Nye, 2017), suggesting a mutual dependency between nations' economies that prevents and deters the worst forms of cyber disruption being visited on adversaries. Fears of escalation in cyber conflict have also been posited as a reason that cyber weapons have not been deployed extensively (Lin, 2012).

If the cyber revolution is indeed a revolution it must also be traced back to technological change that goes back decades. The Arpanet, the world's first internet network, developed by the Advanced Research Projects Agency (ARPA) in the US, was constructed in 1966. The father of modern computing, Alan Turing, also deserves a mention here. Without the intensification of his efforts to create the Turing machine that broke the German enigma codes during World War II, we would not have the modern computer, or at least the science involved would have taken longer to emerge. In this sense, just as with the Manhattan Project, war provided the sense of urgency, which led to scientific breakthroughs which have shaped the modern world, suggesting a mutually constitutive relationship between technology and history.

The idea of a cyber evolution has certainly gained some traction in modern cyber scholarship. Lukas Kello argues that cyber is revolutionary because it creates a sense of urgency around the need to understand the impact of cyber operations on strategy and because policy makers are already falling behind in developing the necessary response. While cyber has not changed the nature of war, he argues, there is a need to understand how a variety of actors can use cyber technologies to expand the range of harm and outcomes between war and peace and to create social and economic harms (Kello, 2013). Cyber weapons are often seen as unpredictable, undetectable, affordable, and diffusible, and this makes them subtly different from new technologies. Nevertheless, the nature of the cyber revolution may not be all it's made out to be. Cyber attribution and digital forensics have come on leaps and bounds over the last decade, and, at least technically, cyber weapons are not that hard to detect. The rate of diffusion should also be questioned. There is little evidence that the more sophisticated cyber weapons are the purview of anyone but the great powers. In this way, the cyber revolutions correspond to the nuclear one in the previous paragraphs. If cyber is a revolution, then it may be a cultural, sociological, or behavioral one. In a reply to the technological determinism of Kello's argument, Lindsay (2013) posits that cyber weapons will be used according to doctrine and organization, and the circumstances of employment of cyber weapons matter more than the existence of the weapons themselves. Why would non-state actors seek to acquire such sophisticated weapons, Lindsay asks, when box cutters (or IEDs or small arms) can cause as much chaos and fear as a cyber-attack? The same note of caution should be ascribed to the use of cyber in actual warfare: Cyber-attacks may amplify and multiply the use of force but have not been used in a decisive way to end any conflict.

The societal impact of cyber-attacks, and the way they have changed the relationship between the state and its citizens is perhaps a more fundamental change. States have used mass surveillance to spy on their citizens, and there is a growing international market in surveillance technology, including e-mail intercepts. This may be a function of the political regime under which they are deployed, as discussed by Simona Soare in Chapter 6 of this book. Social media has laid a tremendous foundation for governments to collect data on their citizenry, and it should also be noted that the impact of cyber-attacks lies predominantly in the societal sector, including banking, health, transport, energy, water, and governance. Offensive cyber capabilities may create security dilemmas between states, as in the traditional realist model, but they also create societal dilemmas between governments and their people, and between different identity groups. Again, the impact of internet technologies may shape great power relations, but, because they are dual use and so widespread among civilian populations, the greater impact may be beneath the state level, at the human behavioral level of analysis, or indeed at the ideational level – i.e. the way we think about technologies and conflict.

Postmodern warfare or warfare in a postmodern era?

If revolutions in military affairs have actually been evolutions, and if we can intellectually separate the character of warfare itself from the broader implication of technologies on strategy more specifically, how are we to judge the latest developments in technology covered in this book? Does the emergence of AI, quantum computing, and automation constitute a new RMA or a revolution in the character of warfare, and should we characterize this technology as "emerging" when AI research, just as with the long history of cyber and nuclear weapons, stretches back decades?

The security environment that these technologies are emerging in is an important starting point for intellectual enquiry. The past decade has seen a resurgence of populism and nationalism connected to the fallout from the 2008 global financial crisis. This has manifested in states' reassertion of their borders, territory, including US President Donald Trump's efforts to build a wall on the US southern border, Russian President Vladimir Putin's annexation of Crimea, and increased tensions over China's militarization in the South and East China Seas. This renationalization of politics (itself a historically recurring process) is reflected too in Brexit, a growing reaction against global free trade, the securitization of the migration crisis, and the institution of more intrusive border technologies. However, while walls appear to be "going up", the process of globalization continues. The COVID-19 crisis is perfectly illustrative of the global connectedness that defines the modern security environment, and there is increased global contestation around other issues, including climate change, the openness of the global internet and communications infrastructure, such as 5G. That is to say that the reassertion of national identity and populism is taking place alongside continuing globalization. In fact, a better characterization is that both these processes – globalization and nationalism, have been a constant feature of global politics that are being magnified by emerging technologies. Great power competition for scarce resources is nothing new, neither is economic protectionism. At the same time, globalization is a process that has deep historical roots, with some analyses arguing that the world was becoming more globalized as early as the 1700s (Beniger, 1986).

Side by side with these macro changes are transitions in the way we see history and knowledge itself, and these changes are being influenced by technology. The move toward a postmodern form of war and conflict and security has been observed by a number of scholars in recent years, and it is worth unpacking the term "postmodern" if we are to understand the current historical and technological transition that we are in.

First, postmodern warfare is framed against modern warfare, the latter being traditional war between nation states, driven by territorial gain, with uniformed combatants, following seemingly rational strategies to achieve strategic gains. War and conflicts in the 21st century rarely conform to these Clausewitzian parameters. In fact, it may be difficult to know and understand exactly what war is in the current postmodern environment (Joenniemi, 2008, p. 233). Mary Kaldor's (2012) conception of new wars captures some of the emerging dynamics. Kaldor highlights that new wars are characterized by the involvement of non-state actors, a criminal war economy that sustains the conflict, the conflict being driven by identity rather than territory, and the use of torture, rape, and child soldiers as weapons of war. These societal conflicts exhibit characteristics that are not entirely new but have intensified and become more prevalent in the post–Cold War era (Kaldor 2013).

Arguably, however, war has undergone a further transition to the postmodern, especially over the last decade, and one that is deeply connected to technological change. This form of postmodern warfare is based on five key frames. First, the erosion of the distinction between war and peace. The war on terror, war on drugs, cyber war, and information wars have all been transplanted from physical conflict between nations to be permanent and constant features of modern international affairs. It is no longer possible to tell when wars start, when they stop, to negotiate a peace, or to even know who to negotiate with.

Relatedly, second, because actual interstate warfare has become so fraught with risk, particularly due to the presence of nuclear weapons, but also because of the fears of dangerous escalations in non-conventional realms, much of the postmodern conflict happens below the threshold of armed force. This dynamic has been captured by a variety of terms, including grey zone conflicts and hybrid warfare, where sabotage, the use of covert actions, information operations, psy-ops, and cyber-attacks are used by nations to achieve influence in ways that facilitate plausible deniability. Conflict below the threshold or the use of force has become a feature of the contemporary environment not least because new technologies provide new and novel ways to achieve strategic effects with limited costs and provide armies and soldiers ways to achieve the goals of their political masters without risk to life or costly arduous counterinsurgency campaigns where they incur mass casualties. The emergence of conflict without physical use of force leads us to think that "warless war" or "wars without warfare" might have emerged (Joenniemi, 2008, p. 234).

The third feature is a questioning of the nature of truth itself. We can see this tendency in a plethora of events in the past few years. China's lack of openness about the initial stages of the coronavirus is illustrative and so too is the continued denial by the Putin government that they are fighting a war in eastern Ukraine, a conflict where 2000 Russians have died, and which continues to be concealed from the Russian public (Nemtsova, 2017). President Trump's insistence that mainstream media outlets disseminate "fake news", and the ongoing politicization of institutions that have traditionally been arbiters of the truth, such as the courts, law enforcement, and scientific experts also fits this paradigm. Hot wars, cold wars, cyber wars, and information wars are taking place in an informational environment that is increasingly becoming twisted and distorted to suit political narratives and interests. In that respect, the leaders of the great powers have been astute at turning the postmodern philosophy to their individual political advantage. They have been aided in their quest by the emergence of social media giants as platforms for misinformation and disinformation and the flattening of the media landscape, to include every type of shock jock, liberal and right wing, and blogs that mirror, replicate, and diffuse whichever political bias consumers happen to hold.

These new actors form part of the fourth feature of the postmodern security environment, which is the emergence of networks or security assemblages that mark a stark contrast from the Cold War national security model in particular.

This suggests that Kenneth Waltz's famous work, *Man, the State and War* (2001), and its delineation of international politics into individual, state, and international system, is aging in its analytic significance. Transnational and subnational groupings are increasingly recognized as having an independent influence on international affairs, but it is the links between them, which are vital to the postmodern approach. The strategic competitions between nations, non-state actors, hackers, cyber criminals, and organized criminal enterprises are endless and unceasing. States are forging complex relationships with criminals and non-state actors – the work done by Russian security services with WikiLeaks in leaking Democratic Party material in the run-up to the 2016 US election is illustrative here. China and Russia maintain networks of patriot hackers and netizens, and North Korea has used cyber-attacks to steal money from international banks to finance the regime. In this respect, postmodern conflict reveals a further blurring of the traditional distinction between the public and private, and the monopoly on the use of force is diffused away from the state (Collier, 2018, p. 13). According to this conception, war will pit a multiplicity of actors against each other, where leaders of governments will be in conflict with rival political parties, religious factions, where there will be a battle over media portrayals of conflict, and in which the financial interests of the markets, financial institutions, and other players will be central to conflict (Lucas, 2010, p. 290).

Finally, fifth, the ways, means, methods, and spatial aspect of postmodern war are profoundly technological. Postmodern warfare is defined by automation, including the use of drones and other unmanned vehicles; advances in AI, which will increasingly take decision-making away from humans and lead to greater battlefield situational awareness; the biological and robotic enhancement of humans as well as human machine teaming, where soldiers will fight alongside machines; and the use of machines to achieve biological, societal, and psychological effects (Burton & Soare, 2019). These technological manifestations will be filtered, framed, and sensationalized through the media and popular culture and will become increasingly performative, as terrorists livestream attacks and states use videos of precision guided munitions to convince publics they are discriminating in their use of force. To those on the winning side, postmodern war will look like and feel like a game, with apps on smartphones harvesting data and predicting troop movements and joystick and video screens being used to control killer robots.

Conclusion

The critique of postmodernist interpretations of modern conflict has been that they have encouraged and facilitated the transition to a post-truth environment, in which it is hard to identify any objective reality and where history itself has been not a series of events but, rather, a series of intersubjective social interpretations that are constantly subjected to reinterpretation (Crilley & Chatterje-Doody, 2019). However, postmodern theory can help shed light on the modern security environment and chimes with many of the recent technological advancements and how they have been used in international affairs.

Have we seen a postmodern revolution in warfare? Due caution should be exercised here, too. Just as with nuclear and cyber, there are elements of continuity. Modern warfare may and likely will still be a feature of the 21st century and will exist alongside postmodern war for the foreseeable future. Propaganda has also long been disseminated to increase the fog of war, to create narratives that deviate from the truth, and to support a nations' strategic goals by blurring the distinction between good and bad, immoral and moral, democratic and non-democratic. Nevertheless, when propaganda can achieve mass effects not through leaflet drops, but by mass manipulation of social media, then the likelihood is we are moving into a new and ill-understood phase of conflict where it is more difficult to determine whether we are being manipulated.

What is history? Can we learn lessons from history that can be applied to the role of technologies in international security? If this chapter has helped clarify these questions, then it has done so in the following ways. First, the transitions that have been brought about in the world have all been influenced in some way by technology. Technology does not appear to have been the deciding factor in any of the examples examined here, but it has been a constant force. Second, we should be careful in describing technological shifts as revolutionary. Quick, unseen, and impactful changes may occur, but they are almost always based on broader historical processes including the slow, steady, and incremental advancement of scientific knowledge. Moreover, our definition of "revolutionary" should be fluid – revolutions start as fundamental change and then track into incremental change. This is connected to the broader cycle of innovation, use, effects, endurance, and, eventually, adaptation. Third, technology should not be seen as having an entirely independent impact on modern war and politics. It is almost always situated in some specific cultural context. To paraphrase Alexander Wendt (1992), technology is what states make of it. In this respect, the conclusion of this chapter is that some common ground may be found between the STS and technological determinist approaches, especially if we recognize that technology and history are mutually constituted.

Where we go from here is another big question. There is something qualitatively different about the current security environment compared to what has existed before. It may not constitute a new generation of conflict, and it may be too soon to tell if we are in the midst of another revolution in military affairs, strategy, or the character of warfare. Caution should be ascribed to statements that overestimate the impact of technology. They often serve political or economic interests, whether to sell newspapers, draw attention to academic writings, or to create a securitized environment in which request for resources, or organizational power is enhanced. Finally, the rate of historical and technological change will always warrant attention. Sometimes technological change will accelerate, sometimes it will decelerate and stagnate. The "AI winter" of the 1970s is a good example. Changes in the security environment, including crises, can stimulate scientific breakthroughs. In this context, the impact of the current

COVID-19 pandemic may lead to a lack of investment in science and technology and a reversion to protectionism and nationalist sentiment. But it could also, as with the nuclear revolution, lead to increased international cooperation and scientific progress.

Note

1 While western states argue that such attacks are prohibited under the Geneva conventions, this is not universally accepted, and the provisions therein are not an easy fit to cyber technologies.

References

Barley, SR (1998) 'What can we learn from the history of technology?', *Journal of Engineering and Technology Management*, 15(4), pp. 237–255.

Beniger, JR (1986) *The control revolution: technological and economic origins of the information society*. Boston: Harvard University Press.

Burton, J and Soare, SR (2019) 'Understanding the strategic implications of the weaponization of artificial intelligence', *11th international conference on cyber conflict (CYCON)*, pp. 1–17.

Cavelty, MD (2012) 'The militarisation of cyberspace: why less may be better', *4th international conference on cyber conflict (CYCON)*, pp. 1–13.

Clark, I (2002) 'International relations: divided by a common language?', *Government and Opposition*, 37(2), pp. 271–279.

Collier, J (2018) 'Cyber security assemblages: a framework for understanding the dynamic and contested nature of security provision', *Politics and Governance*, 6(2), pp. 13–21.

Crilley, R and Chatterje-Doody, P (2019) 'Security studies in the age of "Post-truth" politics: in defence of poststructuralism', *Critical Studies on Security*, 7(2), pp. 166–170.

Jasanoff, S, Markle, GE, Petersen, JC and Pinch, T (1995) *Handbook of science and technology studies*. Thousand Oaks: SAGE Publications.

Joenniemi, P (2008) 'Toward the end of war? Peeking through the gap', *Alternatives: Global, Local, Political*, 33(2), pp. 233–248.

Jones, SH et al (2019) *Science, belief and society: international perspectives on religion, non-religion and the public understanding of science*. Bristol: Policy Press Bristol University Press.

Kaldor, M (2012) *New and old wars: organized violence in a global era*. Cambridge: Polity.

Kaldor, M (2013) 'In defence of new wars', *Stability: International Journal of Security and Development*, 2(1), Art. 4, pp. 1–16.

Kello, L (2013) 'The meaning of the cyber revolution: perils to theory and statecraft', *International Security*, 38(2), pp. 7–40.

Lin, H (2012) 'Escalation dynamics and conflict termination in cyberspace', *Strategic Studies Quarterly*, 6(3), pp. 46–70.

Lindsay, JR and Kello, L (2014) 'Correspondence: a cyber disagreement', *International Security*, 39(2), pp. 181–192.

Lucas, GR (2010) 'Postmodern war', *Journal of Military Ethics*, 9(4), pp. 289–298.

Nemtsova, A (2017) 'How Russia hides its dead soldiers killed in Ukraine', *The Daily Beast*. https://www.thedailybeast.com/how-russia-hides-its-dead-soldiers-killed-in-ukraine, accessed April 29, 2030.

Nye, JS (2017) 'Deterrence and dissuasion in cyberspace', *International Security*, 41(3), pp. 44–71.

Roberts, G (2006) 'History, theory and the narrative turn in IR', *Review of International Studies*, 32(4), pp. 703–714.

Shaw, I (2017) 'Robot wars: US Empire and geopolitics in the robotic age', *Security Dialogue*, 48(5), pp. 451–470.

Wendt, A (1992) 'Anarchy is what states make of it: the social construction of power politics', *International Organization*, 46(2), pp. 391–425.

Part I

The machine and the international system

2 Emerging technologies and the Chinese challenge to US innovation leadership*

James Johnson

Introduction

This chapter considers the intensity of US–China strategic competition playing out within a broad range of artificial intelligence (AI) and AI-enabling technologies (e.g. machine learning, 5G networks, autonomy and robotics, quantum computing, and big data analytics).[1] It describes how great power competition is mounting within several dual-use high-tech fields, why these innovations are considered by Washington to be strategically vital, and how (and to what end) the US is responding to the perceived challenge posed by China to its technological hegemony. The chapter uses the international relations (IR) concept of "polarity" (the nature and distribution of power within the international system) as a lens to view the shifting great power dynamics in AI-related strategic technology (e.g. microchips, semiconductors, big-data analytics, and 5G data transmission networks).[2]

The chapter argues that the strategic competition playing out within a broad range of dual-use AI and AI-enabling technologies will likely *narrow* the technological gap separating great military powers (notably the US and China) and, to a lesser extent, other technically advanced small-medium powers.[3] The chapter builds on the growing body of literature that reinforces the perception in the US that China's pursuit of AI technologies will threaten the first-mover advantage that the US has in a range of dual-use – and military-specific – AI applications (Boulanin, 2019; Johnson, 2019a; Horowitz, 2018; Moore, 2017; Hadley and Nathan, 2017; Allen and Chan, 2017). Because of this perceived threat, Washington will likely consider even incremental progress by China through a military lens and, thus, treat any progress as a national security

* This chapter is derived from an article published in *The Pacific Review*, October 2019, copyright held by the publishers Taylor & Francis, LLC, available online: https://doi.org/10.1080/09512748.2019.1676299

threat (Johnson, 2019c). Certain areas of Chinese progress will, however, likely be considered by the US as more strategically germane than others. For instance, sophisticated Chinese AI-enabled predictive algorithms that can reliably capture the diverse variables, uncertainties, and complex interactions of modern warfare – affording China a full-spectrum multi-domain command and control (C2) capability, hailed as one of the most potentially transformative AI-enabling capabilities – would likely be viewed by the US as a strategic threat (Johnson, 2020b).

What are the implications of US–China defense innovation for the strategic balance and stability and, in particular, for efforts by the US to sustain its first-mover advantages (i.e. locking in a period of military advantages as a result of early technological innovation) in advanced military technology? (Boulanin, 2019; Geist and Lohn, 2018; Ayoub and Payne, 2016; Technology for Global Security, 2019) Why does the US view China's progress in dual-use AI as a threat to its first-mover advantage? How might the US respond to this perceived threat? Will the increasingly competitive US–China relationship dominate world politics creating a new bipolar world order, as opposed to a multipolar one?[4] The chapter is an attempt to acquire greater insight into these questions, to better understand the shifting power dynamics and strategic competition in the development of AI-related and enabling technology, and the implications of these trends for strategic relations between great powers.

The chapter proceeds as follows. First, it summarizes the responses by US decision makers and analysts to the debate about US decline and the rise of the narrative of an imminent shift to a multipolar order.[5] This grand strategic overview will be contextualized with particular reference to the relative decline of the US vis-à-vis China, and the implications of the US being displaced as a global hegemon. It then sets up the debate over rapid advances and proliferation of AI-related technology capabilities, through an exploration of those that view harnessing these capabilities as a central aspect of efforts to maintain Washington's unipolar dominance. Next, it examines the perception of the rise of a bipolar order divided between Washington and Beijing. Again, using the lens of defense innovation, it analyses the credibility of the popular idea that the US has been caught off guard by China's accomplishments in the development of AI-related technologies and that, as a result, the US risks losing its first-mover advantage in the adoption of AI on the future battlefield. Finally, the chapter examines the nature of this particular "arms race" in the context of predictions of a shift toward a multipolar order (Rapoport, 1957, pp. 249–299; Glaser, 2004, pp. 44–84).[6]

This study uses a wide range of open-source Chinese language reports in combination with commercial and defense-centric think-tank and research group reports, to benchmark both China's AI approach and US perceptions of these developments – ostensibly through a national security and military lens (iiMedia, 2017; PwC, 2017; Barton and Woetzel, 2017; China State Council, 2017; Tencent Research Institute et al, 2017; China Academy of Engineering; Chinese Academy of Sciences, 2017; China Institute for Science and Technology Policy at Tsinghua University, 2018; China Economic Net; Council on Foreign Relations, 2017; CB Insights Research, 2018).[7]

The battle for the survival of American technological hegemony

In the post–Cold War era, a preoccupation of US policy makers and analysts has been the nature and implications of US unipolarity. This discourse has centered on two key questions: How long will unipolarity last? Also, is the pursuit of hegemony a viable or worthwhile strategic objective for the US to pursue? The preservation of the US liberal hegemonic role as a unipole has been the overarching grand strategic goal of every post–Cold War administration from George H. W. Bush to Barack Obama (Layne, 2012, p. 2). Having outlined the prominent strands and voices about how the US should (and are able to) respond to the notion of decline and the rise of a multipolar order, the analysis that follows uses IR as a lens to explore how the US is positioning itself vis-à-vis China in AI technology – preparing for bipolarity with China or reluctantly accepting multipolarity?

World leaders have been quick to recognize the transformative potential of AI as a critical component of national security (Work, 2015), in large part driven by the perceived challenges posed by rising revisionist and dissatisfied powers – especially China and Russia (US Department of Defense, 2017). The US Department of Defense (DoD) released, in 2016, a *National Artificial Intelligence Research and Development Strategic Plan* – one of a series of studies on AI machine learning – on the potential for AI to reinvigorate US military dominance (US Department of Defense, 2016). In the context of managing the potential flashpoints in the Taiwan Straits, the South China Seas, and Ukraine, then-US Secretary of Defense, Ashton Carter, stated that Russia and China are the US' "most stressing competitors" and continue to "advance military systems that seek to threaten our [US] advantages in specific areas" (including AI) and in "ways of war that seek to achieve their objectives rapidly, before, they hope, we [the US] can respond" (US Department of Defense, 2016).

In an effort to capitalize on the US comparative advantage in private sector innovation and to circumvent the burdensome military industrial-acquisition process, the DoD also established the Defense Innovation Unit Experimental (DIUx) to foster – albeit with mixed success – closer collaboration between the Pentagon and Silicon Valley (Kaplan, 2016). In a similar vein, the recent summary of the DoD's debut AI strategy, *Artificial Intelligence Strategy*, stated that "China and Russia are making significant investments in AI for military purposes" that "*threaten to erode* our [US] technological and operational advantages", and, in response, the US must "adopt [military-use] AI to *maintain its strategic position*, prevail on future battlefields, and safeguard this [i.e. US-led] order" (US Department of Defense, 2019).

China and Russia have both developed a range of military-use AI technologies as part of a broader strategic effort to simultaneously exploit perceived US military vulnerabilities and reduce their own vulnerabilities.[8] In a quest to become a "science and technology superpower" and catalyzed by AlphaGo's victory (or China's "Sputnik moment"),[9] Beijing launched a national-level AI innovation agenda for "civil-military fusion" – or US Defense Advanced Research Projects Agency (DARPA) with Chinese characteristics (State Council Information Office,

2017). Similarly, the Russian private sector has also benefited from state-directed support of human capital development and early investment in advanced technologies in a broader effort to substitute its continued dependence upon Western technology with indigenous technologies, and despite Russia's weak start-up culture. In short, national-level objectives and initiatives demonstrate recognition by great military powers of the potential military-technological transformative potential of AI for national security and for strategic stability between great military powers.[10]

US analysts and policy makers have suggested a range of possible responses to these emerging security threats to preserve US technological leadership, which harnesses US natural advantages to push back against the rising great military powers in the multipolar order (Hadley and Nathan, 2017; Work and Brimley, 2014; Gesit and Lohn, 2018; White House, 2019). First, the DoD should fund and lead AI-simulated war games and red-teaming creative thinking exercises to investigate existing and new security scenarios involving disruptive AI innovations. Second, the US needs to leverage its world-class think-tank community, academics, AI experts, computer scientists, and strategic thinkers to assess the implications of AI for a range of security scenarios and devise a long-term AI strategic agenda to meet these challenges.

Third, the US should prioritize DoD AI-based R&D to leverage the potential low-cost force multiplier advantages of AI technologies (i.e. autonomy and robotics), and to mitigate potential vulnerabilities and risks. Fourth, the US defense community should actively invest in and establish a commanding position in the nascent development of "counter-AI" capabilities – both offensive and defensive.

Fifth, the US national security policy-making community (e.g. DARPA, the US Intelligence Advanced Research Projects Activity (IARPA), Defense Innovation Board (DIB); the Office of Naval Research (ONR); and the National Science Foundation (NSF)) should seek increased funding for AI-related research to combat the competition for talent and information in AI and actively support university programs to ensure the US retains its relative talent pool advantages – in particular vis-à-vis China. Finally, the Pentagon should fund additional R&D in reliable fail-safe and safety technology for AI systems – especially military AI applications and tools.

America's new Sputnik moment?

As AI military applications have grown in scale, sophistication, and lethality, many in the US defense community have become increasingly alarmed about the implications of this trend for international competition and national security (Hadley and Nathan, 2017, p. 17).[11] In his opening comments at "The Dawn of AI" hearing, Senator Ted Cruz stated, "ceding leadership in developing artificial intelligence to China, Russia, and other foreign governments will not only place the US at a technological disadvantage, but it could have *grave implications for national security*" (Hadley and Nathan, 2017, p. 17). Similarly, Director of US National Intelligence, Daniel Coates, recently opined, "the implications of our adversaries' abilities to use AI are *potentially profound and broad*" (ibid, p. 17).

Given the anticipated national security value US strategic near-peer competitors (notably China and Russia) attach to a range of dual-use (i.e. military and civilian uses) AI-related technologies – notably autonomy and robotics, quantum communications, and 5G networks discussed below – several defense analysts have characterized the exorable pace and magnitude of AI technology as a "Sputnik moment". This could be a harbinger for a military revolution (or perceived as such), triggering a global AI arms race and changing the character (and even nature) of warfare. AI is, however, only one facet of a broader trend toward increasing the speed of modern – conventional and nuclear – war and shortening the decision-making time frame associated with advances in weapon systems, such as cyber-attacks, anti-satellite weapons, and hypersonic missile technology (Wilkening, 2019; Johnson, 2019b; Acton, 2013). These trends could lead to arms race instability between great military powers (especially China, Russia, and the US), as rival states modernize their capabilities to reduce their perceived vulnerabilities (Schelling and Halperin, 1975).

Evidence of exponentially accelerated military-technological competition – in research, adoption, and deployment – of AI-related subset technologies (i.e. 5G networks, IoTs, robotics and autonomy, additive manufacturing, and quantum computing) does not *necessarily* mean an "arms race" is taking place. Rather, framing great power competition (especially US–China) in this way risks the adoption of operational concepts and doctrine that increases the likelihood of arms racing spirals and warfare (Roff, 2019, pp. 1–5). According to the DoD's newly established Joint Artificial Intelligence Center (JAIC) head, Lt. General Jack Shanahan, "its strategic competition, *not an arms race*. They're [China] going to keep doing what we're doing; we [the US] acknowledge that". Shanahan added: "What I don't want to see is a future where our potential adversaries [China] have a fully AI-enabled force and we [the US] do not" (US Department of Defense, 2019).

In response to a growing sense of alacrity within the US defense community, the Pentagon has authored several AI-related programs and initiatives designed to protect US superiority on the future digitized battlefield (e.g. the Third Offset, Project Maven, DARPA's "AI Next Campaign", the establishment of JAIC, the Joint Common Foundation JCF, and the DoD's "AI Strategy"). Taken together, these initiatives demonstrate the perceived gravity of the threat posed to US national security from near-peer states' (especially China and Russia) pursuit of AI-related capabilities – notably autonomy and robotic systems, AI-augmented cyber offense, predictive decision-making tools, and enhancements to intelligence, surveillance, and reconnaissance capabilities (ISR) – to enhance their military power asymmetrically vis-à-vis the US (Johnson, 2020a; Johnson, 2020b). For example, in response to Chinese strategic interest in AI technologies, DIUx proposed greater scrutiny and restrictions on Chinese investment in Silicon Valley companies (Simonite, 2017). This behavior typifies a broader concern that synergies created by China's civil-military fusion strategy could allow the technology, expertise, and intellectual property shared between American and Chinese commercial entities to be transferred to the PLA (Bartholomew and Shea, 2017, p. 507).

Moreover, broader US national security concerns relating to Chinese efforts to catch up (and even surpass) the US in several critical AI-related enabling technologies has prompted Washington to take increasingly wide-ranging and draconian steps to counter this *perceived* national security threat.[12] Against the backdrop of deteriorating US–China relations, responses such as these could accelerate the decoupling of cooperative bilateral ties between these two poles. This could increase the likelihood of strategic competition, mutual mistrust, and negative action–reaction dynamics known as a security dilemma – which continue to manifest in other military-technologies, including missile defense, hypersonic weapons, and counter-space capabilities (Jervis, 1976, chap. 3; Khoo & Steff, 2017; Johnson, 2017, pp. 271–288).

Washington's alarmist tone, and policy responses, to the perceived threat posed by China's bid for technological leadership reveals the following: When we compare the public narratives surrounding the "new multipolarity" thesis with what is happening, two things emerge (Zala, 2017, pp. 2–17). First, the nature of the emerging great power competition in AI suggests that a shift to Sino-American bipolarity (rather than multipolarity) is more likely in the short-medium term. Second, even in the event China surpasses the US in AI (that many experts consider a strong possibility), it still trails the US in several qualitative measures that (for now at least) coalesce to preserve Washington's lead in the development of AI-related technologies (Lee, 2018). The US has the world's largest intelligence and R&D budgets, world-leading technology brands, academic research and innovation (discussed later in the chapter), and the most advanced (offense and defensive) cyber capabilities. Whether these advantages will be enough for Washington to forestall a shift in the military balance of power in the event China catches up or leap-frogs the US in AI – either through mimicry, espionage, or indigenous innovation – and can convert these gains (at a lower cost and less effort than the US) into potentially game-changing national security capabilities is, however, an open question.

China is by some margin Washington's closest peer-competitor in AI-related technology. Beijing's 2017 *Next Generation AI Development Plan* identified AI as a core "strategic technology" and new focus of "international competition". China's official goal is to "seize the strategic initiative" (especially vis-à-vis the US) and achieve "world-leading levels" of AI investment by 2030 – targeting more than US$150 billion in government investment (The State Council Information Office, 2017).[13] Beijing has leveraged lower barriers of entry (discussed later in this chapter) to collect, process, and disseminate data within China to assemble a vast database to train AI systems.

According to a recent industry report, China is on track to possess 20 percent of the world's share of data by the end of 2020, and the potential to have over 30 percent by 2030 (Knight, 2017). These efforts could be enhanced by the synergy and diffusion of a range of disruptive technologies such as machine learning, quantum technology, 5G networks, and electromagnetics. In addition to the availability of vast datasets, comparing the AI-capabilities of US and China also incorporates wider qualitative and quantitative measures such as hardware, high-quality machine-learning algorithms, private–public sector collaboration, and broader technological and scientific initiatives and policies (Ding, 2018).

State-directed Chinese investment in the US AI market has also become increasingly active and, in several instances, Chinese investment has competed directly with the DoD (Kania, 2017). In 2017, for example, a Chinese state-run company Haiyin Capital outmaneuvered the US Air Force's efforts to acquire AI software developed by Neurala in 2017 (Mozur and Perlez, 2017). Incidences such as these are indicative of a broader set of US concerns related to China's willingness (or propensity) to resort to industrial espionage and other means (i.e. forcing foreign partners of China-based joint ventures to divulge their technology) to gain access to US AI technology, in an effort to catch up with, and leap-frog, the US in a range of strategically critical dual-use technologies (e.g. semiconductors, robotics, 5G networks, cyberspace, the internet of things [IoT], big data analytics, and quantum communications).[14] Industrial espionage can, however, only take the Chinese so far. The development of China's national innovation base, expertise, and capacity – even if that foundation builds on industrial and other types of espionage and mimicry – is a broader trend that the DoD also appears to be cognizant of (US Department of Defense, 2019b, p. 96).

Among these critical enabling technologies that could fundamentally change the future of warfare are next-generation data transmission networks. The strategic importance of 5G networks as a critical future military technological enabler was demonstrated during the protracted and ongoing tensions between China's Huawei and Washington. Experts view 5G as a cornerstone technology to increase the speed and stability of data-loads, reduce the latency (i.e. accelerate network response times), and enhance mobile digital communications. According to an AI and telecommunications researcher at the University of Electronic Science and Technology of China, "the 5G network and the internet of things (IoT) enlarge and deepen the cognition of situations in the battlefield by several orders of magnitude and *produce gigantic amounts of data, requiring AI to analyze and even issue commands*" (Liu, 2019).

Against the backdrop of rising tensions in the Sino-American relationship across a plethora of interconnected policy arenas (i.e. trade and geopolitical influence in the Asia-Pacific), the technological race for the access and control of critical enablers that will connect sensors, robotics, autonomous weapons systems, and the exchange of vast volumes of data in real time through AI-machine-learning techniques on the digitized battlefield, will become increasingly intense and strategically motivated (Kania and Costello, 2018, p. 5).

In 2017, Chinese President Xi Jinping explicitly called for the acceleration of the military "intelligentization" agenda, to better prepare China for future warfare against a near-peer adversary, namely the US (Xinhua, 2017). Although Chinese think-tanks and academic discourse are generally poor at disseminating their debates and content, open-source evidence suggests a strong link between China's political agenda related to the "digital revolution", Chinese sovereignty, and national security, and the current public debate surrounding the rejuvenation of the Chinese nation as a great power (Xinhua, 2017). In short, national security is ultimately interpreted by China (and the US) as encompassing economic performance.

President Xi's Belt-and-Road-Initiative (BRI) and its virtual dimension, the "Digital Silk Road", are high-level efforts designed to ensure that the mechanisms, coordination, and state-level support for this agenda will become increasingly normalized (Yuan, 2017). Chinese President Xi Jinping recently stated that AI, "big data", cloud storage, cyberspace, and quantum communications, were amongst the "liveliest and most promising areas for civil-military fusion" (Chin, 2018). Toward this end, Xi has pledged additional state support and resources to enhance China's economic and military dimensions of its national power (Li, 2015; Lee and Sheehan, 2018).[15] While BRI investment is predominantly in emerging markets with comparably low levels of technology maturity, human capital, and military power, the BRI framework supports a broader Chinese agenda to expand (or establish a new) geopolitical sphere of influence to improve its position in the future distribution of power – especially vis-à-vis the US. In the case of quantum technology, the potential convergence between AI and quantum computing could create promising synergies that Beijing intends to leverage to ensure it is at the forefront of the so-called "quantum AI revolution". In 2015, for example, Chinese researchers achieved an apparent scientific breakthrough in the development of a quantum machine-learning algorithm, which may remove several technical bottlenecks (e.g. quantum radar, sensing, imaging, metrology, and navigation) allowing greater independence from space-based systems – where currently China lags the US – enhancing ISR capabilities; potentially creating new vulnerabilities in US space-based GPS and stealth technology in future conflict scenarios – especially in the undersea domain (Kania and Costello, 2018, p. 18). Moreover, Chinese analysts and strategists anticipate that quantum technologies will radically transform future warfare, with a strategic significance equal to nuclear weapons (Li, 2016).

The evidence suggests a strong link between Beijing's pursuit of AI leadership and its broader geopolitical objectives. This link has, in turn, reinforced the narrative within the US defense community that China believes this technological transformation is an opportunity to strengthen its claim on the leadership – and eventual dominance – of the emerging technological revolution, having missed out on previous waves (Godement, 2018, pp. 1–5). In sum, despite the clear economic issues at stake (i.e. the rents to be captured in the data-driven economy), the threat to US technological hegemony is generally interpreted through a military and geopolitical lens.

By contrast, the increasingly strained relationship between the Trump administration and Silicon Valley will likely pose additional challenges to this critical partnership in the development of AI technologies for the US military. Following a recent high-profile backlash from employees at Google, for example, the company recently announced that it would discontinue its work with the Pentagon on Project Maven (White, 2018). Several defense analysts and US government reports have noted the growing gap between the rhetoric and the research momentum (especially in AI and robotics) – and the paucity of resources available, to make the US military more networked and integrated (Harris, 2018).

Specifically, these reports highlight various shortcomings in the US defense innovation ecosystem, such as inadequate funding to sustain long-term R&D, institutional stove-piping that inhibits multidisciplinary collaboration, and an insufficient talent pool to attract and retain top scientists in AI-related fields (US Department of Defense, 2017). In its debut AI Strategy, the DoD committed to "consult with leaders from across academia, private industry, and the international community" and "invest in the research and development of AI systems" (US Department of Defense, 2019a, p. 5). Details of the implementation and funding arrangements for these broad principles remain mostly absent, however. Moreover, the apparent mismatch (even dissonance) between the rapid pace of commercial innovation in AI technologies and the lagging timescales and assumptions that underpin the US DoD's existing procurement processes and practices could exacerbate these bilateral competitive pressures (Kennedy and Lim, 2016, pp. 553–572).

China's pursuit of AI-related (especially dual-use) technologies will fuel the perception (accurate or otherwise) in Washington that Beijing is *intent* on exploiting this strategically critical technology to fulfill its broader revisionist goals (Wohlforth, 2011, p. 37).[16] That is, once the "digital Silk Road" initiative reaches fruition, BRI could enable China's 5G, AI, and precision navigation systems to monitor and dominate the IoT, digital communications, and intelligence of every nation within the BRI sphere of influence as part of Beijing's strategic objective to ensure the leadership of a new international order or a separate China-led bounded order (Mearsheimer, 2001); China's version of the Greater East Asia Co-Prosperity sphere, or Halford Mackinder and Mahan's theories of world control (Beasley, 1991).

In addition to this unique scaling advantage, China's defense AI innovation has also benefited from its approach to AI acquisition: A centralized management system where few barriers exist between commercial, academic, and national security decision-making. While many analysts consider China's centralized approach to the development of AI affords it with unique advantages over the US, others posit that Beijing's AI strategy is far from perfect. Some analysts, for example, have characterized Beijing's funding management as inherently inefficient. These analysts note that China's state apparatus is inherently corrupt and that this approach tends to encourage overinvestment in particular projects favored by Beijing, which may exceed market demand (He, 2017). Moreover, though China has already surpassed the US in the quantity of AI-related research papers produced between 2017 and 2018, the quality of these papers ranks far below US academic institutions (Castro et al, 2019).

Besides, China is currently experiencing a shortage of experienced engineers and world-class researchers to develop AI algorithms; for instance, China has only 30 Chinese universities that produce indigenous experts and research products. As a corollary, industry experts have cautioned that Beijing's aggressive and centralized pursuit of AI could result in poorly conceptualized AI applications that adversely affect the safety of AI-enabled military applications, which increase the potential systemic risks associated with these innovations (Barton and Woetzel, 2017). The comparatively measured pace of US military AI innovation might, therefore, in the longer term result in more capable tools, but without sacrificing safety for speed – even at the cost of falling behind China's AI quantitative lead in the short term. The prioritization by the US of the development of robust, verifiable, and safe military AI technology might, however, come under intense pressure if China's progress in dual-use AI is perceived as an imminent threat to the US first-mover advantages.

A bipolar AI arms race?

As the most powerful nation-states and leaders in the development of AI, the competitive tensions between China and the US have often evoked comparisons with the Cold War–era US–Soviet space race. In response to the global AI arms race, and to sustain US superiority and first-mover advantages in AI, US General John Allen and Spark Cognition CEO Amir Husain have argued that the US must push further and faster to avoid losing the lead to China (and, to a lesser degree, Russia) in the development of AI (Allen and Husain, 2017).

While these depictions accurately reflect the *nature* of the increasingly intense competition in the development of AI-related technologies, the *character* of this particular arms race intimates a more multipolar reality compared to the Cold War–era bipolar space race. Over time, this trend will likely elevate technically advanced small and middle powers (e.g. South Korea, Singapore, Israel, France, and Australia) to become pioneers in cutting-edge dual-use AI-related technology and key influencers shaping the security, economics, global norms, and standards of these innovations in the future world order. While this broader trend will not necessarily run in lockstep with the US–China bipolar contest, it will inevitably be influenced and shaped by the ongoing competition between China and the US (and its allies) on setting global technological standards for AI, nonetheless. For these middle powers, the outcome of this contest and, particularly, how they position themselves on AI technology standards will have a significant impact on their ability to truly become cutting-edge AI innovators – independent of China and the US.

The commercial driving forces underlying AI technology (i.e. hardware, software, and R&D), together with the inherently dual-use nature of AI-related innovations, reduce the usefulness of the space race analogy (Organski and Kugler, 1980; Gilpin, 1981). In short, the particular problem set associated with the Cold War–era bipolar structure of power (i.e. a pernicious obsession with the other sides' military capabilities) is, to date at least, far less intense in the context of contemporary competition in AI. Where primarily commercial forces drive military innovation and, in particular, when imitation is quicker and cheaper than innovation, technology tends to mature and diffuse at a faster pace compared to military-specific applications, such as stealth technology (Horowitz, 2010; Gilli and Gilli, 2019, pp. 141–189). Second-mover catchup possibilities in military-use

AI through imitation are in an adversarial context, therefore, will unlikely be a feasible avenue for states.

Despite the growing sense that the proliferation of AI technologies driven by powerful commercial forces will inevitably accompany (and even accelerate) the shift toward multipolarity, important caveats need to accompany prognostications about the pace and nature of this transformation: The risks associated with the proliferation and diffusion of dual-use AI technologies across multiple sectors and expanding knowledge bases is a very different prospect compared to arms-racing between great power military rivals. Thus, the development of military AI applications based on military-centric R&D would make it much more difficult and costly for smaller (and especially less technically advanced) states to successfully emulate and assimilate (Brooks, 2006; Cavarely, 2007, pp. 598–614).

Moreover, military organizations, norms, and strategic cultural interests and traditions will also affect how AI systems are integrated into militaries, potentially altering the balance of military power (Johnstone, 1995, pp. 65–93). In short, military technologies in isolation are unlikely to alter how militaries prepare for warfare, deter and coerce their adversaries, and fight wars. Instead, the interplay of technology and military power will continue to be a complex outcome of human cognition, institutions, strategic cultures, judgment, and politics (Biddle, 2006). Ultimately, success in sustaining or capturing the first-mover advantages in AI will be determined by how militaries develop doctrine and strategy to seize on the potential comparative benefits afforded by AI-augmented capabilities on the battlefield (Johnson, 2020a).

The pace of military-use AI diffusion to smaller-medium powers (and non-state actors) might also be constrained by three pivotal features of this emerging phenomenon: (1) Hardware constraints (e.g. physical processors) and integrating increasingly complex software and hardware with internal correctness; (2) the algorithmic complexity inherent to AI machine-learning approaches; and (3) the resources and know-how to effectively deploy AI code (Ayoub and Payne, 2016, p. 809).[17]

As a corollary, the advantages China derives from its commercial lead in the adoption of AI and dataset ecosystem will not necessarily be easily directly translated into special-purpose military AI applications (Castro et al, 2019). China's strengths in commercial-use AI (e.g. 5G networks, e-commerce, e-finance, facial recognition, and various consumer and mobile payment applications) will, therefore, need to be combined with specialized R&D and dedicated hardware to unlock their potential dual-use military applications and augment advanced weapon systems. Absent the requisite level of resources, know-how, datasets, and technological infrastructure, therefore, these constraints could make it very difficult for a new entrant to develop and deploy modular AI with the same speed, power, and force as the US or China (Gray, 2015, pp. 1–6).

For example, China and the US are in close competition to develop the supercomputers needed to collect, process, and disseminate the vast amounts of data that traditional computers can handle. While the US possesses more powerful systems, China trumps the US in terms of the number of supercomputers. Thus, military-led innovations could potentially concentrate and consolidate leadership in this nascent field amongst current military superpowers (i.e. China, the US, and, to a lesser extent, Russia), and revive the prospect of bipolar competition (Bostrom, 2014). For now, it remains unclear how specific AI applications might influence military power, or whether, and in what form these innovations will translate into operational concepts and doctrine (Cummings, 2017).

In sum, the degree to which AI alters the military balance of power will depend in large part on the speed of the diffusion of this technology within the military structures of the US and China; as a function of human innovation, political agendas, and strategic calculation and judgment, against the backdrop of a multipolar world (and nuclear) order, and heuristic decision-making (or the propensity for compensatory cognitive shortcuts) associated with decisions taken under compressed timeframes in uncertain and complex environments (Beverchen, 2007, pp. 45–56).

Conclusion

This chapter has made the following central arguments. First, while disagreement exists on the likely pace, trajectory, and scope of AI defense innovations, a consensus is building within the US defense community intimating that the potential impact of AI-related technology on the future distribution of power and the military balance will likely be transformational, if not revolutionary. These assessments have, in large part, been framed in the context of the perceived challenges posed by revisionist and dissatisfied great military powers (China and Russia) to the current US-led international order – rules, norms, governing institutions – and military-technological hegemony.

Second, the rapid proliferation of AI-related military-technology exists concomitant with a growing sense that the US has dropped the ball in the development of these disruptive technologies. If the perception that America's first-mover advantage in a range of dual-use enabling strategic technologies (i.e. semiconductors, 5G networks, and IoT) from rising (especially nuclear-armed) military powers such as China becomes prevalent, the implications for international security and strategic stability could be severe. In response to a growing sense of urgency within the US defense community cognizant of this prospect, the Pentagon has authored several AI-related programs and initiatives designed to protect US dominance on the future digitized battlefield. Further, broader US national security concerns relating to Chinese efforts to catch up – and even surpass – the US in several critical AI-related enabling technologies has prompted Washington to take increasingly wide-ranging and draconian steps to counter this perceived national security threat.

Third, and related to the previous finding, the development of AI evocations of the Cold War-era space race as an analogy does not accurately capture the nature of the evolving global AI phenomena. Instead, compared to the bipolar features of the US–Soviet struggle, this innovation arms race intimates more multipolar characteristics. Above all, the dual-use and commercial drivers of the advances in AI-related technology will likely narrow the technological gap separating great military powers (chiefly the US and China) and other technically advanced small-medium powers. These rising powers will become critical influencers in shaping future security, economics, and global norms in dual-use AI.

In the case of military-use AI applications, however, several coalescing features of this emerging phenomena (i.e. hardware constraints, machine-learning algorithmic complexity, and the resources and know-how to deploy military-centric AI code), will likely constrain the proliferation and diffusion of AI with militaries' advanced weapon systems for the foreseeable future. In turn, these constraints could further concentrate and consolidate the leadership in the development of these critical technological enablers amongst the current AI military superpowers (i.e. China and the US), which could cement a bipolar balance of power and the prospect of resurgent bipolar strategic competition.

Today, the US has an unassailable first-mover advantage in a range of AI applications with direct (and in some cases singular) relevance in a military context. However, as China approaches parity, and possibly surpasses the US in several AI-related (and dual-use) domains, the US will increasingly view future technological incremental progress in emerging technologies – and especially unexpected technological breakthroughs or surprises – through a national security lens. Thus, responses to these perceived threats will be shaped and informed by broader US–China geopolitical tensions (Waltz, 1979).

These concerns resonated in the 2018 US Nuclear Posture Review (NPR). The NPR emphasized that geopolitical tensions and emerging technology in the nuclear domain could coalesce with unanticipated technological breakthroughs in "new and existing innovations to change the nature of the threats faced by the US and the capabilities needed to counter them". (NPR, 2018, p. 14). In sum, against the backdrop of US–China geopolitical tensions, and irrespective of whether China's dual-use applications can be imminently converted into deployable military-use AI, US *perceptions* of this possibility will be enough to justify severe countermeasures.

Notes

1 Recent progress in AI falls within two distinct fields: (1) "Narrow" AI and, particularly, machine learning; (2) "general" AI, which refers to AI with the scale and fluidity akin to the human brain. Most AI researchers anticipate that "general" AI to be at least several decades away.

2 The line between core AI and "AI-related" technology is a blurred one. For the purposes of this study, core AI technology includes: machine-learning (and deep-learning and deep networks sub-set), modeling, automated language and image recognition, voice assistants, and analysis support systems; whereas "AI-related" (and AI-enabling) technology includes: autonomous vehicles, big data analytics, 5G networks, supercomputers, smart vehicles, smart wearable devices, robotics, and the "internet of things", to name a few.

3 There is an important difference, however, between narrowing the gap in certain fields, and gaining an overall lead across all the categories (i.e. talent, research, development, hardware, data, and adoption) in the emerging AI race.

4 Polarity analysis focuses on whether the interstate order is dominated by one (a unipolar order), two (a bipolar order) or three or more (a multipolar order) centers of power. "Multipolarity" in this context implies that no single state is unambiguously in the lead (or polar) in the international order. See, Wohlforth (2011).

5 This includes what might be thought of as "denialists" (those that argue that unipolarity is, in fact, durable and that serious US decline is a myth), "accepters" (those advocating for retrenchment or strategies of "offshore balancing" to navigate the inevitable "rise of the rest"), and "resisters" (those concerned about the rise of peer competitors but who believe that Washington can still see down the challenge and maintain its hegemonic position).

6 The chapter conceptualizes "multipolarity" not simply as an objective description of how capabilities are distributed between great powers, but rather as a multifaceted concept that focuses on great power status, strategic narratives, and, in particular, perceptions.

7 Currently, there are no standardized classifications or metrics for either the commercial or military development of AI technology.

8 For example, Russian focus on AI-enabled land-based infantry robotic systems are considered by analysts to be based on the self-perception that its armored and heavy infantry brigades and divisions are conventionally weaker compared to NATO's.

9 This refers to the shock experienced by Washington when the Soviet Union beat the US into space by launching the world's first satellite, Sputnik 1, in 1957; China experienced a similar shock when Google's DeepMind AlphaGo AI defeated the world's number one Go player, Ke Jie. Sputnik shocked the US into making massive investments in science and technology to compete technologically with the Soviet Union; China responded to AlphaGo by rapidly increasing investment in the field of AI and AI-enabling technologies.

10 There is not space to explore this contest in full in this chapter. For a discussion on the impact of AI on the US–China competitive dyad, see chapter 4. For AI's impact on strategic stability between great powers see, James S. Johnson, "Artificial Intelligence: A Threat to Strategic Stability," *Strategic Studies Quarterly*, Vol. 14, Issue 1 (2020), pp. 16–39; and James Johnson, "Delegating Strategic Decision-Making to Machines: Dr. Strangelove Redux?" *Journal of Strategic Studies* (2020) DOI: 10.1080/01402390.2020.1759038

11 For details on recent developments in the military-use AI – both kinetic and non-kinetic in nature – see, James S. Johnson, "Artificial Intelligence: A Threat to Strategic Stability," *Strategic Studies Quarterly*, Vol. 14, Issue 1, pp.16–39; and James S. Johnson, "The AI-cyber Nexus: Implications for Military Escalation, Deterrence and Strategic Stability," *Journal of Cyber Policy*, Issue 4, Vol. 3, pp. 442–460.

12 For example, Washington's ongoing efforts to stifle China's progress in 5G networks and semi-conductors in the apparent defense of national security.

13 Recent research suggests that China's AI investment targets are falling short of expectations. Some analysts predict the gulf between China and the US will continue to expand, with the US expected to reach a 70% market share in venture capital AI investment within the next few years. See Petrara, 2019.

14 US concerns in this regard relate to a widespread belief that many Chinese companies can readily be recruited (or "forced") by Beijing to transfer technology to the Chinese military to bolster its capabilities; they are adjuncts of the Chinese state. Kinling Lo, "China says US Claims it Uses Forced Technology Transfer to Boost Military are 'Absurd'", *South China Morning Post*, January 16, 2019.

15 The economic gains that China may make through commercial applications such as BRI are not dependent upon dual-use technology or geopolitics alone; gains are also based on geo-economics.

16 A distinction exists between the erosion of US advantages in ancillary applications based on dual-use AI technologies and in military-specific AI applications. Where the US retains an unassailable edge in military capacity and innovation, the actual "threat"

posed to the US in the military-technological sphere is less immediate than in general-use AI. This implies the "threat" narrative is more centered on perceptions of Beijing's future intentions as its military-use AI matures.

17 In Chapter 3 of this volume, Reuben Steff provides a fuller examination of the implications of AI for small states.

References

Acton, JM (2013) *Silver bullet? Asking the right questions about conventional prompt global strike*. Washington, DC: Carnegie Endowment for International Peace.

Advancing Quantum Information Science: National Challenges and Opportunities (2017) *National science and technology council*. Washington, DC: US Department of Defense. July 22.

Allen, G and Chan, T (2017) *Artificial intelligence and national security*. Cambridge, MA: Belfer Centre for Science and International Affairs.

Allen, JR and Husain, A (2017) 'The next space race is artificial intelligence', *Foreign Policy*. November 3. https://foreignpolicy.com/2017/11/03/the-next-space-race-is-artificial-intelligence-and-america-is-losing-to-china/, accessed November 10, 2018.

Ayoub, K and Payne, K (2016) 'Strategy in the age of artificial intelligence', *Journal of Strategic Studies*, 39(5–6), 793–819. doi:10.1080/01402390.2015.1088838

Bartholomew, C and Shea, D (2017) *US–China economic and security review commission – 2017 annual report*. Washington, DC: The US–China Economic and Security Review Commission.

Barton, D and Woetzel, J (2017) *Artificial intelligence: implications for China*. New York: McKinsey Global Institute.

Beasley, WG (1991) *Japanese imperialism 1894–1945*. Oxford: Clarendon Press.

Beverchen, A (2007) 'Clausewitz and the non-linear nature of war: systems of organized complexity', in Strachan H and Herberg-Rothe, A (eds) *Clausewitz in the twenty-first century*. Oxford: Oxford University Press, pp. 45–56.

Biddle, S (2006) *Military power: explaining victory and defeat in modern battle*. Princeton, NJ: Princeton University Press.

Bostrom, N (2014) *Superintelligence: paths, dangers, strategies*. Oxford: Oxford University Press.

Boulanin, V (ed.) (2019) *The impact of artificial intelligence on strategic stability and nuclear risk Vol. I Euro-Atlantic perspectives*. Stockholm: SIPRI Publications.

Brooks, SG (2006) *Producing security: multinational corporations, globalization, and the changing calculus of conflict*. Princeton, NJ: Princeton University Press.

Castro, D, McLaughlin, M and Chivot, E (2019) *Who is winning the AI race: China, the EU, or the United States?* Washington, DC: Center for Data Innovation.

Caverley, JD (2007) 'United States hegemony and the new economics of defense', *Security Studies*, 16(4), 598–614. doi:10.1080/09636410701740825

CB Insights Research (2018) *Advanced search: industry & geography, company attributes, financing & exit*. Retrieved from https://app.cbinsights.com

Chin, J (2018) 'China looks to close technology gap with US April 21', *Wall Street Journal*. https://www.wsj.com/articles/china-looks-to-close-technology-gap-with-u-s-1524316953, accessed April 10, 2020.

China AI Development Report (2018) *China institute for science and technology policy at Tsinghua University*. http://www.sppm.tsinghua.edu.cn/eWebEditor/UploadFile/China_AI_development_report_2018.pdf, accessed April 6, 2020.

China State Council (2017) *State council notice on the new generation artificial intelligence development plan*. http://www.gov.cn/zhengce/content/2017-07/20/content_5211996.htm., accessed April 1, 2019.

Chinese Academy of Sciences (2017) *China's AI business ready to lead the world*. http://english.cas.cn/newsroom/news/201706/t20170602_177674.shtml, accessed April 8, 2019.

Council on Foreign Relations (2017) *Beijing's AI strategy: old-school central planning with a futuristic twist*. https://www.cfr.org/blog/beijings-ai- strategy-old-school-central-planningfuturistic-twist, accessed April 1, 2019.

Cummings, ML (2017) *Artificial intelligence and the future of warfare*. London: Chatham House.

Ding, J (2018) *Deciphering China's AI dream*. Oxford: Future of Humanity Institute, University of Oxford.

Divine, RA (1993) *The sputnik challenge*. New York: Oxford University Press.

Drezner, D (2001) 'State structure, technological leadership and the maintenance of hegemony', *Review of International Studies*, 27(1), 003–025. doi:10.1017/S0260210501000031

Geist, AE and Lohn, J (2018) *How might artificial intelligence affect the risk of nuclear war?* Santa Monica, CA: RAND Corporation.

Gholz, E (2007) 'Globalization, systems integration, and the future of great power war', *Security Studies*, 16(4), 615–636. doi:10.1080/ 09636410701740908

Gilli, A and Gilli, M (2019) 'Why China has not caught up yet', *International Security*, 43(3), 141–189. doi:10.1162/isec_a_00337

Gilpin, R (1975) *US power and the multinational corporation: the political economy of foreign direct investment*. New York: Basic Books.

Gilpin, R (1981) *War and change in world politics*. New York: Cambridge University Press.

Glaser, CL (2004) 'When are arms races dangerous? Rational versus suboptimal arming', *International Security*, 28(4), 44–84. doi:10.1162/0162288041588313

Godement, F (2018) *The China dream goes digital: technology in the age of Xi*. Paris: European Council on Foreign Affairs ECFC, pp. 1–5.

Gray, E (2015) 'Small big data: using multiple datasets to explore unfolding social and economic change', *Big Data & Society*, 2(1), 1–6. doi:10.1177/2053951715589418

Grenoble, R (2017) 'Welcome to the surveillance states: China's AI camera's see all', *Huffpost*. December 12. https://www.huffpost.com/entry/china-surveillance-camera-big-brother_n_5a2ff4dfe4b01598ac484acc, accessed April 4, 2019.

Hadley, D and Nathan, L (2017) *Artificial intelligence and national security*. Washington, DC: Congressional Research Service.

Harris, AHB Jr (2018) 'The integrated joint force: a lethal solution for ensuring military pre-eminence', *Strategy Bridge*. March 2. https://thestrategybridge.org/the-bridge/2018/3/2/the-integrated-joint-force-a-lethal-solution-for-ensuring-military-preeminence, accessed May 4, 2019.

He, Y (2017) *How China is preparing for an AI-powered future*. Washington, DC: Wilson Center.

Horowitz, MC (2010) *The diffusion of military power: causes and consequences for international politics*. Princeton, NJ: Princeton University Press.

Horowitz, MC (2018) 'Artificial intelligence, international competition, and the balance of power', *Texas National Security Review*, 1(3), 37–57.

IIMedia (2017) *China artificial intelligence industry special research report*. April 6. http://www.sohu.com/a/132360429_468646, accessed April 1, 2020.

Ikenberry, J (ed.) (2011) *International relations theory and the consequences of unipolarity*. New York: Cambridge University Press.

Jervis, R (1976) *Perception and misperception in international politics*. Princeton, NJ: Princeton University Press.

Johnson, J (2017) 'Washington's perceptions and misperceptions of Beijing's Anti- Access Area-Denial (A2-AD) "Strategy": implications for military ecalation control and strategic stability', *The Pacific Review*, 30(3), 271–288. doi:10.1080/09512748. 2016.1239129

Johnson, J (2019a) 'Artificial intelligence & future warfare: implications for international security', *Defense & Security Analysis*, 35(2), 147–169. doi:10.1080/14751798.2019. 1600800

Johnson, J (2019b) 'The AI-cyber nexus: implications for military escalation, deterrence and strategic stability', *Journal of Cyber Policy*, 4(3), 442–460. doi:10.1080/2373887 1.2019.1701693

Johnson, J (2019c) 'The end of military-techno Pax Americana? Washington's strategic responses to Chinese AI-enabled military technology', *The Pacific Review*. doi:10.1080/09512748.2019.1676299

Johnson, J (2020a) 'Artificial intelligence: a threat to strategic stability', *Strategic Studies Quarterly*, 14(1), 16–39. doi:10.2307/26891882

Johnson, J (2020b) 'Delegating strategic decision-making to machines: Dr. Strangelove redux?', *Journal of Strategic Studies*. doi:10.1080/01402390.2020.1759038

Johnston, AI (1995) 'Thinking about strategy culture', *International Security*, 19(4), 32–64. doi:10.2307/2539119

Kania, E (2017) Battlefield singularity: artificial intelligence, military revolution, and China's future military power. Washington, DC: Centre for a New American Security.

Kania, EB and Costello, JK (2018) *Quantum hegemony? China's ambitions and the challenge to US innovation leadership*. Washington, DC: Center for a New American Security.

Kaplan, F (2016) 'The Pentagon's innovation experiment', *MIT's Technology Review*. December 16.

Kennedy, A (2016) 'Slouching tiger, Roaring Dagon: comparing India and China as late innovators', *Review of International Political Economy*, 23(1), 65–28. doi:10. 1080/09692290.2015.1105845

Kennedy, A and Lim, D (2017) 'The innovation imperative: technology and US–China rivalry in the twenty-first century', *International Affairs*, 94(3), 553–572.

Khoo, N and Steff, R (2017) *Security at a price: the international politics of US ballistic missile defense*. Weapons of Mass Destruction Series. Rowman & Littlefield, Inc.

Kier, E (1995) 'Culture and military doctrine: France between the wars', *International Security*, 19(4), 65–93. doi:10.2307/2539120

Knight, W (2017) 'China's AI awakening', *MIT's Technology Review*. October 10. https ://www.technologyreview.com/2017/10/10/148284/chinas-ai-awakening/, accessed October 1, 2019.

Layne, C (2012) 'This time it's real: the end of unipolarity and the Pax Americana', *International Studies Quarterly*, 56(1), 203–213. doi:10.1111/j.1468-2478.2011. 00704.x

Lee, K-F (2018) *AI superpowers: China, silicon valley, and the new world order*. New York: Houghton Mifflin Harcourt.

Lee, K-F and Sheehan, M (2018) *China's rise in artificial intelligence: ingredients and economic implications*. Hoover Institution. https://www.hoover. org/research/ch inas-rise-artificial-intelligenceingredients-and-economic-implications, accessed April 5, 2019.

Li, J (2017) *China-US AI venture capital state and trends research report. IT Juzi and tencent institute (full report in Chinese)*. http://voice.itjuzi.com/?p=16960, accessed October 1, 2018.

Li, GS (2016) 'The strategic support force is a key to winning throughout the course of operations', *People's Daily Online*. January 5. http://military.people.com.cn/n1/2016/0105/c1011-28011251.html, accessed January 5, 2020.

Li, R (2015) 'China brain' project seeks military funding as baidu makes artificial intelligence plans', *South China Morning Post*. March 3. https://www.scmp.com/lifestyle/article/1728422/china-brain-project-seeks-military-funding-baidu-makes-artificial, accessed February 1, 2018.

Liu, Z (2019) 'Why 5G, a battleground for US and China, is also a fight for military supremacy', *South China Morning Post*. January 31. https://www.scmp.com/news/china/military/article/2184493/why-5g-battleground-us-and-china-also-fight-military-supremacy, accessed October 10, 2019.

Lo, K (2019) 'China says US claims it uses forced technology transfer to boost military are "absurd"', *South China Morning Post*. January 16. https://www.scmp.com/news/china/military/article/2182402/china-says-us-claims-it-uses-forced-technology-transfer-boost, accessed October 5, 2019.

Mearsheimer, JJ (2001) *The tragedy of great power politics*. WW Norton & Company.

Moore, AW (2017) *AI and national security in 2017*. Washington, DC: Presentation at AI and Global Security Summit. November 1.

Mozur, P and Perlez, J (2017) 'China bets on sensitive US start-ups, worrying the Pentagon', *The New York Times*. March 22. https://www.nytimes.com/2017/03/22/technology/china-defense-start-ups.html, accessed October 11, 2019.

National Science and Technology Council (2016) *The national artificial intelligence research and development strategic plan*. Washington, DC: Executive Office of the President of the United States.

New Intellectual Report (2017) *China academy of engineering: unveiling of artificial intelligence 2.0 era*. http://www.sohu.com/a/126825406_473283, accessed April 1, 2018.

Nye, JJ (2017) 'Deterrence and dissuasion in cyberspace', *International Security*, 41(3), 44–71. doi:10.1162/ISEC_a_00266

Office of the Secretary of Defense (2019) *Annual report to congress: military and security developments involving the people's Republic of China, 2019*. Washington, DC: US Department of Defense.

Office of the Secretary of Defense (2017) *Annual report to congress: military and security developments involving the people's Republic of China, 2017*. Washington, DC: US Department of Defense.

Office of the Secretary of Defense, US Department of Defense (2016) *Remarks by secretary carter on the budget at the economic club of Washington*. February 2. Washington, DC: US Department of Defense.

Office of the Secretary of Defense, US Department of Defense (2018) *Nuclear posture review*. Washington, DC: US Department of Defense.

Office of the Secretary of Defense, US Department of Defense (2019) *Lt. Gen. Jack Shanahan media briefing on A.I.-related initiatives within the department of defense*. Washington, DC: US Department of Defense.

Opinions on Strengthening the Construction of a New Type of Think Tank with Chinese Characteristics (2015) *China Daily*. October 27. https://www.chinadaily.com.cn/china/2014-10/27/content_18810882.htm, accessed April 1, 2017.

Organski, AFK and Kugler, J (1980) *The war ledger*. Chicago, IL: University of Chicago Press.

Paul, TV, Larson, DW and Wohlforth, W (eds) (2014) *Status in world politics*. New York: Cambridge University Press.

Petrara, D (2019) *China's AI ambition gets a reality check as the USA reclaims top spot in global AI investment.* Bloomberg Business Wire. October 30. https://www.bloomberg.com/press-releases/2019-10-30/china-s-ai-ambition-gets-a-reality-check-as-the-usa-reclaims-top-spot-in-global-ai-investment, accessed March 1, 2020.

Porter, P (2019) 'Advice for a dark age: managing great power competition', *The Washington Quarterly*, 42(1), 7–25. doi:10.1080/0163660X.2019.1590079

Posen, BR (2003) 'Command of the commons: the military foundation of US hegemony', *International Security*, 28(1), 5–46. doi:10.1162/016228803322427965

PWC (2017) *Sizing the prize*. https://www.pwc.com/gx/en/issues/ analytics/assets/pwc-ai-analysis-sizing-the-prize-report.pdf, accessed April 1, 2017.

Rapoport, A (1957) 'Lewis F. Richardson's mathematical theory of war', *Conflict Resolution*, 1(3), 249–299.

Roff, HM (2019) 'The frame problem: the AI "arms race" isn't one', *Bulletin of the Atomic Scientists*, 75(3), 1–5.

Schelling, TC and Halperin, MH (1975) *Strategy and arms control*. Washington, DC: Pergamon-Brassey.

Schulman, LD, Sander, A and Christian, M (2015) *The Rocky relationship between Washington & Silicon Valley: clearing the path to improved collaboration*. Washington, DC: CNAS.

Siddiqi, AA (2000) *Sputnik and the Soviet space challenge*. Gainesville, FL: University of Florida Press.

Simmons, JW (2016) *The politics of technological progress*. Cambridge: Cambridge University Press.

Simonite, T (2017) 'Defense secretary James Mattis Envies Silicon Valley's AI ascent', *Wired.com*. November 8. https://www.wired.com/story/james-mattis-artificial-intelligence-diux/, accessed April 1, 2018.

Singer, PW (2009) *Wired for war: the robotics revolution and conflict in the 21st century*. New York: Penguin.

Singer, PW and Friedman, A (2014) *Cybersecurity and cyberwar: what everyone needs to know*. New York: Oxford University Press.

The State Council Information Office of the People's Republic of China (2017) *State council notice on the issuance of the new generation AI development plan*. Beijing, China: State Council of the People's Republic of China.

T4GS and CGSR (2019) 'AI and the military: forever altering strategic stability', *Technology for Global Security*. February 13.

Taylor, MZ (2016) *The politics of innovation: why some countries are better than others at science and technology*. Oxford: Oxford University Press.

Tencent Research Institute (2017) *China academy of information and communications technology, tencent AI lab, and tencent open platform. artificial intelligence: a national strategic initiative for artificial intelligence*. Beijing: China Renmin University Press.

Tomayko, JE (2000) *Computers take flight: a history of NASA's pioneering digital fly-by-wire project*. Washington, DC: National Aeronautics and Space Administration, pp. 24–25, 30.

US Department of Defense, Lt. Gen. Jack Shanahan Media Briefing on AI-Related Initiatives within the Department of Defense (2019) August 20. https://www.defense.

gov/Newsroom/Transcripts/Transcript/Article/1949362/lt- gen-jack-shanahan-media-briefing-on-ai-related-initiatives-within-the-depart/, accessed April 21, 2019.

Waltz, K (1979) *Theory of international politics*. Reading, MA: Addison-Wesley.

White, J (2018) 'Google pledges not to work on weapons after project Maven backlash', *The Independent*. June 7. https://www.independent.co.uk/life-style/gadgets-and-tech/news/google-ai-weapons-military-project-maven-sundar-pichai-blog-post-a8388731.html, accessed January 1, 2019.

White House (2019) *Executive order on maintaining American leadership in artificial intelligence*. February 11. https://www.whitehouse.gov/presidential-actions/executive-order-maintaining-american-leadership-artificial-intelligence/, accessed January 1, 2018.

Wilkening, D (2019) 'Hypersonic weapons and strategic stability', *Survival*, 61(5), 129–148. doi:10.1080/00396338.2019.1662125

Wohlforth, WC 'Unipolarity, Status Competition, and Great Power War', in *International Relations Theory and the Consequences of Unipolarity*, ed. John Ikenberry, pp. 33–65 (Cambridge University Press, 2011).

Work, RO (2015) *Remarks by defense deputy secretary Robert work at the CNAS inaugural national security forum*. Washington, DC: CNAS.

Work, RO and Brimley, SW (2014) *20YY preparing for war in the robotic age*. Washington, DC: Center for a New American Security.

Xi Jinping's report at the 19th Chinese Communist Party National Congress (2017) *Xinhua*. October 27. http://www.xinhuanet.com/english/special/2017-11/03/c_136725942.htm, accessed January 1, 2020.

Yuan, W (2017) 'China's "digital silk road": pitfalls among high hopes', *The Diplomat*. November 3. https://thediplomat.com/2017/11/chinas-digital-silk-road-pitfalls-among-high-hopes/, accessed January 1, 2018.

Zala, B (2017) 'Polarity analysis and collective perceptions of power: the need for a new approach', *Journal of Global Security Studies*, 2(1), 2–17. doi:10.1093/jogss/ ogw025

3 Artificial intelligence

Implications for small states

Reuben Steff

Introduction

While there is a growing body of literature examining the security implications of artificial intelligence (AI) for large states, there is little that specifically considers what it means for the security of *small states*. Given their relative material inferiority, there is a distinct risk they will fall behind larger powers in the race for AI if they do not recognize its implications or lack the material, financial, and technical knowhow to remain at the cutting edge of technology. Relatedly, if they are complacent, they will find themselves in increasingly dependent relationships with larger powers to secure access to critical AI technologies. This would constrain their freedom of action and independence in world affairs and, ultimately, threaten their security and ability to remain economically competitive.

In the short term, AI looks promising for improving a range of simpler processes for small states, including system–user interactions, administration, and logistics. While this may seem like "low-hanging fruit", the benefits to small states with challenging demographics and constrained military budgets could prove considerable by delivering greater efficiency and relieving personnel from repetitive and mundane tasks to take on more cognitively demanding work.

With the above in mind, this chapter makes a contribution to the emerging literature on the security implications of AI, with a focus on small states. It draws upon insights from the nascent literature on AI and interviews conducted by the chapter's author in Washington, DC with experts on AI in April 2018. The first part considers the contours of the small states. The second outlines what AI means for small state security organizations, the prospects of AI diffusion, relevance to alliances, and it considers humanitarian objectives. The third outlines the implications for small-state policy makers. The conclusion identifies areas for further research and warns that complacency about AI poses a grave risk to the security and prosperity of small states in the 21st century; as such, policy makers in smaller states should focus on creating the conditions for them to become "AI powers".

Contours of the small states

A common feature of small states is *weak aggregate structural power* relative to their larger peers. This is a product of (in most cases) having a small territory, a population of less than 15 million people, and limited natural resources (Thorhallsson, 2018). Owing to this, small states face limitations in the resources they have on hand to drive their economies and are unable to field large and technologically diverse militaries. In turn, this challenges a fundamental priority of states: Maintaining their independence and sovereignty. When coupled with a lack of geographic strategic depth and functional buffers, military weakness leaves them vulnerable to coercion, annexation and, at worst, extermination (Thorhallsson and Steinsson, 2017).

As "consumers" of security, working with international organizations and establishing alliances with larger states is an attractive option for small states. It allows them to secure access to resources, military hardware, training, foreign markets, intelligence, infrastructure (such as undersea internet cable), and critical technologies (Thorhallsson and Steinsson, 2017). In the post–war era, the US has traditionally played this role and provided access to state-of-the-art technological capabilities to its partners. In the age of AI there are already signs that the US and other great powers, such as China, will play this role again. Alliances also come with costs for smaller states. Small states fear entrapment – that their interests will become dominated by their larger allies and threaten their autonomy (Burton, 2013, p. 218) – and abandonment – that larger allies will withdraw from alliances if it no longer suits their interests (ibid). It also requires them to engage in foreign military interventions and activities that may be peripheral to their direct national interests but viewed as significant by larger allies. The next section looks at a range of implications AI has for small state military and security organizations.

The implications of AI for small state security

Technological enabler and autonomous vehicles

A key deficiency of small state military forces is the state's limited population. As such, they are often reluctant to deploy forces for missions beyond territorial defense or into dangerous environments. Here, AI contains the potential to act as a technological enabler by enhancing soldier protection and treatment/recovery. This could be facilitated through smart wearable devices that monitor physiology, AI-enhanced protective gear ("smart helmets", protective armor, and exoskeletons) (Shi, Zhao, Liu, Gao and Dou, 2019) and rapid medical interventions (Army Technology, 2019), reducing the risk to personnel in the field. In some cases, greater levels of robotic automation will act as a force multiplier for small states as autonomous vehicles will conduct tasks on the battlefield previously undertaken by existings platforms (for example close air support) and soldiers (such as reconnaissance) with greater levels of precision and reliability, in environs unsuitable for humans, and for durations that exceed human endurance (Scharre and Horowitz, 2018). As such, training, doctrine, recruitment, and organizational structures will need to adjust (Nicholson, 2018).

AI will be fundamental to harnessing and integrating ever-greater amounts of data across air, space, cyberspace, sea, and land domains (Hoadley & Lucas,

2018, p. 11). Without AI, much of these data are virtually useless; with AI filtering it, actionable and time-critical information can be gleaned. This could transform command and control operations, enabling faster decisions and accelerating the speed of conflict (ibid, 27; Gilchrist, 2018). In this context, opponents that rely on human capabilities and judgement may be overwhelmed, unable to understand a rapidly changing battlefield environment, and react fast enough to adversary operations (Hoadley & Lucas, 2018, p. 27). Identifying patterns across large data sets by drones and other automated vehicles will allow improved and faster image recognition for monitoring territory, labeling, and targeting to take place compared to human analysts. The US military through Project Maven is already using AI algorithms for these purposes in Iraq and Syria (Allen, 2017). While a number of these aspects – such as how the speed of military operations will accelerate – are arguably more relevant to larger military forces seeking to gain dominance on the battlefield, understanding these changes is important for small state militaries that will have to coordinate their own forces alongside their larger partners.

Greater efficiency will be possible. For example, in Singapore, wearable gadgets and data analytics are improving targeted training and optimizing soldier performance (Bhunia, 2018b). Additionally, AI is central to Singapore's Smart Air Base as the republic seeks to improve networking between its central base command and its extended airbase system (Bhunia, 2018a). Data analytics and AI will filter data, providing recommendations to decision-makers and speeding up the decision-making cycle and ability to deliver more complex operations (Bhunia, 2018b). Singapore's next-generation smart naval base program seeks to reduce security personnel on its bases by 70% through a new screening system incorporating threat analysis, biometric authentication, and facial recognition (Bhunia, 2018a). Its Maritime Crisis Center is already using AI programs to digest data obtained from multiple sources, producing a single picture of the over 1,500 commercial ships that transit its territory and ports every day (Bhunia, 2017). For its part, New Zealand is using AI for logistics and training, with augmented reality systems used by the Navy to train engineers (Burton and Soare, 2019). In the 1990s, the Israel Defense Forces (IDF) deployed the AI-enabled HARPY drone that hovers over areas and makes independent decisions on whether to attack radar stations and, more recently, rolled out self-driving military vehicles to patrol Israel's border with the Gaza Strip; plans to equip these vehicles with weapons and deploy them in stages to Israel's frontiers; and unveiled two unmanned submarines (Gross, 2017; Fryer-Biggs, 2019). Some militaries are working to empower legacy systems with AI and new capabilities to "pair" older vehicles with newer ones, tasking them with conducting automated tasks to support manned systems by reacting to electronic threats and jamming, conducting reconnaissance, and carrying extra weapons (Hoadley & Lucas, 2018, p. 11). Such a capability, if it proves effective and widely attainable, would further offset the inherent force size limitations facing small states.

Despite immense potential, militaries will need to keep in mind that the bulk of AI development is happening in the commercial sector. This presents unique challenges for military acquisitions with tech companies (and their employees) already showing resistance to cooperating with governments out of concerns their technology will be used to harm humans; there is a trust and culture deficit (Hoadley and Sayler, 2019). While this could exist in small states, they may have an advantage relative to larger powers for two reasons. The first is intimacy – it is easier for individuals from government and the private sector in small states to rapidly interact with one another, and thus find common ground on tech transfers and the rules and norms that should regulate their use. Second, the inherent material vulnerability of small states means that the tech sector may more readily understand the need for greater synergy between their work and the national security imperatives of the state.

It will not be easy to transpose automated vehicles designed for suburban streets and highways where traffic data are plentiful to foreign and unwelcoming terrain. As such, AI-enabled platforms will have to be reconfigured for use in multiple environments – both unstructured (we can imagine unstable, poorly kept, or destroyed roads, off-road environments, and diverse topographies) and structured (for example, urban areas) – and according to higher standards of safety and performance. Furthermore, greater autonomy also presents new tasks, challenges, and unanticipated problems; user errors will be inevitable (Hoffman, Cullen and Hawley, 2016). First principles will require providing a baseline of AI education and literacy across military forces, with AI software offering opportunities to tailor teaching programs, deliver more authentic training environments with advanced simulation of adversary behavior, and greater cognitive skills across the full spectrum of military roles and ranks (Ryan, 2018). As AI forces and command and control processes come online, tech transfers and joint exercises designed to facilitate interoperability between small states forces and their larger partners will be critical to ensure they can operate and deliver value alongside their larger allies.

Military deterrent, technological diffusion and innovative doctrine

The next evolution in military operations could be forces shifting from fighting as a network to fighting as a *swarm*, with large numbers of highly autonomous systems coordinating their actions on the battlefield, enabling greater mass, coordination, intelligence, and speed of operations (Scharre, 2014a; Scharre, 2014b). This offers great asymmetric potential for small states, the ability to offset personnel limitations and to reduce the limitation imposed by economies of scale, as individual soldiers could command relatively low-cost swarms capable of overwhelming comparatively expensive military systems.

For some small states, technological innovations are a critical means to deter larger ones. For example, Singapore views its advanced military capabilities as a means to deter Indonesia and Malaysia (notably, Singapore's Defence Science and Technology Agency (DSTA) has established a partnership with the US Joint Artificial Intelligence Center (JAIC) to enhance their defense operations) (Parameswaran, 2019). Israel has long understood the advantages of leading the battlefield deployment of new military technologies. The IDF was the first to employ elements of the 1980s revolution in military affairs (RMA) by integrating across its networks cutting-edge electronic devices, unmanned aerial vehicles (UAVs), and precision-guided munitions (PGMs), deploying them to devastating effect against a larger state, for example, against the Syrian army in Lebanon (ibid).

First-mover advantages could prove fleeting in the age of AI, with smaller states gaining relative to larger ones since, as noted, development is largely being driven by the private sector which has incentives to sell to the global market rather than just to the government in specific nations in which they reside. This could lead to the rapid diffusion of military applications with immense implications for the balance of power at the global level and in regional environments. Emily O. Goldman and Richard B. Andrew make a telling point when they write that "It's far easier and quicker to assemble very advanced software with a small team of experts than it ever was to amass a modern mass-production industrial base in the late nineteenth and early twentieth centuries" (Goldman & Andrew, 1999). This increases the chances that states will be able to "leap frog" one another in the technological pecking order and small states that leverage new military advances in AI or generate breakthroughs on their own could benefit relative to larger ones.

The history of technological revolutions in warfare suggests the winner is not the state/s that gets the technology first but those that are able to innovate and use it in novel ways. Given the advent of forces like autonomous swarms, there is no reason why innovative *uses* (on the operational and doctrinal side) of these technologies could not come from small states; indeed, the military bureaucracies in small states may prove more nimble and culturally receptive to technological innovations than larger ones, able to expedite the testing of new doctrines in the field with AI-enabled military units before larger powers. Here there are grounds for caution. Algorithms are vulnerable to theft, hacking, data manipulation, and spoofing. This makes it essential that high cyber security standards are set. This is especially relevant for small states that could be considered "weak links" – "back doors" into the networks of their allies – in cyber security. This could have immense implications for alliances; investment in AI-based automated network defenses to strengthen small state's cyber security would thus seem a critical area of investment to minimize their vulnerabilities.

Emergency response and policing

AI can enhance the ability of small states' emergency services to react to and prepare for national emergencies (Meier, 2015), enabling planners and responders to analyze real-time population and seismic data, physical infrastructure schematics, risk assessments, and response plans. This information is combined with data from citizens posting on social media and first responder's feedback to enable command and control personnel to make informed decisions. By being continuously updated with new information, algorithms can provide a constant picture of the changing needs on the ground. This capability has already been successfully put into action through the Artificial Intelligence for Digital Response (AIDR) open-source platform that utilized machine learning and human intelligence ("crowdsourcing" by identifying informative tweets) during earthquakes in Pakistan, Nepal, and Chile (Imran, Castillo, Lucas, Meier & Vieweg, 2014).

AI systems can also support police work. Recent research on crime around sports stadiums reveals that social media data can enhance the police's ability to preemptively identify trouble hotspots and thus allocate resources more effectively (Ristea, Kurland, Resch, Leitner & Langford, 2018). AI could also enhance police force's abilities to predict terrorist attacks and to track the tens of thousands of people on terrorist watch lists (Loeb, 2018). Dubai has shown how AI can be used to create efficiencies at the "back end" of police organizations and operations. This includes a new money transfer service for prisoners; an interactive voice response system for payments by civilians related to non-emergency issues; "smart suits" to allow police to operate in difficult environments; a high-altitude pseudo satellite for surveillance; self-driving motorcycles; and police chatbots and robots to serve clients. Other systems include: Smart security patrol system, virtual smartphone plugin, smart hand clock for emergency situations, 3D printer model, and an "AI clinic" (Agarib, 2018). Ultimately, their ambition is for AI to reduce the time it takes to capture criminals, decreasing it from thousands of hours to just one; police have already used a smart facial recognition camera system to locate and return kidnapped children (Malek, 2018).

Police use of AI also comes with ethical and practical concerns. What if flawed data lead to the surveillance of people not involved in criminal behavior? These issues have surfaced in the US, where predictive policing and offender management algorithms have been criticized for unfairly targeting African Americans and those from disadvantaged backgrounds (Robitzski, 2018). "Feedback loops" might be created, whereby increased police resources in a particular area generate increased crime identification and an eventual over-allocation of resources (Loeb, 2018), or governments use law enforcement to repress political opposition or minorities.

Advancing human security

Improved situational awareness afforded by AI could facilitate humanitarian objectives and the UN Sustainable Development Goals. This should appeal to small states, which spend more of their time on "human security" activities compared to large-scale military operations. As populations in the developing world secure access to the internet and smart devices, ever-increasing amounts of data will provide aid and multilateral organizations the ability to comb through information to improve targeting and delivery of aid to the most needy, reducing the severity of famines and improving the ability to respond to droughts and other natural disasters. Data could identify early warning conflict indicators to prevent humanitarian crimes, such as ethnic cleansing and genocide, and respond to them by empowering stakeholders, domestic groups, and regional organizations (Cummings et al, 2018). The Early Model-Based Event Recognition using Surrogates (EMBERS) represents a successful nascent example. It combines data produced from social media platforms, new articles, blog posts, Google search and Wikipedia with meteorological and event data, economic and financial indicators, satellite imagery, and even restaurant reservations. It forecasts events up to eight days in advance with reportedly 94% accuracy to notify users in real time, a capability that vastly outstrips traditional political science efforts (ibid, p. 23). The time required to put together UN peacekeeping and humanitarian responses to crises can be reduced.

With the above in mind, small states could take a lead in investigating and promoting the ethical and social aspects of AI, or "equitable AI", thus facilitating access to AI software and education in the developing world (Cummings et al, 2018). Promoting cheap or free transfer of AI technologies from larger states will, in turn, provide economic opportunities and immense creative potential to address local problems, resource shortages, and critical threats from climate change set to hit the developing world the hardest. AI also offers great potential healthcare benefits for the developing world, improving efficiency, reducing misdiagnoses, and overall outcomes, especially in rural areas (Guo & Li, 2018). This would align with the tradition of many small states to promote humanitarian objectives. Indeed some, such as the Nordic states, exert disproportionate international influence and soft power in this way, allowing them to make both a positive contribution to global security while creating expertise and a reputation that distinguishes themselves from larger states (Thorhallsson and Steinsson, 2017).

In the shadow of the great technological powers

History shows that while technological innovations do not always diffuse from a core set of powerful countries, (Goldman & Andrew, 1999) at present the world's greatest powers, the US and China, stand ahead of the pack by a significant margin on AI (Horowitz, 2018a). Small states cannot ignore this fact, nor can they neglect their material weaknesses which have historically required many of them to seek out alliances with larger powers. Indeed, like other technologies relevant to military and economic development and thus to the balance of power, the great powers will likely seek to influence, impede, and proffer the proliferation of AI technologies/AI-supporting technologies, infrastructure, and expertise to allies and adversaries. While there is the potential that some small states will be able to acquire AI rapidly through the open market and some develop their own cutting-edge algorithms, enhancing their sovereignty and independence in the process, some small states may not achieve this. As such, with AI coming to play a defining role in what states consider to be in their national interests throughout the 21st century, strategic alignments will be influenced by this.

The recent international controversy over China's multinational telecommunications company, Huawei, is a sign of things to come. Huawei has offered the cheapest rates to roll out 5G (the next generation of mobile networks) in the developed and developing world. This network will dramatically improve internet and data speed. This has immense applications for the development of advanced AI and, thus, for international security affairs, military operations, and economic competition. Like all networks, 5G is susceptible to hacking, and data flows could be siphoned off and manipulated, giving corporations that lay these networks immense power over critical state functions. This led to a collision between the US and China: Washington asserts that there is little that separates the Chinese government from its international corporations. These concerns have been heightened recently by a 2018 research paper that claims China used its telecom company network "point of presence" in North America to redirect and copy information-rich traffic (Demchak & Shavitt, 2018). As such, the US has made it clear to its allies that they must reject Huawei's advances, revealing how great powers will look to leverage critical AI-supporting infrastructure in their competition with one another. Notably, while some of America's larger allies have resisted the US on this issue, a number of other smaller allies, for example New Zealand, are under pressure to balance their strategic alignment with the US with their economic interests that emphasize trade with China (Steff & Dodd-Parr, 2019) leading them to ban Huawei from rolling out 5G.

The fact that AI development is largely driven by the private sector means that there are – beyond military and security reasons – vital economic and commercial factors that could deepen small state dependency on larger powers; governments will need to sustain close ties with larger powers to ensure their businesses retain access to advancements on AI software. But this may cut the other way – small states that foster commercial AI "national champions" could be a moderating force, providing them with more independence and turning the tables by inducing larger powers to seek to import key AI technology from smaller states.

Finally, as mentioned above, small states also fear abandonment. In the age of AI this concern will increase for those small states that are not high tech and high capital. If they are overly cautious, late adoption will threaten their militaries' ability to remain interoperable with their allies' armed forces, calling into question their reliability on the battlefield and value as an ally. According to Kenneth Payne, failure on this score means that allies "might be useful legitimizers for action, without contributing themselves, reducing them to client-state status, effectively demilitarizing states content to function under a Pax AI" (Payne, 2018, p. 25). One way to alleviate this is for small states to focus on making *niche AI contributions* and local optimization of AI that support allied nations' intelligence and defense capabilities. Additionally, a flexible bureaucracy and national security sector that embraces change could lead to fast track adoption of AI technologies, generating operational insights, innovative applications, and doctrine that is valued by larger allies.

Implications

This section outlines three key tasks for small state decision makers in the AI age.

Pragmatic multilateralism and international cooperation

A key issue is balancing moral concerns over lethal autonomy with strategic imperatives. For example, lethal autonomous weapon systems (LAWS) that independently identify targets and fire weapon systems with little or no human control are advancing rapidly (Hoadley and Sayler, 2019). Many states will soon be able to deploy them. At present, leaders in this field, such as the US (and Germany) have regulations preventing deployment of fully autonomous LAWS on moral grounds, and debates are taking place at the UN through the Convention of Certain Conventional Weapons (UN CCW) over what norms and rules should govern autonomous weapons and vehicles. The only major agreement among states so far is on the requirement of guaranteeing meaningful human control in the use of these weapons (Rickli, 2018).

Small states will need to bear in mind that a complete and verifiable ban over the lethal use of AI, including LAWS, will be difficult, especially as the competition for military AI takes on the trappings of an "arms race" (Geist, 2016; Horowitz, 2018b). As AI comes to outperform humans across a range of increasingly complex military tasks, military necessity will increase pressure for adoption of LAWS, especially if strategic rivals deploy them, leaving non-adopters at a disadvantage and fearing that even if international agreements to restrain their use are signed, rivals cannot be trusted to stick to them. The advent of "security dilemmas" – where states with defensive intentions (i.e. that have no aggressive designs toward one another) engage in a tit-for-tat cycle of deployment that ultimately lowers the security of both states – is plausible, making the threat of war more likely.

To restrain the AI arms race, bilateral and multilateral engagement will be required to update arms control and non-proliferation treaties as AI (and other emerging technologies) challenge existing orthodoxies that grew out of Cold War efforts focused on nuclear weapons. Rather than aiming to ban the full category of "military AI", banning specific military uses would seem more credible, for example handing over command and control of full-scale military operations to AI or operations of drone fleets (Horowitz, 2018b). Efforts toward moderating this could include developing a "security culture" to manage the race, akin to that which developed between the Soviets and Americans during the Cold War. This should include "Track II" diplomacy and establishing the basis for what realistic arms controls can do in the future (Geist, 2016). Collective agreement to prevent proliferation to rogue states and terrorists is an issue where a broad international consensus should be possible and could form the basis for broader confidence building.

Given the corporate investment in AI, the private sector will need to become a central partner in this. International efforts should start to identify "red lines" to ensure their use within the bounds of international norms and laws. If these efforts fail, they could have grave consequences, inaugurating an era where strategic and crisis stability is fluid, regional security dilemmas proliferate, and risk of great power confrontation grows that could draw in small states or makes them the battleground for proxy wars (Lieber & Press, 2017; Altmann & Sauer, 2017; Saalman, 2018; Geist & Lohn, 2018; Unal & Lewis, 2018). Given this, is it essential that small states expand their expertise on these issues to credibly contribute to and, in some areas, lead debates in major international forums over lethal AI technologies.

Efforts between small states to pool resources together and work with middle-tier powers will improve their chances of remaining relevant in the race for cutting-edge artificial intelligence. This kind of collaboration is already evident, albeit in regional settings. For example, in May 2018, Denmark, Estonia, Finland, the Faroe Islands, Iceland, Latvia, Lithuania, Norway, and Sweden, signed the *Declaration on AI in the Nordic-Baltic Region* (Nordic Council of Ministers, 2018).

Recalibrating the diplomatic toolkit

The existing diplomatic toolkit, at both the capital and embassy ends, will need to be adjusted to monitor foreign developments in tech firms, governments, and their military forces. Free trade agreements should include provisions specifically guaranteeing access to AI technologies to open doors for domestic economic interests. Indeed, acquiring a sense of debates taking place in foreign capitals about AI and how they intend to use it would be informative. Researchers from government, the military, and academic sectors must be funded to engage with their foreign counterparts and pursue fact-finding missions to identify collaborative opportunities. Furthermore, methods to combat disinformation campaigns by malicious actors seeking to impede civilian–government trust will be necessary, alongside setting standards for reliable open communication channels, and educating the public about AI-assisted fake news and deepfake campaigns.

Efforts by foreign affairs ministries to tackle and adapt to AI in a proactive manner can be informed by their experience with cyber security. While many ministries now have units and personnel focused on cyber issues, they had to play catch up (Scott, et al, 2018). States that allow this to occur could miss opportunities to establish trading, diplomatic, and military arrangements with foreign partners, inhibiting the development of their AI expertise and capabilities. Efforts must be taken to educate personnel in foreign policy units (including those whose portfolios do not explicitly include AI) on what AI is and why it is poised to play a decisive and broad role across society and to the economic and security fortunes of states.

Furthermore, much as the speed of conflict will increase as AI increasingly enters the battlefield, so too will AI impact on how diplomacy functions, with policy makers seeking to make data-informed decisions quicker in response to unfolding global events. Foreign and strategic-political affairs would be "accelerated", and, as a consequence, foreign ministries and diplomats may increasingly find their jobs emphasize the need to assimilate, combine, refine, and go through large amounts of data to get relevant decision-making advice to political actors as quickly as possible.

Internationally, AI could have profound implications on the balance of power, requiring deft diplomacy by small states (Horowitz, 2018a). At present, it is an open question whether liberal democracies or authoritarian political and economic systems will prove better at harnessing AI (Wright, 2018; also see Simona Soare's chapter in this volume). Some, such as China and Russia, that have greater access to their citizen's data may gain an advantage relative to democracies that develop stringent regulations around private data, or that are unable to forge coherent AI strategies. Diplomatic reporting will need to place greater emphasis on this to ensure decision-makers can make informed strategic decisions vis-à-vis the larger geopolitical competition underway.

Assessing the foundations of AI power

A global race between the US and China to acquire AI talent (Kiser and Mantha, 2019) is already underway. This threatens to further polarize AI talent in favor of the world's most powerful states, with negative repercussions for the ability of small states to incorporate AI. On the other hand, an idealist view is that it is possible the metrics of international influence, economic success, and military power could be altered by AI and come down to the quality of a state's digital systems, sensors, high-speed networks, data protocols, and ability to create a strong technological and science base supported by interdisciplinary teams of researchers. In principle, there is no reason why innovative uses of AI could not come from small states. As such, small state policy makers should assess how their own country is positioned to generate and harness "AI power". While geography, population, and industrialization are still important for states to generate power, in the information and AI age the foundations of success are multidimensional. They include:[1]

- *Large quantities of effective data tailored for the right applications.* Countries and organizations with large high-quality sets of data will gain advantages in developing AI applications
- *Foreign AI talent and human capital.* AI programmers are scarce and salaries for new PhDs large. As such, competing for existing top-tier talent in the global marketplace will be difficult and could require small states to employ innovative marketing strategies. For example, those with stable political systems, a high quality of life, and a culture that welcomes immigrants could use these factors to entice AI technicians from abroad
- *Domestic AI talent and human capital.* Creating domestic expertise by improving education systems is essential. This should include establishing a "baseline" of knowledge across society and prioritizing training in relevant industries. On this point, Finland has recently announced a program to educate 1% of its population in the basics of AI (Delcker, 2019), a model for other small states to consider
- *Computing hardware and software.* Large high-tech computing resources and hardware shortens the time it takes to train AI algorithms, to conduct experimentation, prototyping, and reduces power consumption and related costs. Some cutting-edge AI applications will require significant computing resources and custom chips to enable their use in small devices
- *Organizational strategies and incentives to adopt AI.* Government agencies and non-government organizations can play a critical role by educating industries and forging AI "ecosystems" (groups of like-minded organizations with

a stake in AI that share insights and expertise). The United Arab Emirates (UAE) became a leader in this field in October 2017 when it appointed the world's first Minister of State for Artificial Intelligence

- *Public-private cooperation.* As noted, many key AI developments are taking place in the private sector. Therefore, effective public–private cooperation will be critical in the years to come in order to regulate practical applications of AI to facilitate transfer of private innovations into the government sector
- *Government strategies and effective regulations.* A number of small states (including Singapore, Finland, Denmark, Sweden, Tunisia, and UAE) have official AI strategies (Dutton, 2018) (to the author's knowledge, no small state has an unclassified military-related AI strategy. Rather, small states' AI strategies focus on economic and commercial activities and contribute to economic security interests, rather than military or other traditional security objectives). While every one of these strategies is unique, other small states can use them as reference points for drafting their own. They should, however, be mindful that strategies can operate as forms of light regulation. In the AI sphere this takes on added importance, since regulations will need to balance a range of competing interests. This includes privacy concerns and ensuring automated vehicles and drones are safe while allowing access to data for economic competitiveness and incentivizing innovation
- *Resilience and safety.* To ensure their ability to generate, access, and harness AI power is not impeded, small state policy makers must consider how to deal with the inherent limitations of "smallness" (limited AI talent pools, smaller data sets, and the challenge of adapting algorithms to national data sets) – issues likely to affect them on a greater scale relative to larger powers. This requires devising best policy practices (and potentially collaborating with other small states in this issue and sharing data), educating their population and military forces, expanding training opportunities, running practices and simulations, and adopting a whole-of-society approach to ensure populations are informed of the challenges posed by AI.

Conclusion

Many of the issues touched on in this chapter are ripe for further research. This includes crafting best practice AI policy principles for small state decision-makers; embedding AI in diplomatic operations; considering how, and in what specific areas, small states can collaborate with one another to develop their domestic AI expertise; how near-term applications of AI can improve the efficiency and functioning of security organizations; considering how AI can strengthen the asymmetric warfighting capabilities of small states to strengthen their ability to deter coercion and credibly contribute to allied and multinational and peacekeeping operations.

Ultimately, small states cannot passively assume critical AI technologies will be cheaply sourced and readily accessible on the international market. While this is possible, it is also plausible that the strongest states will come to view AI in zero-sum terms, denying market access to other countries and classifying the most advanced developments. In this world, the first-mover advantages for states like China and the US will increase, resulting in greater inequalities between large and small states, and the autonomy of the latter reduced (Horowitz, 2018a). As such, decision-makers, civilians, and public and private sector organizations in small states must awaken to the challenges and opportunities presented by AI and take decisive steps to create AI ecosystems that bring together domestic stakeholders, society, and industry. This will maximize their prospects of developing AI domestically, allow them to credibly engage in multilateral forums, collaborate on joint AI projects with international partners, and contribute to allied nations' intelligence and defense capabilities, ensuring they remain a valued and credible partner.

Note

1 For a deeper dive into these aspects of "AI power", see Horowitz et al, 2018.

References

Agarib, A (2018) 'Dubai police unveil artificial intelligence projects', *Smart Tech.* March 12. https://www.khaleejtimes.com/nation/dubai/dubai-police-unveil-artificial-intelligence-projects-smart-tech, accessed February 1, 2019.

Allen, GC (2017) *Project Maven brings AI to the fight against ISIS*. December 21. https://thebulletin.org/2017/12/project-mahttps://thebulletin.org/2017/12/project-maven-brings-ai-to-the-fight-against-isis/ven-brings-ai-to-the-fight-against-isis/, accessed July 31, 2018.

Altmann, J and Sauer, F. (2017) 'Autonomous weapon systems and strategic stability', *Survival*, 59(5), pp. 117–142.

Army Technology (2019) *US DoD awards contracts for autonomous trauma care system*. May 24. https://www.army-technology.com/news/us-dod-awards-contracts-for-autonomous-trauma-care-system/, accessed November 1, 2019.

Bhunia, P (2017) *Building next gen Singapore armed forces: cyber defence, analytics, artificial intelligence and robotics*. October 28. https://www.opengovasia.com/building-next-gen-singapore-armed-forces-cyber-defence-analytics-artificial-intelligence-and-robotics/, accessed December 3, 2018.

Bhunia, P (2018a) *Fact sheet: the republic of Singapore air force's smart airbase of the future*. March 2. https://www.mindef.gov.sg/web/portal/mindef/news-and-events/latest-releases/article-detail/2018/march/02mar18_fs4, accessed July 31, 2018.

Bhunia, P (2018b) *How MINDEF Singapore is using technology to enhance quality and ability of soldiers*. March 3. https://www.opengovasia.com/how-mindef-singapore-is-using-technology-to-enhance-quality-and-ability-of-soldiers/, accessed July 31, 2018.

Burton, J (2013) 'Small states and cyber security: the case of New Zealand', *Political Science*, 65(2), pp. 216–238.

Burton, J and Soare, S (2019) 'Understanding the strategic implications of the weaponization of artificial intelligence', *11th international conference on cyber conflict.* http://ccdcoe.org/uploads/2019/06/Art_14_Understanding-the-Strategic-Implications.pdf, accessed December 2, 2019.

Cummings, ML, Roff, HE, Kenneth Cukier, K, Parakilas, J and Bryce, H (2018) *Artificial intelligence and international affairs: disruption anticipated*. Chatham House Report. https://www.chathamhouse.org/sites/default/files/publications/research/2018-06-14-artificial-intelligence-international-affairs-cummings-roff-cukier-parakilas-bryce.pdf, accessed February 2, 2019.

Delcker, J (2019) 'Finland's grand AI experiment', *Politico*. January 2. https://www.politico.eu/article/finland-one-percent-ai-artificial-intelligence-courses-learning-training/, accessed January 18, 2019.

Demchak, CC and Shavitt, Y (2018) 'China's maxim – leave no access point unexploited: the hidden story of China telecom's BGP Hijacking', *Military Cyber Affairs*, 3(1), pp. 1–9.

Dutton, T (2018) *An overview of national AI strategies*. June 28. https://medium.com/politics-ai/an-overview-of-national-ai-strategies-2a70ec6edfd, accessed February 1, 2019.

Elsa, BK (2018) 'The pursuit of AI is more than an arms race', *Defense One*. April 19. https://www.defenseone.com/ideas/2018/04/pursuit-ai-more-arms-race/147579/, accessed June 8, 2018.

Fryer-Biggs, Z (2019) 'Coming soon to a battlefield: Robots that can kill', *Atlantic*. September 3, https://www.theatlantic.com/technology/archive/2019/09/killer-robots-and-new-era-machine-driven-warfare/597130/

Geist, EM (2016) 'It's already too late to stop the AI arms race—we must manage it instead', *Bulletin of the Atomic Scientists*, 72(5), pp. 318–321.

Geist, EM and Lohn, AJ (2018) *How might AI affect the risk of nuclear war?* RAND Corporation. https://www.rand.org/pubs/perspectives/PE296.html, accessed June 8, 2018.

Gilchrist, M (2018) 'Emergent technology, military advantage, and the character of future war', *The Strategy Bridge*. https://thestrategybridge.org/the-bridge/2018/7/26/emergenttechnology-military-advantage-and-the-character-of-future-war, accessed January 12, 2019.

Goldman, EO and Andrew, RB (1999) 'Systemic effects of military innovation and diffusion', *Security Studies*, 8(4), pp. 79–125.

Gross, JA (2017) *Unmanned subs, sniper drones, gun that won't miss: Israel unveils future weapons*. September 5. https://www.timesofisrael.com/unmanned-subs-and-sniper-drones-israel-unveils-its-weapons-of-the-future/, accessed 31 July 2018.

Guo, J and Li, B (2018) 'The application of medical artificial intelligence technology in rural areas of developing countries', *Health Equity*, 2(1). https://www.liebertpub.com/doi/full/10.1089/heq.2018.0037, accessed February 2, 2019.

Hoadley, DS and Lucas, NJ (2018) *Artificial intelligence and national security*. Congressional Research Service. April 26. https://www.hsdl.org/?abstract&did=810166, accessed November 1, 2018.

Hoadley, DS and Sayler, KM (2019) *Artificial intelligence and national security*. Congressional Research Service. January 30. https://fas.org/sgp/crs/natsec/R45178.pdf, accessed October 31, 2018.

Hoffman, RR, Cullen, TM and Hawley, JK (2016) 'The myths and costs of autonomous weapon systems', *Bulletin of the Atomic Scientists*, 72(4), pp. 247–55.

Horowitz, MC (2018a) 'Artificial intelligence, international competition, and the balance of power', *Texas National Security Review*, 1(3). https://tnsr.org/2018/05/artificial-intelligence-international-competition-and-the-balance-of-power/, accessed June 8, 2018.

Horowitz, MC (2018b) *The algorithms of August*. September 12. https://foreignpolicy.com/2018/09/12/will-the-united-states-lose-the-artificial-intelligence-arms-race/, accessed February 6, 2019.

Horowitz, M, Elsa, BK, Allen, GC, and Scharre, P (2018) *Strategic competition in an era of artificial intelligence*. July 25. https://www.cnas.org/publications/reports/strategic-competition-in-an-era-of-artificial-intelligence#fn14, accessed January 5, 2019.

Imran, M, Castille, C, Lucas, J, Meier, P and Sarah, V (2014) *AIDR: artificial intelligence for disaster response*. https://dl.acm.org/citation.cfm?id=2577034, accessed February 1, 2019.

Kiser, G and Mantha, Y (2019) *Global AI talent report 2019*. https://jfgagne.ai/talent-2019/, accessed October 31, 2019.

Lieber, KA and Press, GD (2017) 'The new era of counterforce: technological change and the future of nuclear deterrence', *International Security*, 41(4), pp. 9–49.

Loeb, J (2018) *AI and the future of policing: algorithms on the beat*. https://eandt.theiet.org/content/articles/2018/04/ai-and-the-future-of-policing-algorithms-on-the-beat/, accessed July 31, 2018.

Malek, C (2018) *AI will help police fight crime more efficiently, Abu Dhabi forum hears*. March 7. https://www.thenational.ae/uae/government/ai-will-help-police-fight-crime-more-efficiently-abu-dhabi-forum-hears-1.711149, accessed February 1, 2019.

Meier, P (2015) *Digital humanitarians: how big data is changing the face of humanitarian response*. Boca Raton: CRC Press.

Military Technology (2019) *US DoD awards contracts for autonomous trauma care system*. May 24. https://www.army-technology.com/news/us-dod-awards-contracts-for-autonomous-trauma-care-system/, accessed October 31, 2019.

Nicholson, BPW (2018) *Singer: adapt fast, or fail*. July 7. https://www.realcleardefense.com/articles/2018/07/07/pw:singer_adapt_fast_or_fail_113585.html, accessed January 1, 2019.

Nordic Council of Ministers (2018) *AI in the Nordic-Baltic region*. May 14. https://www.regeringen.se/49a602/globalassets/regeringen/dokument/naringsdepartementet/20180514_nmr_deklaration-slutlig-webb.pdf, accessed January 18, 2019.

Parameswaran, P (2019) 'What's in the new US-Singapore artificial intelligence defense partnership?', *The Diplomat*. July 1. https://thediplomat.com/2019/07/whats-in-the-new-us-singapore-artificial-intelligence-defense-partnership/, accessed November 3, 2019.

Payne, K (2018) 'Artificial intelligence: a revolution in strategic affairs?', *Survival*, 60(5), pp. 7–32.

Rickli, J (2018) 'The economic, security and military implications of artificial intelligence for the Arab Gulf countries', *EDA Insight*. November. https://www.gcsp.ch/download/8671/197214, accessed November 20, 2018.

Ristea, A, Kurland, J, Resch, B, Leitnerm, M and Langford, C (2018) 'Estimating the spatial distribution of crime events around a football stadium from georeferenced tweets', *ISPRS International Journal of Geo-Information*, 7(2), pp. 1–25.

Robitzski, D (2018) *The LAPD's terrifying palantir-powered policing algorithm was just uncovered and yes it's basically "Minority Report"*. May 10. https://futurism.com/lapd-documents-show-their-policing-algorithms-continue-to-target-minorities-and-past-offenders/, accessed July 31, 2018.

Ryan, M (2018) 'Intellectual preparation for future war: how artificial intelligence will change professional military education', *War on the Rocks*. July 3. https://warontherock

s.com/2018/07/intellectual-preparation-for-future-war-how-artificial-intelligence-will-change-professional-military-education/, accessed January 1, 2019.

Saalman, L (2018) 'Fear of false negatives: AI and China's nuclear posture', *Bulletin of the Atomic Scientists*. April 24. https://thebulletin.org/landing_article/fear-of-false-negatives-ai-and-chinasnuclear-posture/, accessed June 8, 2018.

Scharre, P (2014a) 'Robotics on the battlefield: part I: range, persistence and daring', *Center for a New American Century*. September 25. https://www.cnas.org/publications/reports/the-coming-swarm-the-cost-imposing-value-of-mass, accessed July 31, 2018.

Scharre, P (2014b) 'Robotics on the battlefield part II: the coming swarm', *Center for a New American Century*. October 15. https://www.cnas.org/publications/reports/robotics-on-the-battlefield-part-ii-the-coming-swarm, accessed July 31, 2018.

Scharre, P and Horowitz, M (2018) *Artificial intelligence: what every policymaker needs to know*. June 19. https://www.cnas.org/publications/reports/artificial-intelligence-what-every-policymaker-needs-to-know, accessed July 31, 2018.

Scott, B, Heumann, S and Lorenz, P (2018) *Artificial intelligence and foreign policy*. January. https://www.stiftung-nv.de/sites/default/files/ai_foreign_policy.pdf, accessed November 5, 2018.

Shi, H, Zhao, H, Liu, Y, Gao, W and Sheng-Chang, D (2019) 'Systematic analysis of a military wearable device based on a multi-level fusion framework: research directions', *Sensors*, 19(12), pp. 1–22.

Steff, R and Dodd-Parr, F (2019) 'Examining the immanent dilemma of small states in the Asia-Pacific: the strategic triangle between New Zealand, the US and China', *The Pacific Review*, 32(1), pp. 90–112.

Thorhallsson, B (2018) 'Studying small states: a review', *Small States & Territories*, 1(2), pp. 17–34.

Thorhallsson B and Steinsson, S (2017) *Small state foreign policy*. Oxford Research Encyclopedia of Politics. http://oxfordre.com/politics/view/10.1093/acrefore/9780190228637.001.0001/acrefore-9780190228637-e-484, accessed January 1, 2019.

Unal, B and Lewis, P (2018) *Cybersecurity of nuclear weapons systems: threats, vulnerabilities and consequences*. Chatham House. https://www.chathamhouse.org/publication/cybersecurity-nuclear-weapons-systems-threatsvulnerabilities-and-consequences, accessed January 2, 2019.

Wright, N (2018) 'How artificial intelligence will reshape the global order: the coming competition between digital authoritarianism and liberal democracy', *Foreign Affairs*, July 10.

4 Artificial intelligence and the military balance of power

Interrogating the US–China confrontation

Reuben Steff and Khusrow Akkas Abbasi

Introduction

Throughout the 21st century, artificial intelligence (AI) will decisively alter relative military power balances. Like the advent of electricity in the late 19th century, AI is not a weapon in itself – rather, it is a broad technological enabler of electronic and digital programs and platforms. To military organizations able to innovate and successfully incorporate AI, it will offer cost savings at the 'back end' of operations (logistics and human resources); improve command and control, weapon system designs, and lethality; offer a range of tactical and doctrinal innovations and, potentially, strategic breakthroughs. As such, it will play a significant role in the future of warfare across all domains, including for operations short of war, providing military advantages for those states that can best harness it.

By broadly enabling military forces, AI is set to alter the balance of military power between the existing status quo superpower, the US, and its great power challenger, the People's Republic of China (PRC).[1] This chapter considers this topic in four stages. The first section outlines the historical relationship between technology and the balance of military power. The second addresses AI's implications for the nature of conflict. The third considers where the US and China stand in their quest for AI supremacy by looking at the recent national efforts by Beijing and Washington to advance their AI strategies, and presents quantitative data to assess where the US and China stand on (military and general) AI. It also discusses the challenges to effective AI adoption Beijing and Washington face. Finally, the conclusion ponders how the injection of AI into military affairs could increase uncertainty between the US and China and suggests that a range of factors could intersect with their mounting competition to enhance or reduce one or the other's military power.

Artificial intelligence and the balance of power

Historically, technology has been a fundamental building block of national power, critical in determining a state's position in the global balance of power, their relative influence, level of security, and ability to emerge victorious in conflict

(McNeill, 1982). For example, the advent of machine tools and harnessing of steam by Great Britain placed it at the forefront of the First Industrial Revolution in the late 18th century. This led it to become the world's leading technological innovator and shift the balance of military power in its favor by generating a preeminent navy, allowing it to secure a global trading empire. Two more industrial revolutions followed, with the second in the 19th century driven by electricity and new mass production techniques, and a third in the early 1970s as electronics and information technology further automated production. In each revolution, entire industries, societies, and warfare were altered, in turn altering the balance of national and military power between states.

How does AI relate to this? In the broadest sense, AI will prove instrumental to powering the Fourth Industrial Revolution. This is characterized by intensification and increased dependency on virtual networks, access to ever-expanding amounts of data, cloud-based systems, and net-centric mobile applications and electronics that will dramatically change economies and industries, society, and warfare (Schwab, 2015). The renewed excitement over AI results from the confluence of advancing computing power and robotic technologies and ever-growing data which act as the 'fuel' to train AI; companies and organizations use data to train their algorithms to anticipate and mimic human behavior. The level and richness of national and global data will exponentially accelerate in coming years, owing to the expansion of the 'Internet of Things (IoT)', in turn increasing efficiency and bringing down the cost of training algorithms. Indeed, some 2.2 billion gigabytes of data were generated globally every day in 2018 (Goujon, 2018). According to International Data Corporation, the global datasphere will expand from 33 zettabytes in 2018 to 175 by 2025 (Reinsel, Gantz, Rydning, 2018). As such, political leaders, tech CEOs, defense organizations, and academics are virtually unanimous in asserting that AI will be fundamental to the future fortunes of states and the balance of power. For example, in 2017, the US Department of Defense (DoD) declared AI to be "the most disruptive technology of our time," (Department of Defense, 2017, p. 8) and said in its 2018 AI strategy that it will "impact every corner of the Department" (Department of Defense, 2018, p. 8), while the Belfer Center declared AI has the "potential to be a transformative national security technology, on a par with nuclear weapons, aircraft, computers, and biotech" (Allen and Chan, 2017). Furthermore, on July 16, 2019, the Trump administration's Secretary of Defense, Mark Esper, said the first priority for the DoD's technology modernization should be:

> artificial intelligence. I think artificial intelligence will likely change the character of warfare, and I believe whoever masters it first will dominate on the battlefield for many, many, many years. It's a fundamental game changer. We have to get there first.
>
> (Shanahan, 2019)

Kenneth Payne asserts that securing battlefield advantage from AI comes down to a competition of optimizations, "between the best algorithm and the rest ...

marginal quality might prove totally decisive: Other things being equal, we can expect higher-quality AI to comprehensively defeat inferior rivals" (Payne, 2018, p. 24). In other words, AI systems that provide marginal improvements in terms of decision-making and command and control, speed of operations, and ability to employ concentrated force and with greater accuracy will provide dominance on the battlefield relative to states with lesser AI-enabled forces. In addition, even if forecasts over the benefits of AI on the battlefield prove to be exaggerated, Washington and Beijing are operating on the assumption that they cannot fall behind the other; complacency risks the other state acquiring escalation dominance and shifting the military balance of power in their favor. Furthermore, it could prove difficult for either state to assess the relative capabilities of rival AI across one another's forces without seeing them in action on the battlefield. This will ensure worst-case analysis remains at the forefront of their strategic planning, propelling their efforts to acquire a position of AI military supremacy.

AI and the nature of conflict

During the 1991 Gulf War, the US military decimated Iraqi forces through the employment of new precision-guided weapons and stealth technologies and by harnessing 'informatized' warfare (the circulation of information through networks to improve communications, targeting, surveillance, reconnaissance and the speed of combined operations). In the process, the US cemented itself as the preeminent global military power, and, in the conflict's wake, scholars emphasized the growing importance of sophisticated military technologies in international security affairs with the term, "revolution in military affairs" (RMA), coming into common use (Collins & Futter, 2015). Rather than augur a new RMA, AI is primed to advance the previous evolution by delivering qualitative improvements in military technology and a paradigm shift from digitized/'informatized' warfare to 'intelligentized' warfare, offering innovations at the tactical and potentially strategic levels. According to Elsa Kania, China's People's Liberation Army (PLA) strategists and officials believe this will

> result in fundamental changes to military units' programming, operational styles, equipment systems, and models of combat power generation, even leading to a profound military revolution… At that point, as warfare occurs at machine speed, keeping humans 'in the loop' for the employment of weapons systems or even certain aspects of decision-making could become a liability, rather than an asset. Consequently, AI could take on a greater role in command and control [C2].
>
> (Kania, 2017a)

Along these lines, Jean-Michel Valantin says AI in C2, at its fullest extent, would "imply managing operations by AI-led air, ground, and sea vehicles, by AI-led entire units, as well as by AI cyber warfare units" (Kania, 2017b). Michael Horowitz reinforces this point, stating that algorithms will be used to coordinate different military systems for purposes such as swarming and generating and learning adversarial networks (Horowitz, 2018). AI will evolve existing capabilities by improving existing battlefield concepts and doctrine, accelerating the speed of combat, enhancing situational awareness, decision-making, lethality, and the concentration of military power while improving logistics and maintenance (reducing costs); and will replace mundane and repetitive jobs, allowing manpower to be redirected to more cognitively demanding and valuable tasks.

AI is already in use or slated for use to automate turrets, drones, aircraft, spacecraft, and submarines; to improve missile targeting and missile defense systems; for cyber and electronic warfare, modelling, simulations, and wargaming; and intelligence collection and analysis (Hawley, 2017; Harper, 2019; Roth, 2019; Davis, 2019). Ultimately, the cumulative effects of tactical level AI could become a strategic-level game changer, with decisive advantage on the battlefield accruing to militaries that are at the forefront of the 'intelligentized' AI-led RMA, relative to those who fall behind.

AI offers the potential for military operations to evolve from forces fighting as a network to forces fighting as a 'swarm'. For example, large numbers of highly autonomous systems, such as micro-drones, will coordinate their actions on the battlefield, enabling greater mass, coordination, intelligence, and speed of operations (Scharre, 2014). As such, individual soldiers could command small relatively low-cost fleets of swarming drones capable of overwhelming comparatively expensive military systems. Swarms of synchronized and widely distributed drones operating over large areas will allow more efficient allocation of assets, with systems cooperating for sensing, deception, and attack, able to 'self-heal' in response to enemy attacks, and able to conduct missions too dangerous for soldiers and thus at greater ranges. This could prove a game-changer on the battlefield; while the state that is most able to find innovative uses of them could benefit enormously. We can envision swarms becoming critical tools in states' Anti Access/Area Denial (A2/AD) capabilities. For example, coming into play during a crisis over the status of Taiwan or the South China Seas, low-cost swarms launched by China[2] could overwhelm key elements of US naval power, like aircraft carriers, auguring a paradigm shift in naval strategy and tactics (and making expensive, multi-mission platforms obsolete). Such a development would have immense implications for the military balance of power and for global security by undermining US security guarantees and its alliance networks. On the other hand, since swarms can conduct missions too dangerous for soldiers and thus at greater ranges, they could potentially address concerns over the growth of A2/AD capabilities by China and Russia in the Western Pacific, East Europe, and other former Soviet territories, providing the US with a military edge.

Indeed, China's lead in domestic adoption of AI technologies suggests they may be better placed to establish a lead in the military adoption of AI across a range of forces in the battlefield relative to the US. On this note, the US Congressional Research Service asserts that China's lack of active combat in recent decades may actually benefit it by making it more likely to develop unique AI combat tactics.

Moreover, while the US has prioritized AI to solve tactical-level problems and improve existing processes, China is using AI-generated war games to overcome its lack of combat experience (Hoadley & Lucas, 2018, p. 18). Further, while China may have reservations, similar to the US, about trusting AI-generated decisions owing to its centralized authority, it may have fewer moral qualms about employing lethal autonomous weapon systems (LAWS).

AI could also have a decisive effect on the balance of nuclear power. While analysis of AI's implications for nuclear deterrence and stability are, at present, speculative and theoretical, the potential implications are vast (Altmann and Sauer, 2017; Geist and Lohn, 2018; Saalman, 2018). It could transform C2 operations (AI could provide 'advice' on decisions or replace human decision making wholesale), enabling faster decisions and targeting of other states' mobile second-strike forces. For much of the Cold War the ability of nuclear arsenals to survive a first strike and deliver a retaliatory strike to ensure deterrence seemed feasible, as the superpowers' arsenals were vast, mobile, hardened, and hidden. However, the internet revolution – especially the advent and confluence of cyber capabilities, AI and big data analytics (and eventually quantum computing) – could upset nuclear deterrence in a number of ways in favor of the US (Lieber & Press, 2017; Kania & Costello, 2017; Futter, 2018). First, they could improve the *accuracy* and re-targeting capabilities of nuclear missiles, increasing counterforce capabilities and making the prospect of low-casualty nuclear strikes plausible, thereby reducing the political and normative barriers to a nuclear strike. Secondly, the big data revolution, and the ability of AI to harness and integrate it, will allow a more systematic approach to *searching* and identifying mobile systems, as well as probing large data sets comprised of signals intelligence and imagery to identify patterns and connections. Thirdly, the use of cyber weapons (which in turn will be increasingly automated via AI algorithms) during the early parts of a crisis to sow confusion in the networks of nuclear states and impede C2 systems could reduce political and military leaders' faith in their ability to assess the situation and control escalation. Finally, these capabilities could be combined with AI-improved missile defense systems to shoot down retaliatory second strikes (Hawley, 2017; Harper, 2019). The combination of the above factors risks upsetting the nuclear balance of power between the US and China, providing the US with *unidirectional deterrence*. This would be a one-way deterrent capability that denied others the ability to deter the US (Knopf, 2010, p. 3). In turn, this would give Washington greater freedom of action and, with it, enhance its ability to bend the outcome of future crises in its favor. This scenario is intolerable to Beijing (and Moscow) compelling their elites to take measures, including using AI, to advance their own nuclear deterrence and defense arsenals.

Marshaling AI power: Comparing the US and China in AI development

Although much of the data in this section are not specifically military-related (we turn to that toward the end), it provides us with an idea of the balance of technological and AI power between the US and China and how the trend lines are shifting between them. As previously noted, owing to its role as a general purpose technology, AI is akin to technologies that played an important role in previous industrial revolutions in determining the balance of military and economic power. Consider that the development of the microchip, which occurred in the US, has been a core technology out of which numerous subsequent technologies have emerged; in turn creating new military capabilities and enhancing national power. In the emerging Fourth Industrial Revolution, the state/s with the most advanced technological power will likely be best at translating AI into military might and global influence.

National efforts

In July 2017, China announced its first national strategy on AI – the *New Generation Artificial Intelligence Development Plan* (Webster, Creemers, Triolo & Kania, 2017) – and a *Three-Year Action Plan* in December 2017. The short-term objective is to catch up with America's lead in AI-based technology and applications by 2020. Medium-term objectives include making major achievements in AI applications in the field of manufacturing, defense, regional, and extra-regional security assessment, medicine, agriculture, and rules and regulations on AI by 2025. Long-term objectives encompass the establishment of China as the world leader in AI by 2030 (ibid). Another key element of China's effort is the reduction of barriers between the government and key tech giants (including Baidu, Alibaba, and Tencent), enabling Beijing to channel AI developments and accelerate technology transfer from the private sector to the military with the help of a Military-Civil Fusion Development Commission created in 2017 (Horowitz, Kania, Allen & Scharre, 2018).

By contrast, the US approach has been hesitant and erratic. In 2014, US Secretary of Defense, Chuck Hagel, articulated the *Third Offset Strategy* that sought to draw upon modern technologies (like AI) to offset growing US disadvantages against A2/AD and to "sustain and advance America's military dominance for the 21st century" (Dombrowski, 2015, p. 4). This was followed in 2015 when the Obama administration began efforts to build government–private sector partnerships and to incorporate commercial technology into its security organization. This included the Department of Homeland Security and DoD (the Defense Innovation Unit) opening offices in Silicon Valley (Fang, 2019). In the last days of the Obama administration, two government reports released in October 2016 (the first titled the *National AI R&D Strategic Plan*, the second *Preparing for the Future of Artificial Intelligence*) sought to elevate the importance of AI in US government research and positioned the US to fast track an AI strategy (Agrawal, Gans & Goldfarb, 2017). However, after the unexpected victory of Donald Trump in the 2016 November elections, Obama's strategy was shelved as science and technology was placed on the backburner as the administration dithered over next steps.[3]

It was not until 2019 that substantive moves were recommenced. In February 2019, President Donald Trump released an *Executive Order* (EO) *on Maintaining*

American Leadership in Artificial Intelligence (the 'AI Initiative'). This directed the government to play a key role in driving "technological breakthroughs in AI"; "promote scientific discovery, economic competitiveness, and national security" and to "sustain and enhance the scientific, technological, and economic leadership position of the United States in AI R&D and deployment" (White House, 2019). Another key objective, consistent with aspects of the US-initiated trade war with China, is to "protect the United States advantage in AI and AI technology critical to United States economic and national security interests against strategic competitors and adversarial nations" (ibid). This was followed by the *National AI R&D Strategic Plan* in June 2019 (an updated version of the 2016 plan that added the priority of enhancing public–private partnerships). It explained that the AI Initiative "implements a whole-of-government strategy in collaboration and engagement with the private sector, academia, the public, and likeminded international partners", while the Plan laid out a set of R&D priorities to ensure federal "investments can achieve their full potential" and the US remain the "world leader in AI" (Executive Office of the President, 2019).

In short, the strategic plans of China and the US contain clear statements of intent to secure a lead in AI and emphasize the need to establish links between their commercial, academic, and military industrial sectors to deliver benefits to their military establishments. This exists against the backdrop of their broader AI 'ecosystems', which in turn provide a structural if fluid basis out of which AI military power can be forged. It also raises the question of which state is currently in the lead. This topic is considered below.

Comparing the US and China

According to STRATFOR, in August 2018, the US was ahead of China in AI development in the critical areas of "hardware, research, and development", owing to its "dynamic commercial AI sector" (Goujon, 2018). As of August 2019, the US remained the leading AI power, according to the Center for Data Innovation (CDI) (Castro, McLaughlin, and Chivot, 2019). It leads in four of six categories (talent, research, development, and hardware), with China catching up and currently leading in two categories (adoption and data). For more data, Table 4.1 below provides a breakdown of where the US and China stand across a range of metrics. While we cannot easily identify a direct line between these statistics and the military advantages they will or could offer, they represent elements that could be harnessed to generate AI military power, while AI and other military technologies in turn will influence the general balance of power.

In addition to the information above, the most prestigious institutes for AI education and R&D (such as Carnegie Mellon University, University of California, Berkeley, Stanford University, Massachusetts Institute of Technology, and Cornell University) are located in the US (Chuvpilo, 2018). These attract people from diverse backgrounds in AI, leading to a diverse AI pool of information. Furthermore, according to Sarah O'Meara, "Most of the world's leading AI-enabled semiconductor chips are made by US companies such as Nvidia,

Table 4.1 Comparing the US and China in AI development[4]

Main driver in AI	*Proxy measures*	*China*	*US*
Hardware	International market share of semiconductor products (2015)	4% of world	50% of world
	Financing for FPGA chipmakers (2017)	$34.3 million (7.6% of world)	$192.5 million (5.5% of world)
Data	Mobile users (2016)	1.4 billion (20% of world)	416.7 million (5.5% of world)
Research and algorithms	Number of AI researchers (2017)	18,232	28,536
	Percentage of AAAI Conference Presentations (2015)	20.5% of world	48.4% of world
Commercial AI sector	Proportion of world's AI companies (2012–2016)	23%	42%
	Total investments in AI companies (2012–2016)	$2.6 billion (6.6% of world)	$172 billion (43.4% of world)
	Total global equity funding to AI startups (2017)	48% of world	38% of world
	Number of 'unicorns' (private startups valued at over $1 billion)	64	114
	Total valuation of unicorns	$280 billion	$392 billion
	Number of firms in top 15 for semiconductor sales (2019)	1	6
	Number of firms designing AI chips (2019)	26	55
	Number of supercomputers ranked in top 100 (2019) (top500.org)[5]	22	37

Intel, Apple, Google, and Advanced Micro Devices" (O'Meara, 2019). This gap is primarily because China is new to the chip industry, with estimates that China requires five to ten more years to develop expertise in the cutting-edge technology to design and manufacture integrated circuits and semiconductors (Soo, Sai, and Jing, 2019). China also has to deal with the US-initiated trade war and efforts by Washington to keep Beijing at arm's length from sensitive American hardware and technologies, efforts consistent with the February EO's objective of "protecting the American AI technology base from attempted acquisition by strategic competitors and adversarial nations" (White House, 2019).

The number (in 2017) of AI start-ups in the US also outstripped China, standing at 1,393 to 383, while (also in 2017) 1,727 US AI firms received more than $1 million in funding compared to 224 firms in China (Castro et al, 2019). On the other hand, there are more Chinese workers working in firms adopting and piloting AI than there are American workers (ibid), which is likely attributable to China's larger population. In the categories of mobile users and access to data, China outstrips the US by a wide margin owing to its vast population. Notably the founder of Sinovation (a Chinese venture capital firm), Kai-fu Lee, has stated "in the age of AI, data is the new oil, and China is going to be the new Saudi Arabia" (Lee, 2018). A larger population and low barriers to data collection owing to the absence of liberal values regarding personal privacy and the personal ownership of data, provides an immense pool of data to feed and train AI algorithms in China and opportunities to pilot and experiment with AI on the domestic population. In this area, China leads in comparison to the US (Castro, et al. 2019). Chinese researchers also publish more AI papers than their American counterparts do (in 2018 China's output was 27,231 compared to 23,965 for the US) – and the quality of research is expected to catch up to US publications by 2020 (Simonite, 2019). Additionally, according to a *Global Times* report, "China accounts [in July 2018] for 60% of global AI investment" (Sheng, 2018).

China also leads the US in public–private partnerships that further enhance its ability to develop, pilot, and employ AI, with the state taking an equity in relevant companies (Goujon, 2018). This military–civil fusion allows the Chinese military to directly benefit from developments in the private sector (Simonite, 2019). Lower barriers across the Chinese system mean the government can potentially operate "like a grand symphony" on AI.[6] China has also found a willing partner in Russia, as the two Eurasian giants expand their military and technological cooperation that includes AI, 5G, and robotics (Bendett and Kania, 2019). Joint development of dual-use technologies to augment their power is part of a broader deepening of their strategic partnership that, in turn, is driven by their intensifying rivalries with the US and Russia's policies to limit their engagement with the global technological ecosystem. Relatedly, for the US (and plausibly for Russia, and China in the future), ensuring partner and allied militaries are AI-enabled and remain interoperable with their own forces has been identified as a priority (Department of Defense, 2018). It will ensure they can contribute on the battlefield in support of their aims rather than become liabilities or de facto neutral in future conflicts.

Comparing trends and the Center for Security and Emerging Technology estimates

Here, firstly, we consider the trends lines of the US and Chinese economies and defense budgets, which provides a rough indication of where Washington and Beijing stand and how AI might change the balance of military power between them (notwithstanding distortions based on the difficulty of actually integrating AI into military capabilities, discussed in the next section). We then consider statistics related to how much each is estimated to spend specifically on military AI.

Relevant statistics illustrate the growth in China's relative power: In 1994, the US accounted for roughly one-fourth of world GDP and 40% of global military spending; by 2015, it accounted for 22.4% of world GDP and 33.8% of global military spending. In comparison, China accounted for 3.3% of the world GDP in

1994 and 2.2% of global military spending; by 2015, these figures rose to 11.8% of global GDP and 12.2% of world military spending (Brands, 2018, pp. 11–12). Estimates for 2019 indicate the US economy accounted for 24.8% of world GDP compared to China with 16.3% and US military spending in 2018 was 36% of the global total and China's 14% (IMF, 2019; SIPRI, 2018, p. 2). In short, the US remains ahead but China is rapidly catching up.

A common figure that makes the rounds in Washington, DC suggests China spent $12 billion on AI in 2017, with projections this will rise to $70 bn in 2020. However, estimates from a December 2019 report by the Center for Security and Emerging Technology (CEST) (Acharya and Arnold, 2019)[7] suggests China is spending far less than the aforementioned figures, with Kareo Hao explaining that most of Beijing's expenditure is directed toward "non-military-related research, such as fundamental algorithm development, robotics research, and smart-infrastructure development". She continues, "By contrast, the US's planned spending for fiscal year 2020 allocates the majority of its AI budget to defense, which means it could possibly outspend China in that category" (Hao, 2019).

Showing the large uncertainty over what the real figures are, CEST estimates that China's *specific expenditure on military AI* in 2018 was at the low end, $300 million, or $2.7 billion at the high end. This compares to the approximately $4.9 billion the US military will spend on military AI in 2020. In terms of total spending on non-military *and* military AI, China (in 2018) spent between $2 billion and $8.4 billion, against the projected 2020 figure for the US of $5.9 billion (Acharya and Arnold, 2019, p. 13). What the precise figure of Chinese spending on AI by 2020 will be, and how it will compare against America's spending, is unknown and likely will remain so, given the lack of authoritative data; but spending by both may very well be close to one another, suggesting a relative dead-heat.

Marshaling AI power: A brief discussion

Based upon the aforementioned data, it appears the US is in the lead in terms of AI power. At the macro level, its economy remains larger and more dynamic; its military is better funded and retains a broader technological edge, and has more experience over decades of developing, fielding, and deploying advanced weapons systems (Imbrie, Kania, and Laskai, 2020). Additionally, in many of the categories considered by STRATFOR and the CDI (captured in Table 1 above), the US is ahead of China. This is not to dismiss China's increasing level of investments in AI and proximate industries, the seeming seamlessness of public–private partnerships (although this may be exaggerated, as the next section discusses), or its ability to generate and harness a larger amount of data than the US. These are all important, but even the question of who has access to more data is more complex than it first appears. At the domestic level, China clearly has more data to draw upon given its larger population; yet the amount of data the US has on hand, owing to its global extent of military installations and sensors, likely outstrips China by an order of magnitude at the global level.

We should note that there is also an active debate over whether (as a general and dual-use technology) it is meaningful to separate spending on military AI from considerations over general AI expenditure given algorithms that can augment, say, a smart phone's applications could be applied and repurposed to improve missile targeting. In this view, whether funds are specifically earmarked for civilian, commercial, or military R&D is not particularly meaningful, and the most significant take away for considering the military balance of power, and AI's role in influencing it, is the relative levels of general expenditure on AI. However, it is difficult to authoritatively determine the level of relative spending on general AI; the consensus is that China is ahead, and this may be the case, yet it may not be ahead by much. Furthermore, if, as Payne notes, future military efficacy and victory on the battlefield depends on a competition of optimizations (who has the best algorithms) then we have to admit we do not know which state has a qualitative advantage, and we may not know until we see capabilities actually used on the battlefield. All things being equal, given the above data and the US lead across a range of metrics, we guess the US will, in the short term, have the best algorithms on hand that that can be transferred and applied to its military systems.

Challenges to adoption

Data and calculated speculation can only tell us so much, as both the US and China face a range of challenges to successful AI-military adoption. The expectation that AI will accelerate the speed of battlefield decision making suggests that delegating more control and authority to lower levels of the military command will maximize tactical optimization, a notion that conflicts with the PLA's culture of consolidating control at higher levels. Indeed, the existing information-led RMA has led PLA commanders to increase their micromanagement rather than encourage initiative at lower levels (Wright, 2019, p.159). Furthermore, as AI technologies become increasingly complex, it may offset the "advantage of backwardness" (the ability of states to imitate and replicate technological developments of other countries that made first-mover investments) and impede the diffusion effects of globalization and communications, making it more difficult for the PRC to copy US advances (Gilli & Gilli, 2019). Trust will also be a critical factor for both the US and Chinese militaries, necessitating rigorous testing and exercises of AI-enabled forces prior to deployment, as well as high levels of cyber security and methods to ensure the reliability of data.

Moreover, it is not technology itself but the way it is used that can have a decisive advantage on the battlefield (Horowitz, Kania, Allen & Scharre, 2018). For example, while the British were first to develop aircraft carriers in 1918, they only envisioned their use to carry airplanes as 'spotters' for their naval destroyers. Instead, it was the navies of the ascendant powers – Japan and the US – that innovated them as mobile airfields, revolutionizing naval warfare in the 20th century (ibid, p. 19; Horowitz, 2018). Additionally, Prussia was the first European state to use railways as a military tool, playing a role in the defeat of France during the 1870 Franco-Prussian War (Lorimer, 2006), while the Nazi 'blitzkrieg' of

Europe utilized existing mobile forces in a new way to make rapid gains. All these instances showed how revisionist powers were more willing and able than status quo states to innovate and gain asymmetric advantages on the battlefield. In short, the ability of states to innovate – to be *creative* – in their use of AI in military affairs may be as important or possibly more important to gain a decisive edge in future conflicts than it was in earlier eras.

Militaries accustomed to research and development and deployment of innovations, will more readily incorporate AI than others (the US has a lot of historical experience on this score); innovation must be matched with successful integration. However, military bureaucracies may resist technological changes since the assimilation of AI into preexisting military platforms is expected to alter, and in the process disrupt, existing military force structures and processes ranging from training, doctrine, and strategies to logistics and human resources. For example, the US Navy was reluctant to finance the X-47B (an autonomous fighter/bomber aircraft) despite successful sea trials, because it threatened the most prestigious "'tip-of-the-spear' jobs" (Cummings, 2017, p. 9). The implementation of the Algorithmic Warfare Cross-Functional Team (Project Maven) faced significant resistance within the Pentagon mainly because it was deemed disruptive and not certain to confer any immediate advantages to the US military (Sayler, 2019).

The US military has also experienced difficulty dealing with Google – a private sector actor – during Project Maven. The revelation that Google joined the project in March 2018 led to protests and a petition by employees against Google's participation, leading Google to not renew the contract in March 2019 (Fang, 2019). This speaks to a key point: At present, AI is primarily driven by the private sector; where once the federal government took the lead in funding ambitious high-risk high-reward research such as nuclear, GPS, and the internet – providing some military advantage to the US, at least initially – now it is private companies pushing emerging technologies forward. As such, the US DoD often has to adapt commercially developed technologies for military use after its development and commercialization. Much of this AI is available on the open market, reducing or even obviating the ability of the US military to secure first-mover advantages over new algorithms. However, it must be recognized that the US tech culture is not a monolith as other companies, such as Anduril Industries, do not share Google's concerns. Not only is Anduril actively engaged in moving ahead with involvement in Pentagon projects (for example by using virtual reality (VR) to assist soldiers with identifying targets and direct unmanned vehicles in combat and to deploy VR for battlefield-management systems) but their founder, Palmer Luckey, is outspoken in his critique of other tech companies unwillingness to work with the Pentagon, asserting that it impedes the US national interest and gives advantage to China (ibid).

Nevertheless, China also faces challenges on this front and the effectiveness of China's vaunted civil–military fusion could be exaggerated, as the motivation behind it is, in part, driven by the fact that its state-owned enterprises comprising China's defence industrial base are separated from the economy; its booming civilian tech sector is not easily able to translate its innovations into the military.

Lorand Laskai explains, "the [PRC] defense sector remains bureaucratic, secretive, and vertically-integrated. This strongly contrasts with horizontal, 'move fast and break things' workstyle of entrepreneurs and innovators both in [the] US and China" (Laskai, 2018). He further warns that, unless greater state-owned enterprise reforms take place, many private firms will not be able to contribute effectively and, as such, "Chinese military planners would largely prefer an American-style defense industry that promotes the efficient allocation of resources through private sector collaboration and procurement" (ibid).

Conclusion

As a broad technological enabler, AI will play an increasingly important role in defining the global balance of military power between China and the US. The implications are immense, but also uncertain. For example, the speed of AI technological advance makes it a fast-changing capability – if either state secures an operational military advantage that outstrips the systems of the other, the military balance could rapidly change and with it a shift at the systemic level of power (unipolar, bipolar, multipolar, etc.). Advances by one state in AI may also provide "windows of opportunity" by shifting the (real or perceived) military balance of power in their favor (Lebow, 1984). Into this space, revisionist states may be more tempted to engage in limited interventions in regional settings or expand their use of 'salami tactics' (gradual actions to incrementally shift the status quo in their favor) where the balance of power favors them. Recent efforts by China to expand its military presence in the South China Seas is a pertinent example. Furthermore, the proliferation of sensors worldwide and ever-increasing amounts of data could be harnessed by AI to improve the ability to calculate the relative balance of conventional military power, at the same time that uncertainty would remain in assessing the relative capabilities of rival AI across those very same forces. The prospects of brinkmanship and miscalculation would grow.

Additionally, there is a range of factors that could intersect with the mounting competition, directing it in an unpredictable direction or decisively accelerating AI developments in Washington or Beijing at the other's expense. This could come from the first and second order effects of the ongoing US trade war; related efforts driven by Washington to technologically and economically decouple from China (and compel US allies to do the same) – efforts which appear all-but-certain to be accelerated due to the fallout from COVID-19 (Pamuk and Shalal, 2020); the emergence of a domestic political crisis in either nation; and/or the intersection between AI with other disruptive emerging technologies that facilitates one nation's offensive or defensive capabilities relative to the other. It is also an open question whether the US's more decentralized approach to AI development or China's 'symphonic' state-led strategy will provide a better foundation for developing military power. Indeed, the general consensus is that the combined effect of the cultural differences between Silicon Valley tech companies and the US government/military – and desire by businesses to profit off the international market – make it difficult to imagine that US tech companies will ever cooperate with the US government to the degree China's tech-giants will work with the PRC. At the same time, there are questions over whether Beijing's efforts at civil-military fusion will adequately lower boundaries between China's tech sector and military industrial base to accelerate the speed of transfer of AI innovations to the military. Ultimately, the answer to these questions, as well as the true implications of AI for the balance of power between the US and China, will only become apparent in the years and decades to come.

Notes

1 US and Chinese efforts to harness AI take place against the regional backdrop of a number of states announcing their own AI strategies (Singapore announced its plan in May 2017; Japan in March 2017; while Taiwan, Australia, Malaysia and New Zealand announced theirs in 2018, and Russia and the US in 2019). Word count considerations do not provide space to consider their efforts in this chapter; for chapters that cover some of these states in this volume see Chapters 2, 3, 5, 6, 8, and 9.
2 In June 2017, the China Electronic Technology Corporation overtook the US record by launching the world's largest fixed-wing drone swarm (Feng and Clover, 2017).
3 The Trump administration's slow approach on AI was based on an exaggerated belief that the private sector would drive US advances in AI; resistance to Obama policies; and a culture gap between the administration and Silicon Valley (Dr Reuben Steff interviews with think tank officials in Washington, DC).
4 This table draws upon information provided by the Future of Humanity Institute and Center for Data Innovation. See Goujon, 2018 and Castro, McLaughlin and Chivot, 2019.
5 In the top 10, US computers rank 1st, 2nd, 5th, 7th, and 10th; Chinese computers rank 3rd and 4th.
6 Interview: Dr Steff and think tank expert, Washington DC, April 2018.
7 The report caveats its analysis by explaining that any assessment of China's spending on AI generally – and specifically on military AI – is inherently speculative, owing to significant data gaps.

References

Acharya, A and Arnold, Z (2019) 'Chinese public AI R&D spending: provisional findings', *Center for Security and Emerging Technology Issue Brief*. December. https://cset.georgetown.edu/wp-content/uploads/Chinese-Public-AI-RD-Spending-Provisional-Findings-1.pdf, accessed April 9, 2020.

Agrawal, A, Gans, J and Goldfarb, A (2017) 'The Obama administration's roadmap for AI policy', *Harvard Business Review*. December 21. https://hbr.org/2016/12/the-obama-administrations-roadmap-for-ai-policy, accessed November 13, 2019.

Allen, G and Chan, T (2017) 'Artificial intelligence and national security', *Belfer Center Study*. July. https://www.belfercenter.org/sites/default/files/files/publication/AI%20NatSec%20-%20final.pdf, accessed November 13, 2019.

Altmann, J and Sauer, F (2017) 'Autonomous weapon systems and strategic stability', *Survival*, 59(5), pp. 117–142.

Bendett, A and Kania, E (2019) 'A new Sino-Russian high-tech partnership', *Australian Strategic Policy Institute*. October 29. https://www.aspi.org.au/report/new-sino-russian-high-tech-partnership, accessed April 9, 2019.

Brands, H (2018) *American grand strategy in the age of Trump*. Washington: Brookings Institution Press.

Castro, D, McLaughlin, M and Chivot, E (2019) *Who is winning the AI race: China, the EU or the United States?* Center for Data Innovation. August 19. https://www.datainnovation.org/2019/08/who-is-winning-the-ai-race-china-the-eu-or-the-united-states/, accessed November 13, 2019.

Collins, J and Futter, A (2015) *Reassessing the revolution in military affairs transformation, evolution and lessons learnt*. London: Palgrave Macmillan.

Chuvpilo, G (2018) 'Who's ahead in AI research?', *Medium*. August 8. https://medium.com/@chuvpilo/whos-ahead-in-ai-research-insights-from-nips-most-prestigious-ai-conference-df2c361236f6, accessed November 13, 2019.

Cummings, M (2017) *Artificial intelligence and the future of warfare*. Chatham House. January. https://www.chathamhouse.org/sites/default/files/publications/research/2017-01-26-artificial-intelligence-future-warfare-cummings-final.pdf, accessed November 13, 2019.

Davis, ZS (2019) *Artificial intelligence on the battlefield: an initial survey of potential implications for deterrence, stability, and strategic surprise*. Center for Global Security Research Lawrence Livermore National Laboratory. March. https://cgsr.llnl.gov/content/assets/docs/CGSR-AI_BattlefieldWEB.pdf, accessed December 6, 2019.

Department of Defense (2017) *The operational environment and the changing character of future warfare*. https://community.apan.org/wg/tradoc-g2/ace-threats-integration/m/documents/266218, accessed November 13, 2019.

Department of Defense (2018) *Summary of the 2018 department of defense artificial intelligence strategy*. https://media.defense.gov/2019/Feb/12/2002088963/-1/-1/1/SUMMARY-OF-DOD-AI-STRATEGY.PDF, accessed December 5, 2019.

Dombrowski, Peter (2015) 'America's third offset strategy: new military technologies and implications for the Asia Pacific', *RSIS*. June 2015. https://www.files.ethz.ch/isn/191706/PR150608Americas-Third-Offset-Strategy.pdf

Executive Office of the President (2019) *The national artificial intelligence research and development strategic plan: 2019 update*. https://www.nitrd.gov/pubs/national-AI-RD-Strategy-2019.pdf, accessed November 13, 2019.

Fang, L (2019) *Defense tech startup founded by Trump's most prominent silicon valley supporters wins secretive military AI contract*. Intercept. March 10. https://theintercept.com/2019/03/09/anduril-industries-project-maven-palmer-luckey/, accessed November 13, 2019.

Feng, F and Clover, C (2017) 'Drone Swarms vs. conventional arms: China's military debate', *Financial Times*. August 25. https://www.realcleardefense.com/2017/08/25/drone_swarms_vs_conventional_arms_chinarsquos_military_debate_296204.html

Futter, A (2018) *Hacking the bomb: cyber threats and nuclear weapons*. Washington: Georgetown University Press.

Geist, E and Lohn, AJ (2018) *How might AI affect the risk of nuclear war?* RAND. https://www.rand.org/pubs/perspectives/PE296.html, accessed November 13, 2019.

Gilli, A and Gilli, M (2018–19) 'Why China has not caught up yet', *International Security*, 43(3), pp. 141–189.

Goujon, R (2018) 'AI and the return of great power competition', *Stratfor*. August 2. https://worldview.stratfor.com/article/ai-and-return-great-power-competition, accessed November 13, 2019.

Hao, K (2019) 'Yes, China is probably outspending the US in AI—but not on defense'. *MIT's Technology Review*. December 5. https://www.technologyreview.com/2019/12/05/65019/china-us-ai-military-spending/, accessed April 9, 2020.

Harper, J (2019) 'Just in: Pentagon contemplating role of AI in missile defense', *National Defense*. October 7. https://www.nationaldefensemagazine.org/articles/2019/10/7/pentagon-contemplating-role-of-ai-in-missile-defense, accessed April 9, 2020.

Hawley, JK (2017) 'Patriot wars: automation and the patriot air and missile defense system', *Center for a New American Century*. January. https://css.ethz.ch/content/dam/ethz/special-interest/gess/cis/center-for-securities-studies/resources/docs/CNAS-Patriot%20Wars.pdf, accessed April 9, 2020.

Hoadley, D and Lucas, NK (2018) *Artificial intelligence and national security*. April 26. https://www.hsdl.org/?abstract&did=810166, accessed November 13, 2019.

Horowitz, M (2018) 'Artificial intelligence, international competition, and the balance of power', *Texas National Security Review*, 1(3), pp. 37–57.

Horowitz, M, Kania, E, Allen, G and Scharre, P (2018) 'Strategic competition in an era of artificial intelligence', *Center for a New American Security*. July 25. https://www.cnas.org/publications/reports/strategic-competition-in-an-era-of-artificial-intelligence, accessed November 13, 2019.

Imbrie, A, Kania, E and Laskaim, L (2020) 'The question of comparative advantage in artificial intelligence enduring strengths and emerging challenges for the United States', *CEST Policy Brief*. January. https://cset.georgetown.edu/wp-content/uploads/CSET-The-Question-of-Comparative-Advantage-in-Artificial-Intelligence-1.pdf, accessed April 9, 2020.

IMF. Projected GDP Ranking (2019–2024) *Statistics Times*. http://statisticstimes.com/economy/projected-world-gdp-ranking.php, accessed November 12, 2019.

Kania, E (2017a) 'Go and beyond: the Chinese military looks to future "Intelligentized"', *Lawfare*. June 5. https://www.lawfareblog.com/alphago-and-beyond-chinese-military-looks-future-intelligentized-warfare, accessed November 13, 2019.

Kania, E (2017b) *Battlefield singularity: artificial intelligence, military revolution, and China's future military power*. November. https://s3.amazonaws.com/files.cnas.org/documents/Battlefield-Singularity-November-2017.pdf?mtime=20171129235804, accessed November 13, 2019.

Kania, E and Costello, JK (2017) 'Quantum technologies, U.S.-China strategic competition, and future dynamics of cyber stability', *2017 international conference on cyber*, Washington, DC.

Knopf, JW (2010) 'The fourth wave in deterrence research', *Contemporary Security Policy*, 31(1), pp. 1–33.

Laskai, L (2018) 'Civil-military fusion: the missing link between China's technological and military rise', *Council on Foreign Relations*. January. https://www.cfr.org/blog/civil-military-fusion-missing-link-between-chinas-technological-and-military-rise, accessed December 6, 2019.

Lebow, RN (1984) 'Windows of opportunity: do states jump through them?', *International Security*, 9(1), pp. 147–186.

Lee, K (2018) *AI superpowers: China, silicon valley, and the new world order*. New York: Houghton Mifflin Harcourt.

Lieber, KA and Press, D (2017) 'The new era of counterforce: technological change and the future of nuclear deterrence', *International Security*, 41(4), pp. 9–49.

Lorimer, JG (2006) 'Why would modern military commanders study the Franco-Prussian war?', *Defence Studies*, 5(1), pp. 108–123.

McNeill, WH (1982) *The pursuit of power: technology, armed force, and society since A.D. 1000*. Chicago: University of Chicago Press.

Morgenthau, H (1960) *Politics among nations: the struggle for power and peace* (3rd ed.). New York: Alfred Knopf.

O'Meara, S (2019) *Will China overtake the U.S. in artificial intelligence research?* August 24. https://www.scientificamerican.com/article/will-china-overtake-the-u-s-in-artificial-intelligence-research/, accessed November 13, 2019.

Pamuk, H and Shalal, A (2020) *Reuters*. May 4. https://uk.reuters.com/article/us-health-coronavirus-usa-china/trump-administration-pushing-to-rip-global-supply-chains-from-china-officials-idUKKBN22G0BZ, accessed May 6, 2020.

Payne, Kenneth (2018) 'Artificial intelligence: a revolution in strategic affairs?', *Survival*, 60(5), 7–32.

Reinsel, D, Gantz, J and Rydning, J (2018) 'The digitization of the world from edge to core', *International Data Corporation*. November. https://www.seagate.com/www-content/our-story/trends/files/idc-seagate-dataage-whitepaper.pdf, accessed December 6, 2019.

Robles, P (2018) 'China plans to be a world leader in artificial intelligence by 2030', *South China Morning Post*. October 1. https://multimedia.scmp.com/news/china/article/2166148/china-2025-artificial-intelligence/index.html, accessed November 13, 2019.

Roth, M (2019) *Artificial intelligence in the military – an overview of capabilities*. February 22. https://emerj.com/ai-sector-overviews/artificial-intelligence-in-the-military-an-overview-of-capabilities/, accessed December 6, 2019.

Saalman, L (2018) 'Fear of false negatives: AI and China's nuclear posture', *Bulletin of the Atomic Scientists*. https://thebulletin.org/landing_article/fear-of-false-negatives-ai-and-chinasnuclear-posture/, accessed November 13, 2019.

Sayler, K (2019) *Artificial intelligence and national security*. January. https://fas.org/sgp/crs/natsec/R45178.pdf, accessed November 13, 2019.

Scharre, P (2014) *The coming swarm: the cost-imposing value of mass*. https://www.jstor.org/stable/resrep06138, accessed November 13, 2019.

Schwab, L (2015) 'The fourth industrial revolution what it means and how to respond', *Foreign Affairs*. December 12. https://www.foreignaffairs.com/articles/2015-12-12/fourth-industrial-revolution, accessed November 13, 2019.

Shanahan, J (2019) 'Artificial intelligence – a game changer and decisive edge', *Armed Science*. September 11. https://science.dodlive.mil/2019/09/11/artificial-intelligence-a-game-changer-and-decisive-edge/, accessed March 15, 2019.

Sheng, Y (2018) *China attracts 60% of global AI investment*. July. http://www.globaltimes.cn/content/1110703.shtml, accessed November 13, 2019.

Simonite, T (2019) 'China is catching up to the US in AI research—fast', *Wired*. March 13. https://www.wired.com/story/china-catching-up-us-in-ai-research/, accessed November 13, 2019.

Soo, Z, Sai, S and Jing, M (2019) 'Lagging in semiconductors, China sees a chance to overtake the US with AI chips as 5G ushers in new era', *South China Morning Post*. https://www.scmp.com/tech/enterprises/article/3027775/lagging-semiconductors-china-sees-chance-overtake-us-ai-chips-5g, accessed November 13, 2019.

Top 500 List (Supercomputers) (2019) June. https://www.top500.org/, accessed November 13, 2019.

Trends in World Military Expenditure 2018 (2018) *SIPRI*. April 2019. https://www.sipri.org/publications/2019/sipri-fact-sheets/trends-world-military-expenditure-2018bra

Webster, G, Creemers, R, Triolo, P and Kania, E (2017) *China's plan to "Lead" in AI: purpose, prospects, and problems*. New America Foundation. August 1. https://www

.newamerica.org/cybersecurity-initiative/blog/chinas-plan-lead-ai-purpose-prospects-and-problems/, accessed November 13, 2019.

White House (2019) *Executive order on maintaining american leadership in artificial intelligence*. February 11. https://www.whitehouse.gov/presidential-actions/executive-order-maintaining-american-leadership-artificial-intelligence/, accessed November 11, 2019.

Wright, N (2019) *Artificial intelligence, china, russia, and the global order*. Air University Press. October. https://www.airuniversity.af.edu/Portals/10/AUPress/Books/B_0161_WRIGHT_ARTIFICIAL_INTELLIGENCE_CHINA_RUSSIA_AND_THE_GLOBAL_ORDER.PDF, accessed December 6, 2019.

5 Mitigating accidental war

Risk-based strategies for governing lethal autonomous weapons systems

Aiden Warren and Alek Hillas

Introduction

The continual advancement of new technologies in theaters of conflict poses distinct challenges to global security. While concerns surrounding drones have been debated extensively over the last decade, it is the advent of lethal autonomous weapons systems (LAWS) which is intensifying the ethical, legal, governance, and strategic debates pertaining to the use (and misuse) of unmanned military power. The actions that LAWS can, will, or should be allowed to perform is a major policy question. In describing these "machines with the power and discretion to take lives without human involvement", the UN Secretary-General, António Guterres, claims that LAWS are "politically unacceptable, morally repugnant, and should be prohibited by international law" (UN News, 2019). While the UN Group of Governmental Experts (GGE) has not agreed on a definition for LAWS, the major difference compared to that of drones is that humans would not directly control the use-of-force by LAWS. While no one can predict with certainty the actions LAWS might perform, experts from a variety of fields – including roboticists, military personnel, politicians, lawyers and philosophers – must consider policies to mitigate accidental conflict involving LAWS, particularly if considered to be a device used in a humanitarian or strategic context.

In attempting to add much needed connective tissue to such complex debates, this chapter will begin with a brief review of the context that decision-makers faced during major crises during the Cold War, and how crises could play out very differently today. Second, it considers regulation to enforce limits on the weapons carried by LAWS in a bid to minimize the consequences of present-day and future conflicts. And lastly, the chapter puts forward suggestions for finding areas of transparency and communication for development in artificial intelligence (AI) algorithms to minimize the likelihood of accidental war breaking out.

Decision-makers and crisis response

When political leaders decided to use force or de-escalate tensions in the twentieth century, their decision-making often took into account the notion that adversaries were (usually) rational actors with defined interests. Imagine how events could have changed if robots had been on hand for the totalitarian powers to use in the twentieth century, or if robots had coding errors and it was unclear whether they were acting under the direction of civilian or military authorities or non-state actors. A thought experiment to consider is the Kennedy administration's calculations during the Cuban Missile Crisis in 1962. The flashpoint involved several near misses but fortunately never escalated to armed hostilities or nuclear war. Consider if an American vessel or aircraft such as the downed U-2 reconnaissance plane piloted by Major Rudolf Anderson Jr. had been struck – potentially by a LAWS – and Havana or Moscow denied attribution. President Kennedy and advisers would have been required to factor into their response whether the "actor" using lethal force had acted outside of the chain of command. In fast forwarding to the twenty-first century, as the rapid pace and intensification of technology advances, it is likely that political leaders confronted with such situations in the future will need to make some of their strategic decisions based on ambiguities posed by LAWS. This chapter will consider such scenarios and the implications.

Proposals to address risks posed by new technologies are helpful in understanding historical precedents. A year after the Cuban Missile Crisis, John Phelps from the Institute for Defense Analyses considered contingency strategies for technology that could lead to an accidental war (Phelps was also the Arms Control Committee Chairperson of the Federation of American Scientists). He identified false radar signals, miscommunication, and misunderstood orders as possible causes and sources of accidents. Phelps was concerned with more than machine error. Human error existed in the form of "miscalculations by statesmen and military decision-makers" and would "present a greater danger than do specific man-machine failures" (Phelps, 1963, pp. 176–179).

Indeed, the establishment of a direct communications link between the superpowers in 1963 (updated again in 1971, 1984, and 1988) indicated that political leaders wanted to reduce misunderstandings and avoid unnecessary military escalation. The US Department of State (c. 2009–2017a) assesses that the hotline "proved its worth since installation", for example, by providing the ability to "prevent possible misunderstanding of US fleet movements in the Mediterranean" during the 1967 and 1973 Arab-Israeli wars. Nevertheless, according to Phelps, communication cannot always lead to peaceful resolutions. This includes when "one side faces the prospect of having its retaliatory forces (its deterrent) substantially destroyed unless it launches these forces immediately" (Phelps, 1963, p. 178). When, in 1983, this almost occurred in a time of heightened tensions between the superpowers, a Soviet officer, Lieutenant Colonel Stanislav Petrov, correctly assessed that early warning system signals emanating from "incoming" US intercontinental ballistic missiles were a false alarm. Petrov made the decision not to "retaliate" based on his knowledge of previous computer system malfunctions and the nuclear first-strike scenario he had trained for being different to this "attack". Given the ambiguity decision-makers will encounter in future crises where LAWS exist, when assessing a potential near miss, accident, or deliberate use-of-force, it is imperative they understand the plethora of variables and the wider context before finalizing their judgment.

While AI may assist with situational awareness and reducing uncertainty – for example, monitoring unexpected troop or vessel movements in peacetime – once a conflict involving LAWS breaks out, decision-makers will need to be prepared to engage their counterparts, when possible, to avoid miscalculation. The UN Charter gives states an inherent right to individual or collective self-defense against an armed attack. Yet, some legal scholars, such as Ugo Pagallo, have claimed it is uncertain "whether military robotics technology somehow changes this 'inherent right'" (Pagallo, 2011, p. 315). However, in August 2018 the GGE's Possible Guiding Principles (for emerging technologies in the area of LAWS) "affirmed that international law, in particular the UN Charter and international humanitarian law", was *non*-binding (United Nations 2018). Peter Asaro (2008, p. 55) has noted, "it may be difficult to distinguish a genuine intention from a technical error", but found that states may retaliate, provided they do not launch or expand the scenario into a larger war. While states could certainly return "fire with fire", human judgement is required before scaling up hostilities and LAWS could not necessarily make these decisions independently.

When states have an offensive doctrine, they can raise the risk of war breaking out. These states prefer to "shoot first, [and] ask questions later" in uncertain conditions (Paul, 1995, p. 272). Some political leaders have used accidental attacks as a justification to expand into wider hostilities, regardless of the legality (Asaro, 2008, p. 55). On the other hand, calculated belligerence can, in some instances, lead to de-escalation. Here, ideas and tactics related to brinkmanship are relevant. Brinkmanship is "the ability of an aggressor to undertake an observable action that will lead with positive probability to war or some other mutually undesirable outcome" (Schwarz & Sonin, 2008, p. 165). The calculated use of brinkmanship, near-misses, and shows of force can lead to concessions from the enemy. For example, US shows of force have succeeded against North Korea, including securing the release of the crew of the USS *Pueblo* in 1968; moving to DEFCON 3 after two US servicemen were killed in the DMZ in August 1976 (compelling Kim Il-sung to express regret); and preparing a military response to a crisis over North Korea's nuclear program in 1993, leading to the resumption of talks in Geneva (Lee, 2011, p. 13). In such cases, perhaps the learnings or experiences from past disputes acted as a framework for the escalation and subsequent de-escalation of the next conflict (Senese & Vasquez, 2008, p. 182). Following missile and nuclear tests and the death of American citizen Otto Warmbier, President Trump's unprecedented rhetoric that North Korea "will be met with fire, fury, and, frankly, power, the likes of which this world has never seen before" arguably contributed to creating the conditions for the aspirational Singapore Summit in 2018 (Trump, 2017; Trump, 2018). However, brinkmanship tactics, as a high-risk approach, historically have had mixed and often undesirable results.

As the chapter will elucidate further, the spread of autonomy and AI in existing military technology – without effective international regulation or common understanding between states pertaining to their appropriate use – is establishing an ambiguous standard for its role in the use-of-force, a situation with broader global implications. Our preference is to canvas a robust set of measures aimed at reducing the likelihood and consequences associated with the unintended use-of-force – specifically, the use-of-force by LAWS.

Minimizing the consequences: Enforcing limits on the weapons carried by LAWS

Our first focus is mitigating or minimizing the consequences of unintended use-of-force. We examine existing regulations on the types and yields of weapons, and then consider their applicability in relation to LAWS.

Most agree that LAWS should not carry payloads of chemical, biological, radiological, or nuclear (CBRN) weapons of mass destruction (WMD). Yet given the uncertainty of state adherence to the Nuclear Non-Proliferation Treaty (NPT), it could be possible that some states develop LAWS that one day carry nuclear weapons. In 2018, the US *Nuclear Posture Review* identified Moscow's development of a "new intercontinental, nuclear-armed, nuclear-powered, undersea autonomous torpedo" (US Department of Defense 2018, p. 9). During his presidential address to the Federal Assembly, Vladimir Putin (2018) announced Russia would "engage various targets, including aircraft groups, coastal fortifications and infrastructure". If, as Putin claims, the drone will "move at great depths … at a speed multiple times higher than the speed of submarines, cutting-edge torpedoes and all kinds of surface vessels" and with "hardly any vulnerabilities for the enemy to exploit", it would be difficult to disengage. Therefore, it is likely to require fully autonomous capabilities. A component of the chapter's argument – strategies to address or mitigate the unintended use-of-force by LAWS – must span from strategic to tactical weapons and explore various arms control agreements, acknowledging that a one-size-fits-all arms control agreement would not be fit for purpose. In relation to nuclear weapons, historically, the US and USSR/Russia agreed to place mutual constraints on their nuclear doctrine (including arsenal and policies) following an arms race. In the case of the *Poseidon*, intended as a second or third-strike nuclear weapon to "serve as a Russian response to concerns about the US withdrawal from the ABM [Anti-Ballistic Missile] Treaty and US advances in ballistic missile defenses" (Woolf, 2019, p. 23), diplomacy may be the most effective strategy to contain its use. On the other hand, less destructive categories of weapons would require a different set of responses. Below are some examples of the different weapon categories, each necessitating a tailored arms control regime to address distinct risks.

An advantage of LAWS would be their ability to go to areas beyond the reach of human soldiers – over land, under water, into the subterranean (such as caves) or even navigate outer space. There are different combat operation requirements for each component of the armed forces. For some missions, LAWS may only need to carry weapons with minimal firepower. For example, there is a lower risk of collateral damage if LAWS need only to fire low-caliber bullets into the soft tissue of the human eye (Russel, 2015, p. 20). To some extent, this helps to address the compliance areas of International Humanitarian Law (IHL) that relate to the principle of proportionality. Therefore, to adopt a risk-based approach to mitigate the likelihood and consequences for each category of LAWS to an acceptable tolerance level, reducing collateral damage and maximizing compliance with IHL, the political question becomes how to: (a) agree on respective limits and (b) ensure that any transfers of LAWS do not exceed these agreed-upon limits.

The history of arms control diplomacy provides some examples of how to successfully develop these limits. Paul Scharre, a Senior Fellow at the Center for a New American Security, notes it could be counterproductive to define technical limitations for LAWS. Previous arms control agreements demonstrate that "simple rules have the best track record of success". These focus more on the intent behind the weapon, than specific rules (Scharre 2017, pp. 29–30). This approach addresses the risk of improvements in technology circumventing the agreement in the future and does not incentivize states to look for loopholes in the arms control regime. There are several existing arms control agreements that can be drawn upon to prevent LAWS from exceeding category transfer limits. We explore three below:

- The Missile Technology Control Regime
- Wassenaar Arrangement on Export Controls for Conventional Arms and Dual-Use Goods and Technologies
- Convention on Prohibitions or Restrictions on the Use of Certain Conventional Weapons

The Missile Technology Control Regime (MTCR) is "an informal and voluntary association of countries which share the goals of nonproliferation of systems capable of delivering weapons of mass destruction (other than manned aircraft), and which seek to coordinate national export licensing efforts aimed at preventing their proliferation" (MTCR Annex Handbook, 2017, p. ii). Members apply the MTCR Guidelines, as an export policy, to items listed under the MTCR Equipment, Software, and Technology Annex (2019). Since its formation by the G-7 in 1987, membership has expanded to 35 countries including Russia, Turkey, India, and South Korea. China and Israel, among the world's top arms exporters, are not partners; Iran is also not a member. The MTCR keeps pace with technological advances via regular updates. It has been relatively successful in curbing or restraining the missile programs of some states (Mistry, 1997, pp. 71–73; de Klerk, 2016, pp. 2–4). In arms control, the "growing popularity [of soft law] is due to advantages such as the relative ease by which they can be adopted and updated, and the broader roles they create for stakeholders to participate in their substantive formulation" (Marchant et al, 2011, p. 306). Perhaps this type of model could be suitable for regulating weapons that are in the early stages of development, because any technical aspects that have not reached full maturity would remain subject to change.

The MTCR currently limits the payload and distance of certain types of weapons including the transfer of rocket systems or drones, and related technologies, capable of carrying WMD or any payload of at least 500 kg (1102 lb) to a distance of at least 300 km (186 miles). The MTCR may already be applicable in relation to airborne LAWS because these could be considered as a type of drone or capable of carrying the payload restricted by the agreement. The MTCR would, therefore, address some of the risks related to the proliferation of LAWS. While there are no formal verification provisions, a state can pass laws under the national legislature that allow for the unilateral punishment of MTCR violations, in effect giving more rigor to the arms control agreement. For example, US legislation allows for sanctions against non-compliant members (Center for Arms Control and Non-Proliferation, 2017). Moreover, the regime incentivizes responsible behavior as members cannot undercut each other. A member which has refused export of a prohibited item must be consulted before another member can sell a similar item. This is an important point, as most of the world's top-ten highest-spending militaries and exporters of military components and arms are MTCR members.

While the MTCR addresses some risks of WMD proliferation, the Wassenaar Arrangement (WA) focuses on conventional arms and dual-use goods. The WA was established in 1996 and has 42 states; a list of members similar to the MTCR. Chertoff (2018, pp. 2, 7) has proposed to use the WA as "a ready platform for the near-term creation of a new export control on LAWS and critical LAWS components" to "reduce the risk of transfer to malicious non-state actors". The WA has export control lists for munitions and dual-use goods and technologies (Wassenaar Arrangement, 2018). Given military technology can find its way into the hands of civilians, this consideration of dual-use technology is generally a good approach, but it is important to note potential unintended consequences arising from the regulation of dual-use technologies. As Chertoff (2018, p. 7) posits, an amendment made to the WA in 2013 (and later addressed in 2017) had "accidentally criminalized many of the necessary tools for stopping malware". While both agreements are voluntary, another difference between the WA and the MTCR is that, for very sensitive items, WA members are required to notify the Secretariat only after they have completed an export that has been previously denied by another member. This contrasts with the need to consult with the other members before finalizing an export under the MTCR (Arms Control Association, 2017). Bearing this in mind, the MTCR model has fewer incentives for exporters to undercut each other, while the WA is more tailored for dual-use technologies. Noting the limited membership of the MTCR and WA, however, in the long term any form of regulation outside of the UN system must draw on the combined strengths of both these arms control agreements, unless the purpose of the regulation is to ban LAWS altogether.

The Convention on Certain Conventional Weapons (CCW) contributes to the UN's disarmament agenda by prohibiting or restricting the use of an entire class of weapons which are indiscriminate or cause excessive injury. It can be updated with new protocols to "respond to new developments in weapons technology and challenges in armed conflicts" (United Nations, 2014a, p. 2). From 2014 to 2016, the UN Office at Geneva convened annual informal Meetings of Experts on LAWS comprised of the contracting parties to the CCW. Its initial mandate was "to discuss the questions related to emerging technologies in the area of lethal autonomous weapons systems, in the context of the objectives and purposes of the Convention", including an agenda that covered technical, ethical, sociological, legal, and military considerations (United Nations, 2014b, p. 1). From 2017 onwards, it became an open-ended Group of Governmental Experts (GGE) on emerging technologies in the area of LAWS. Given these existing institutional efforts and the fact the CCW contains protocols that take into account new technological developments, the CCW is considered by many to be the best prospect for a protocol on LAWS. It will be essential that any agreement is able to take into account new technological advances which, given the speed of technological change in the field of AI (and thus LAWS), are likely to be numerous. However, advocates for a ban are growing impatient. After the August 2019 GGE, the Campaign to Stop Killer Robots (2019), a coalition of NGOs seeking to preemptively ban LAWS, criticized participants' "diplomatic talk and no regulatory action" which had "moved even further away from a legally binding instrument or any credible outcome for their work". The Campaign asserted that, "If the CCW cannot produce a credible outcome, alternative pathways must be pursued", though no specifics were proposed.

Provided the unlikelihood, at least in the near future, of restrictions coming into effect on certain types or usages of autonomous weapons, the CCW negotiations may benefit from drawing on the technology transfer components of the MTCR and WA. This could assist because there would be some overlap or similarities between some of the hardware and software required for LAWS and non-military robots. The ability to identify and locate a target – whether it is a target to be killed in combat or a person to be saved in an emergency, like a natural disaster – has dual applications. The listing of items prohibited under the MTCR and WA is another approach to minimize consequences from the misuse of LAWS if an outright ban is not implemented as an additional protocol to the CCW.

Given the length of time it generally takes to finalize large arms control agreements, an alternative approach could be for technologically advanced states to exercise the option of developing and implementing their own export control regime on LAWS. This "pilot" could influence the negotiation for other potential agreements needing broader membership, such as the CCW. Of course, the lack of consensus to date can be perceived as a reflection of the ongoing challenge in developing a substantive governance mechanism that, as the title of this chapter denotes, works toward mitigating "accidental war".

Minimizing the likelihood of accidental war: Finding areas of transparency during development

States have an opportunity to alleviate and redefine a security environment marked by increased complexity and heightened tensions between great powers, by finding areas of transparency and communication for the development of AI algorithms and protocols. Indeed, for LAWS to comply with Rules of Engagement (ROE) and IHL, lawyers and coders will need to collaborate to translate existing or future regulations into the language of a computer algorithm. This is a difficult task. According to Anderson and Waxman (2013, p. 13), "'Programming the laws of war' at their conceptually most difficult (sophisticated proportionality, for example) is a vital research project over the long run, in order to find the greatest gains that can be had from machine decision-making within the law". That said, to minimize the likelihood of a programming error, we suggest using an open source approach where possible. Open source is "freely available with its source code" and based on "a belief that the cooperative approach it seeks to foster is the best way to create high-quality software" (Daintith and Wright, 2014). Open source site GitHub, acquired by Microsoft in 2018, is an example of collaboration across projects and repositories encompassing the public and private sector; it brings together over 50 million software developers, providing them with a platform to share and build software together (Github, 2020). In the future, this model for the development of accurate and trustworthy algorithms could demonstrate whether LAWS have compliance with IHL and ROE. If so, it would also lead to a lower margin of error during testing prior to use in the field.

Of course, governance arrangements for states intending to resolve technical issues in relation to programming the laws of war, should be limited in scope and not release information that may harm persons or damage national security. For example, a risk that will need to be managed is prospect of non-state actors (or other adversaries) reprogramming commercially available drones into LAWS. Each state thus will need to develop policies on addressing collaboration requirements with trusted partners, whether NGOs, other states, tertiary research institutions, or private providers, such as IT contractors. The assessment of a cost-benefit ratio will inform the extent to which states choose to participate in open source. The "cost" is the loss of sensitive information and foreseeable consequences arising from this action. The "benefit" is a reduction in the likelihood that there will be unintended consequences arising from conflicts involving LAWS. Fortunately, certain types of models offer flexibility. While many disarmament treaties prohibit reservations, other treaties allow reservation "space". It is preferable for a state to join a treaty with some form of reservation than not participate at all (Rosenboom, 2009). Alternatively, states may declare an intention to unilaterally abide by a treaty while remaining outside of a binding instrument. For example, while article 19 of the Mine Ban Treaty prohibits reservations and the US is not a signatory, the Obama administration announced a Presidential Policy in 2014 to align the US with the Treaty everywhere except for the Korean Peninsula (though the US Government's international legal obligations had not changed when the policy was rescinded in 2020 under the Trump administration). States that want to pursue some of the benefits of open source code without losing their sovereignty may favor limited participation in open source projects through a reservation basis and revisiting the policy as changing circumstances require.

A benefit of open source coding is that it is possible for independent identification of oversights that were left out during the initial self-assessment. For example, the 2001 Code of Conduct for Canadian Forces Personnel (ICRC, 2019a) recognizes some non-governmental organizations "do not benefit from international legal protection" compared to the special role of the International Committee of the Red Cross as an independent humanitarian institution under the Geneva Conventions. Yet, the Code requires NGOs to be respected "upon recognition that they are providing care to the sick and wounded". An algorithm should always respect the important role of non-combatants on the battlefield. Open source code would provide states and NGOs an opportunity to develop internationally recognized protective markings for use by humanitarian workers who feel that they may not be safe from LAWS, due to robots' lack of contextual awareness compared with humans (of course, this system could be abused; though any resort to perfidy would amount to a violation of IHL, as is the case today). These types of technologies could offer other benefits to enhance compliance, such as capabilities for operations where NGOs work in partnership with militaries – for example, humanitarian operations during ceasefires or natural disasters.

In addition, the creation of a no-strike list would reduce the likelihood of LAWS attacking a protected site by all participating militaries. An "off-the-shelf" list of global navigation satellite system (GNSS) coordinates and agreed-upon broadcast methods may allow humanitarian organizations to identify sites they occupy outside the main battlefields, reducing the likelihood of accidental harm (Henckaerts & Doswald-Beck, 2009, p. 127). To minimize the loss of civilian casualties when ROE processes are not followed, non-combatants need a way to request human overseers put the LAWS attack on hold, pending a rapid review of the situation. Deliberate falsification of signals "would arguably amount to perfidy" under the Geneva Conventions (Homayounnejad, 2017, p. 22), and as emphasized in the above, such penalties already exist under IHL.

A shared understanding of algorithm equivalents of ROE among states will reduce the risk of escalation. Being able to estimate more accurately whether LAWS would respond proportionately to conflicts and whether they would execute the requisite physical damage or casualty count that political and military leaders deem to be necessary is a desirable policy outcome. This would promote mutual understanding and reduce some of the unpredictability of engagements between robots. The adoption of open source common standards and legal interpretations (Davies, 2013) should reduce the margin of error or, at the very least, function as a political process to clarify the distinctions in how the LAWS of each nation would process IHL differently. This may build on the Transparency Action Plan advocated by the Stockholm International Peace Research Institute during the 2015 CCW Meeting of Experts on LAWS (Anthony, 2015, p. 3). Alternatively, such an approach may provide for greater interoperability between allied partners in joint efforts. Ideally, collaboration through open source would improve practices in areas that all responsible states agree on, such as observing the principles of distinction and proportionality. The above measures would be possible to achieve without exposing the sensitive coding of the LAWS because it brings sources publicly known, such as through the ICRC (2019b) Customary IHL Database, into a translated format. This spans treaties and other instruments across international organizations and judicial bodies, through to national legislation, military manuals, case law, and practices.

Unfortunately, even if the algorithms behind LAWS can successfully demonstrate compliance with IHL and ROE, there is a risk that robots would not have enough contextual awareness, particularly in relation to symbolic actions that are important for military signaling and political messaging. Mock attacks by the Soviets were once intended to "enable US commanders to recognize these attacks during crises … in effect, simulated attacks (gave) the Soviet navy an 'action language' for signaling their US adversaries and American political leaders" (Lynn-Jones, 1985, p. 159). Humans have learned to signal each other in such ways. Unfortunately, robots may not learn this. Even the use of "human–machine teaming" – which sounds like a ready-made solution – would not necessarily enable robots to comprehensively understand action language. Human-machine teaming is a process designed for robots to help humans make decisions in real time, with humans only being able to "help" robots during the earlier development stages which are time-consuming and not suitable once in the field.

With this in mind, it is important to note that the informal development of such "rules of the road" has arisen, historically and in recent times, because of dangerous events, not through proactive diplomacy. For example, during the late 1960s there were countless near-misses, warnings of attack, and actual collisions. According to the US Department of State (c. 2009–2017b), the 1972 bilateral Treaty on the Prevention of Incidents On and Over the High Seas, which drew on the International Maritime Organization's Convention on the International Regulations for Preventing Collisions at Sea adopted at the UN, was brought about by frequent occurrences of "planes of the two [superpower] nations passing near one another, ships bumping one another, and both ships and aircraft making threatening movements against those of the other side". A decade after implementation, the treaty was dubbed a "quiet success for arms control" by one commentator (Lynn-Jones, 1985, p. 159). It is plausible that similar near-misses involving LAWS will occur in the years to come, leading to pressure to establish agreements to prevent accidents. For example, LAWS might be excluded from flashpoint areas until they develop a foundational understanding of contextual awareness. It would, however, be ideal to develop arrangements to prevent these incidents from occurring in the first place, rather than running the risk of accidents escalating to crises and conflict.

During a crisis today, the US still relies on bombers to "be forward deployed to help deter regional aggression and assure distant allies … providing effective signaling for deterrence and assurance, including in times of tension" (US Department of Defense, 2018, p. 47). In the coming years, the US and China will likely have increased visual contact around maritime areas. In 2014, a Chinese fighter performed unsafe "Top Gun-style" maneuvers against a US patrol aircraft. The US "apparently assessed that the seriousness of the incident was below a threshold that would demand an escalatory military response" and used diplomacy to eventually agree on Rules of Behavior for the Safety of Air and Maritime Encounters (Green et al, 2017, pp. 224, 234). How would a robot view such a scenario? Rules of Engagement have steps to escalate after communicating a threat to use force or a display of hostile intent (Cole et al, 2009, pp. 22–23, 72). Even with open source code, states may have different interpretations of the law and would consider their operations as legal. Our primary concern is that the LAWS may not "comprehend" or have awareness of the purpose of signaling. As both states have an interest in avoiding an accidental conflict that leads to unintended escalation, humans must be able to communicate with each other in volatile and ambiguous situations to avoid a crisis. This practice would ideally extend already well-established customs (Cole et al, 2009, pp. 79–80). Conversely, we should not rule out that humans may need to understand the "action language" of LAWS, just as we would need LAWS to be able to interpret human maneuvers intended to convey warnings. Availability of limited, unclassified open source code could help to bridge this gap while keeping other information secure in accordance with classified security protocols.

To achieve this, it may be necessary to implement a system akin to a "virtual red phone" – a form of direct communications link between humans who oversee the operation of LAWS, for example, like how "seconds" communicated with each other during a duel. The important consideration here is that humans, as adversaries, should retain the capacity to communicate with their opposite number when neither of them physically may be present in the field. In terms of how this process could be achieved technologically, one human sends a message to their LAWS, which, being in proximity to another LAWS, then "bridges" that message to the human on the other side (open source coding would ensure this does not compromise security protocols). In a situation where there is a misunderstanding between the LAWS and the human, which has the effect of making the LAWS "misunderstand" the order, the human will immediately need to signal the unintentional nature of the attack to their opposite number so that they can prevent escalation or mitigate the magnitude of retaliation. As stated above, damage can be limited to below catastrophic thresholds by ensuring LAWS do not carry weapons capable of delivering irreparable harm. Unless this action is taken, the other state could consider itself under attack. Adoption of the above measures should reduce the likelihood of responding to an unintentional use-of-force by LAWS and prevent unnecessary political or military escalation.

Of course, the difference between intelligence, surveillance and reconnaissance (ISR) and being combat-ready is significant. The former pertains to evaluating options on the use-of-force while the latter requires accepting the political consequences of fighting. Despite the United States' withdrawal, the Treaty on Open Skies between NATO and the former Warsaw Pact countries exemplifies this distinction. As signified by Bell and Wier (2019, pp. 14–15), while the agreement restricts surveillance technology in which the "cameras just need to be good enough to distinguish a tank from a truck", it is markedly easier to use pictures obtained via Open Skies Treaty flights instead of "sensitive satellite sources to gain diplomatic advantage … [or] without the usual delays or concerns that accompany sensitive intelligence declassification." Like any other arms control agreement, timing is key. The concept was first envisaged by President Eisenhower in 1955, but Moscow only came on board in 1992 following the dissolution of the Soviet Union, to "prove their commitment to real change". Although there were always challenges relating to implementation and compliance, the demise of the Open Skies Treaty was based more on political leadership (or lack of), as Moscow and Washington questioned each other's broader commitment to such mechanisms amid deteriorating bilateral relations. Notwithstanding such developments, how can this relate to LAWS? Something that incoporates the successful aspects of the Treaty on Open Skies could be a "quick win" for advocates of open code, but leaders would need to get the timing right. As alluded to earlier, this may follow on from a near-miss incident, but proactive diplomacy is of course preferable. An open code solution could make it possible for a state to disclose whether their robot is carrying a weapon or just conducting surveillance. If autonomous vehicles remain in continuous use for months, states should consider a provision that demonstrates whether they carry weapons or not, particularly if doing so reduces tensions and builds mutual confidence.

Even if states agree to this in principle, development of the infrastructure required for such a communications link would be difficult in certain environments, such as underwater. As Andrew Krepinevich (2014, p. 82) illustrates, it seems conceivable that, granted the cost difference between a small UUV [Unmanned Underwater Vehicle] or AUV [Autonomous Underwater Vehicle] and a modern submarine, "maritime competitors could arm and deploy them as undersea 'kamikaze' devices or delivery systems for ASW [Anti-Submarine Warfare] munitions." Yet the maritime communication infrastructure required for people to monitor such devices would be difficult to build and to decide who pays the bill. Unlike ground or airborne systems, underwater operations are usually unable to use GNSS. Instead, they can utilize less accurate techniques such as:

- Acoustic beacons, which are limited by having a small range
- Dead reckoning, which is limited by factors that cannot be as easily measured
- Location approximation, such as the Simultaneous Localization and Mapping program developed by the US (Finn & Scheding 2010, pp. 16, 21, 99–102, 138)

Anthony Finn and Steve Scheding (p. 100) suggest that data collected from multiple underwater drones operating in a network could be, "fused together to form an aggregate, global map". This would mean that underwater LAWS could remain covert but stay within a defined location. To communicate with people, the systems could travel near to a specific location and, while remaining underwater, communicate with data gateway buoys that have been set up on the ocean surface to facilitate communication from underwater to elsewhere. This type of communication would rely on various pieces of maritime and airborne communication infrastructure (Melnick, 2011, pp. 100, 102). The alternative, fiber-optic cables, is more expensive. Without the maritime infrastructure proposed above, communication with underwater LAWS would be far less economical. This would mean states would have less oversight over their maritime robotic systems, making it difficult to utilize some of the proposals we have outlined above in relation to minimizing the likelihood or mitigating the consequences of an unintended attack.

Conclusion

One of the purposes of developing LAWS is to counter threats posed by adversaries or for states to multiply their force size or projection capabilities. Nevertheless, autonomous systems carrying out unintended attacks and provoking retaliation would likely create greater political instability and overall (in) security. In our consideration of this problem, we have proposed several methods to address risks associated with the use of LAWS. The desired outcome of our risk-based approach is the minimized likelihood and consequences of unintentional attacks. Of course, to be ready for adoption by policy makers, each recommendation will require further development beyond the exploratory stage presented here.

First, we examined the MTCR, WA, and CCW – all voluntary models – to place a limit on the unnecessarily damaging effects of LAWS and to limit the transfer of sensitive technology to non-state actors. States would also need to consider a mechanism to enforce limits on the types and yields of weapons carried by different classes of LAWS. Secondly, we outlined some benefits from greater transparency and communication. The adoption of open source code, making clear which acts are prohibited, would place further constraints on the use-of-force by LAWS and provide greater protection to non-combatants. This could enable prominent decision-makers to resolve ambiguous situations, ultimately providing a pathway toward de-escalation strategies for political and military leaders alike, and ultimately mitigating the specter of accidental war and violence.

Political calculations will decide whether it is worth cooperating with an adversary in this area. History suggests most countries cooperate on "common sense" measures to reduce accidents, such as "rules of the road". Of course, LAWS would add a level of complexity regarding the admission of wrongdoing. We acknowledge that a state may not always cooperate if it can take advantage of the potential anonymity afforded by LAWS to evade attribution, such as after attacking non-combatants or using a prohibited weapon. Yet, the introduction of LAWS does not fundamentally change such calculations. Even today, few leaders go public about ordering the use of chemical weapons against civilians, but the world continues to be shocked by the available evidence. Our primary concern is that states should find a way to cooperate, to reduce the likelihood and consequences of unintended attacks, when it is clearly in their mutual interest to do so. The core risks of regulating LAWS amount to more than humanitarian challenges. The rules-based international system, underpinned by the UN Charter, hands responsibility to five permanent members of the Security Council for determining an act of aggression and making recommendations to restore international peace and security. This concert relationship between China, France, Russia, the UK, and the US creates incentives to support increased measures and, as the title of this chapter signifies, a potential platform in mitigating accidental war by adopting risk-based strategies for governing LAWS.

References

Anderson, K and Waxman, MC (2013) 'Law and ethics for autonomous weapons systems: why a ban won't work and how the laws of war can', *American University Washington College of Law Research Paper 2013–11.*

Anthony, I (2015) 'Transparency and information sharing', *Meeting of experts on LAWS from April 13–17.* United Nations. http://web.archive.org/web/20190915072202/https://www.http://unog.ch/80256EDD006B8954/(httpAssets)/24986665C2487E35C1257E2D002E0314/$file/2015_LAWS_MX_Anthony_PP.pdf

Arms Control Association (2017) *The Wassenaar arrangement at a glance.* http://web.archive.org/web/20191104112718/, https://www.http://armscontrol.org/factsheets/wassenaar

Asaro, PM (2008) 'How just could a Robot war be?' in Briggle, A, Waelbers, K and Brey, PAE (eds) *Current issues in computing and philosophy.* Amsterdam: IOS Press, pp. 50–64.

Bell, A and Wier, A (2019) 'Open skies treaty: a quiet legacy under threat', *Arms Control Today*, 49(1), pp. 13–19.

Campaign to Stop Killer Robots (2019) *Russia, United States attempt to legitimize killer robots.* August 22. http://web.archive.org/web/20190920045909/, https://www.http://stopkillerrobots.org/2019/08/russia-united-states-attempt-to-legitimize-killer-robots/

Center for Arms Control and Non-Proliferation (2017) Fact sheet: missile technology control regime (MTCR). Washington, DC. http://web.archive.org/web/20190915220835/, https://armscontrolcenter.org/missile-technology-control-regime-mtcr/

Chertoff, P (2018) *Perils of lethal autonomous weapons systems proliferation: preventing non-state acquisition.* Geneva: Geneva Centre for Security Policy. http://web.archive.org/web/20190915220303/, https://dam.gcsp.ch/files/2y10RR5E5mmEpZE4rnkLPZwUleGsxaWXTH3aoibziMaV0JJrWCxFyxXGS

Cole, A, Drew, P, McLaughlin, R and Mandsager, D (2009) *Rules of engagement handbook.* Sanremo: International Institute of Humanitarian Law. http://web.archive.org/web/20190915072903/, http://iihl.org/wp-content/uploads/2017/11/ROE-HANDBOOK-ENGLISH.pdf

Daintith, J and Wright, E (2014) *A dictionary of computing* (6th ed.). Oxford: Oxford University Press. http://dx.doi.org/10.1093/acref/9780199234004.001.0001

Davies, TR (2013) 'Governing communications', in Harman, S and Williams, D (eds) *Governing the world? Cases in global governance.* Oxon: Routledge, pp. 114–127.

de Klerk, P (2016) *The missile technology control regime: successful international co-operation, with limits.* http://web.archive.org/web/20190920025328/, http://mtcr.info/wordpress/wp-content/uploads/2016/07/160228-Presentation-MTCR-for-AECS2016.pdf

Finn, A and Scheding, S (2010) *Developments and challenges for autonomous unmanned vehicles: a compendium.* Berlin: Springer. http://dx.doi.org/10.1007/978-3-642-10704-7

Githib (2019) https://github.com/

Green, M, Hicks, K, Cooper, Z, Schaus, J and Douglas, J (2017) *Countering coercion in maritime Asia: the theory and practice of gray zone deterrence.* Washington, DC: Center for Strategic & International Studies. http://web.archive.org/web/20190915093922/, https://csis-prod.s3.amazonaws.com/s3fs-public/publication/170505_GreenM_CounteringCoercionAsia_Web.pdf?OnoJXfWb4A5gw:n6G.8azgEd8zRIM4wq

Henckaerts, J and Doswald-Beck, L (2009) *International committee of the red cross customary international humanitarian law volume 1: rules*. Cambridge: Cambridge University Press. http://web.archive.org/web/20190915105735/, https://www.icrc.org/en/doc/assets/files/other/customary-international-humanitarian-law-i-icrc-eng.pdf

Homayounnejad, M (2017) 'Ensuring Lethal autonomous weapon systems comply with international humanitarian law', *TLI Think!* Paper 85/2017. http://dx.doi.org/10.2139/ssrn.3073893

International Committee of the Red Cross (2019a) *Canada: practice relating to rule 25. Medical personnel*. http://web.archive.org/web/20191207033324/, https://ihl-databases.icrc.org/customary-ihl/eng/docs/v2_cou_ca_rule25

International Committee of the Red Cross (2019b) *Sources*. http://web.archive.org/web/20191208020555/, https://ihl-databases.icrc.org/customary-ihl/eng/docs/src

Krepinevich, AF (2014) *Maritime competition in a mature precision-strike regime*. Washington, DC: Center for Strategic and Budgetary Assessments. http://web.archive.org/web/20191207035230/, https://csbaonline.org/uploads/documents/MMPSR-Web.pdf

Lee, S (2011) *The ROK-US joint political and military response to North Korean armed provocations*. Washington, DC: Center for Strategic and International Studies. http://web.archive.org/web/20190915111436/, https://csis-prod.s3.amazonaws.com/s3fs-public/legacy_files/files/publication/111006_Lee_ROKUSJointResponse_web.pdf

Lynn-Jones, SM (1985) 'A quiet success for arms control: preventing incidents at sea', *International Security*, 9(4), pp. 154–184. http://dx.doi.org/10.2307/2538545

Marchant, GE, Allenby, B, Arkin, R, Barrett, ET, Borenstein, J, Gaudet, LM, Kittrie, O, Lin, P, Lucas, GR, O'Meara, R and Silberman, J (2011) 'International governance of autonomous military robots', *Columbia Science and Technology Law Review*, 12, pp. 272–315. http://web.archive.org/web/20190915111931/, http://stlr.org/download/volumes/volume12/marchant.pdf

Melnick, A (2011) 'Instant infrastructure: Interview with Dan Erdberg, director of business development, World Surveillance Group Inc', *Anthropology Now*, 3(3), pp. 98–104. https://doi.org/10.1080/19492901.2011.11728337

Missile Technology Control Regime. *Guidelines for sensitive missile-relevant transfers*. Government of Canada on behalf of Missile Technology Control Regime Partners. http://web.archive.org/web/20190920022831/, https://mtcr.info/guidelines-for-sensitive-missile-relevant-transfers/

Missile Technology Control Regime Annex Handbook (2017) *United States Government*. http://web.archive.org/web/20190920012532/, http://mtcr.info/wordpress/wp-content/uploads/2017/10/MTCR-Handbook-2017-INDEXED-FINAL-Digital.pdf

Missile Technology Control Regime Equipment, Software and Technology Annex (2019) http://web.archive.org/web/20190920023222/, http://mtcr.info/wordpress/wp-content/uploads/2019/07/MTCR-TEM-Technical_Annex_2019-05-29.pdf

Mistry, D (1997) 'Ballistic missile proliferation and the MTCR: a ten-year review', *Contemporary Security Policy*, 18(3), pp. 59–82. http://dx.doi.org/10.1080/13523269708404169

Pagallo, U (2011) 'Robots of just war: a legal perspective', *Philosophy & Technology*, 24(3), pp. 307–323. http://dx.doi.org/10.1007/s13347-011-0024-9

Paul, TV (1995) 'Time pressure and war initiation: some linkages', *Canadian Journal of Political Science*, 28(2), pp. 255–276. https://doi.org/10.1017/S0008423900018837

Phelps, J (1963) 'Causes of accidental war', *Survival: Global Politics and Strategy*, 5(4), pp. 176–179. https://doi.org/10.1080/00396336308440409

Putin, V (2018) *Presidential address to the federal assembly*. Kremlin. March 1. http://web.archive.org/web/20190920034727/, http://en.kremlin.ru/events/president/news/56957

Rosenboom, A (2009) 'Reservations and declarations in multilateral treaties', *Capacity-building workshop on treaty law and practice and the domestic implementation of treaty obligations*. Wuhan, China, October 13–17. United Nations. http://web.archive.org/web/20190920042129/, https://treaties.un.org/doc/source/training/regional/2009/13-17October-2009/reservations-declarations.ppt

Russell, S (2015) 'Artificial intelligence: implications for autonomous weapons', *Meeting of experts on LAWS from* April *13–17*, 2015. United Nations. http://web.archive.org/web/20190915222425/, https://www.unog.ch/80256EDD006B8954/(httpAssets)/36AF841749DE9819C1257E2F0033554B/$file/2015_LAWS_MX_Russell+bis.pdf

Scharre, P (2017) 'A security perspective: security concerns and possible arms control approaches', *UNODA occasional papers no. 30, November 2017: perspectives on lethal autonomous weapons systems*. New York: United Nations Office of Disarmament Affairs, pp. 19–33. http://web.archive.org/web/20190915222204/, https://www.unog.ch/80256EDD006B8954/(httpAssets)/6866E44ADB996042C12581D400630B9A/$file/op30.pdf

Schwarz, M and Sonin, K (2008) 'A theory of brinkmanship, conflicts and commitments', *Journal of Law, Economics & Organization*, 24(1), pp. 163–183. https://doi.org/10.1093/jleo/ewm038

Senese, PD and Vasquez, JA (2008) *The steps to war: an empirical study*. Princeton: Princeton University Press.

Trump, DJ (2017) *Remarks by President Trump before a briefing on the opioid crisis*. White House. August 8. http://web.archive.org/web/20190915223316/, https://www.whitehouse.gov/briefings-statements/remarks-president-trump-briefing-opioid-crisis/

Trump, DJ (2018) *Press conference by President Trump*. White House. June 12. http://web.archive.org/web/20190915223423/, https://www.whitehouse.gov/briefings-statements/press-conference-president-trump/

United Nations (2014a) *Convention on certain conventional weapons booklet*. Geneva: United Nations. http://web.archive.org/web/20190920084430/, https://unoda-web.s3-accelerate.amazonaws.com/wp-content/uploads/assets/publications/more/ccw/ccw-booklet.pdf

United Nations (2014b) 'Report of the 2014 informal meeting of experts on lethal autonomous weapons systems (LAWS)', *Meeting of the high contracting parties to the CCW from November 13–14, 2014*. United Nations. [UN document system: CCW/MSP/2014/3.]

United Nations (2018) '10 possible guiding principles…', tweet by @CCW_UNODA. September 1. http://web.archive.org/web/20190920050454/, https://twitter.com/CCW_UNODA/status/1035938908482232321?s=20

UN News (2019) *Autonomous weapons that kill must be banned, insists UN chief*. March 25. http://web.archive.org/web/20190915222759/, https://news.un.org/en/story/2019/03/1035381

US Department of Defense (2018) *Nuclear posture review*. http://web.archive.org/web/20190920051805/, https://media.defense.gov/2018/Feb/02/2001872886/-1/-1/1/2018-NUCLEAR-POSTURE-REVIEW-FINAL-REPORT.PDF

US Department of State c (2009–2017a) *Memorandum of understanding between the United States of America and the Union of Soviet socialist republics regarding the establishment of a direct communications link*. http://web.archive.org/web/20190920051308/, https://2009-2017.state.gov/t/isn/4785.htm

US Department of State c (2009–2017b) *Agreement between the government of the United States of America and the government of the Union of Soviet socialist republics on the prevention of incidents on and over the high seas*. http://web.archive.org/web/20190920051506/, https://2009-2017.state.gov/t/isn/4791.htm

Wassenaar Arrangement on Export Controls for Conventional Arms and Dual-Use Goods and Technologies. List of Dual-Use Goods and Technologies and Munitions List (vol. 2). Wassenaar Arrangement Secretariat. http://web.archive.org/web/20190920044319/, https://www.wassenaar.org/app/uploads/2019/consolidated/WA-DOC-18-PUB-001-Public-Docs-Vol-II-2018-List-of-DU-Goods-and-Technologies-and-Munitions-List-Dec-18.pdf

Woolf, AF (2019) *Russia's nuclear weapons: doctrine, forces, and modernization*. Washington, DC: Congressional Research Service. http://web.archive.org/web/20191207020334/, https://crsreports.congress.gov/product/pdf/R/R45861

Part II

Emerging technologies, the state, and the changing character of conflict

6 Politics in the machine

The political context of emerging technologies, national security, and great power competition

Simona R. Soare

Introduction

Technological competition in the area of emerging technologies is at the heart of renewed great power competition, especially between the US and China. Artificial intelligence (AI), quantum computing (QC), Internet of Things (IoT), 5th generation telecommunications (5G), biotechnology, robotics and autonomy, directed energy, and advanced materials are shaping international power and are vital for national and international security. In 2019, the US Department of Defense (DoD) Strategy on Artificial Intelligence emphasized that Chinese and Russian investments in military AI "threaten to erode our technological and operational advantages and destabilize the free and open international order" (DoD, 2019, p. 5). Russian President Vladimir Putin argued, "whoever becomes the leader in this sphere [of AI] will rule the world" (CNN, 2017), while Chinese President Xi Jinping noted that "under a situation of increasingly fierce international military competition, only the innovators win" (Kania, 2020, p. 2). The 2017 Chinese pledge to invest $150 billion in securing global leadership in AI by 2030 led the European Union (EU) to announce an estimated 20-fold increase in public research and development (R&D) spending on emerging technologies (Soare, 2020, p. 3). Over four dozen countries are pursuing AI-enabled military capabilities, fueling concerns about an ongoing "AI arms race" (Auslin, 2018) and an increased interest in the implications of AI for national security (Sayler, 2019).

However, emerging technologies are not just an external variable of great power competition. Many present-day dilemmas about emerging technologies are inherently political and presented in the framework of democratic and authoritarian political regimes. Washington regularly portrays the strategic competition with Beijing in ideological terms, calling for a democratic way of AI that is rooted in liberal values (Imbrie et al, 2020; Shahbaz, 2018; Wright, 2018). Chinese and Russian transfers of emerging technologies to bolster other authoritarian regimes (Feldstein, 2019) and perceptions of the increased aggressiveness of informational authoritarians (Rosenbach & Mansted, 2019; Kendall-Taylor et al, 2020) have become a strategic concern in the democratic West. Scholars are investigating the role of digital emerging technologies in undermining democratic institutions in the context of the "democratic disintegration" (Freedom House, 2020, p. 1) and the global downward trend in civil and political liberties (Economic Intelligence Unit, 2020). Others question who is ahead in this "race for global technological dominance" (Schneider-Petsinger et al, 2019, p. 4) and warn it is not clear the democratic narrative is winning (Watts et al, 2020).

What role do political regimes play in our understanding of the development, adoption, and use of emerging technologies? And how do authoritarianism and democracy[1] shape the use of emerging technologies? In answering these questions, I explore the role of politics in the machine – a play on Ryle's 1949 "ghost in the machine". I argue that the development and use of emerging technologies is shaped by political context, which in turn redefines the scope and the means through which different political regimes pursue perennial political interests. The chapter is structured in three parts. The first part explores the relationship between technology and political contexts and highlights three theoretical approaches within international relations (IR) literature – technological determinism, sociopolitical determinism, and science and technology studies. In the second section, I turn to a comparative analysis of the different democratic and authoritarian uses of digital emerging technologies. The final section highlights the role of political dynamics in the debate about whether democratic or authoritarian states are performing better in the adoption of emerging technologies.

Theory, technology, and politics

Technology is a central variable in the mainstream IR theories – realism, liberalism, and constructivism. For realists, including neoclassical realists, technology is critical to national power and security (Morgenthau, 1988). Technologies shift the balance of power between international actors and alter the structure of the international system. They do so by shaping economic and military power, shrinking time and distance, increasing the frequency of international interactions and enhancing security dilemmas (Waltz, 1979). Military technologies may transform the conduct and effectiveness of warfare (Murray, 2011). Conversely, liberalism acknowledges the role of these technologies in facilitating and fostering cooperation (Ikenberry, 2018), whereas social constructivism is primarily concerned with the relationship between technology and identity (Wendt, 1995, p. 73). These mainstream approaches see technology as exogenous to international politics. They treat technological impact either as independent of political regimes (technological determinism), predetermined by sociopolitical structures (sociopolitical determinism), or in a generic (technology, not technologies), instrumentalist (i.e. as politico-military tools) and ahistorical manner.

However, technologies do not exist "in a cultural or political vacuum" (Black, 2013, p. 3), but rather shape and are shaped by political contexts and human agency. They shape society, politics, and institutions by molding political authorities, redefining governmental tools, including their reach, scope, and efficiency (Herrera, 2003, p. 560). Technological systems, such as the smartphone and the tablet, are highly networked (e.g. global supply chains) and institutionalized (e.g. commercial relations between parent company, outsourcing company, and service providers). These "complex socio-technical systems" include the knowledge (articles, books, patents), technical standards, manufacturing, retailer networks, users, social and political rules, and practices of use (Herrera, 2003, p. 572). Changes in these norms and values are as consequential as technological change itself.

Understanding how technological impact is politically structured can be challenging. International actors regularly use different tools to shape the uses of technologies, such as wargaming exercises and "AI champions" who promote innovative uses of AI to solve existing military problems. The European Commission (EC) recently finished a year-long project that tested the implementation of its trustworthy AI guidelines on a cross-section of 500 companies active in the European market in different sectors of activity (from digital and health services to manufacturing). A measure of political and technological opportunism is at play among military and political elites who seek to use emerging technologies to improve the performance and output of their existing strategies. In doing so, they generally (and mostly unconsciously) follow technological determinist, sociopolitical determinist, or evolutionary science and technology studies (S&T) approaches, all of which have important limitations.

In 2000, former US President Bill Clinton argued: "The genie of freedom will not go back into the bottle… In the new century, liberty will spread by cell phone and cable modem". This *determinist* view inferred political impact from the technological features of the internet – a distributed network of actors with self-regulating behaviors. Clinton believed the global spread of the internet and computer technologies would profoundly challenge the status quo for authoritarian regimes for whom trying to control the internet would be "sort of like trying to *nail Jell-O to the wall*" (*New York Times*, 2000). Seen as the teleological driver of political progress, a phenomenon perfectly encapsulated by the designation of the internet as "liberation technology" (Diamond, 2010), democracy was perceived as having an inherent innovation advantage over other regimes (Kroenig, 2020; Rosenbach & Mansted, 2019). This advantage helped bring about what Fukuyama famously called the "end of history".

In hindsight, scholarly research has proven authoritarian regimes are just as interested in adopting digital technologies but with a temporal delay, in which they learn to control online content and communication to curb possible democratizing effects (Stier, 2017). Over the last decade, despite the progressive expansion in the use of the internet, mobile phones, and other IoT devices, the number of democratic states has declined and civil and political liberties have atrophied in many states, including former democratic regimes, leading to what Freedom House (2020, p. 5) calls "a decade of democratic deficits". Moreover, not all technologies impact politics in the same way. There is no scholarly consensus on which point in their life cycles (emerging or mature) technologies have greater political or military impact. Network-centric capabilities produced the largest impact upon their first operational use in the Iraq War (1990/91) and not during their development in the 1980s. Similarly, the effectiveness of offensive cyber capabilities is dependent upon an attack surface that is created by the wide diffusion and adoption of digital technologies. General-purpose technologies – such as electricity, the combustion engine, digital technologies, and AI – have a whole-of-society impact. In comparison, military technologies – such as stealth and precision – have an impact on the military sector (Black, 2013; Herrera, 2003, p. 578). Moreover, technologies with force multiplier effects – such as AI and QC – are expected to have a differentiated impact on great powers and smaller international actors.

In an interview for *Wired* magazine (2000), former Polish leader Lech Walesa adopted a more *sociopolitical determinist* view of technologies, arguing:

> Communism is a monopolistic system, economically and politically. The system suppresses individual initiative, and the 21st century is all about individualism and freedom. The development of technology supported these directions. (…) To control the free flow of information, the Communists would have to increase the secret police by a factor of four. It would be a huge effort for police to control the channels you get on TV or the phone numbers you are allowed to dial. So technology helped end communism by bringing in information from the outside.

Walesa's view was cognizant of the different constraints and opportunities that technology creates for political regimes but failed to recognize their non-linear character. In the 1980s, information technologies exposed the limits of traditional political policing and eventually led to communism's demise in the Soviet Union and Eastern Europe. But by 2019, new technological opportunities were enabling Beijing to monitor and police 210–280 million text messages daily (Statista, 2019). AI-enabled mass surveillance and big-data analysis are fueling China's "social credit" system, which is rapidly becoming an efficient tool to control individual behavior without threatening to make China any less communist. These same technologies enable authoritarian regimes in other parts of the world to perpetuate their rule (Imbrie et al, 2020), challenge democratic institutions (Rosenbach & Mansted, 2019), and threaten to force a democratic retreat (Economic Intelligence Unit, 2020).

Finally, in 2018, Google's former CEO, Eric Schmidt articulated a view on the role of the internet more in line with the *science and technology studies* approach, saying: "I think the most likely scenario now is not a splintering but, rather, a bifurcation into a Chinese-led internet and a non-Chinese internet led by America [by 2028]" (Hamilton, 2018). This view encapsulated the evolutionary and mutually constitutive relationship between political regimes and technologies and highlighted their international and geopolitical consequences. While AI, 5G, and other emerging technologies will not change the nature of international politics, they do change frameworks, issues, and even actors in international politics, while also being "determined largely by familiar, national-based political processes in which science and technology are considerations that must be taken into account" (Skolnikoff, 2002, p. 29).

Democracy, authoritarianism, and emerging technologies

Scholars expect emerging technologies such as AI to profoundly shape global order through the competition between digital authoritarians and democracies

(Wright, 2018). But how do authoritarianism and democracy shape emerging technologies today? This section offers a comparative analysis of the impact of democratic and authoritarian regimes on the adoption and use of emerging technologies. I argue political regimes play a key role in shaping technological development based on a mix of domestic and international constraints, opportunities, and perceptions about their role in aiding their respective political interests.

Digital authoritarians: Automating information control and mass disruption?

A fundamental political imperative for the use of emerging technologies in authoritarian regimes is domestic, namely regime survival (Talmadge, 2016). This is increasingly dependent upon controlling the information environment, companies, and infrastructure. China's Great Firewall has been effective in preventing Chinese citizens from freely accessing information on the internet deemed undesirable by the regime. A Russian plan to ensure that, by 2020, 95% of its domestic internet traffic never leaves Russia's borders is part of Moscow's strategy to control its domestic vulnerability to external information (Rosenbach & Mansted, 2019). There is no consensus among scholars whether Russia and China are indeed trying to export their nascent ideological projects, constructed as alternatives to the Western liberal model (Watts et al, 2020; Weiss, 2020), though the lack of intention does not negate the consequences of the replication of their models beyond their national borders. While recent research suggests there is an increasing synergy between Moscow and Beijing's information operations (Kliman et al, 2020), the Russian and Chinese models also diverge, not least because they are informed by different levels of ambition and technological capability (Weber, 2019). There are similarities too. Both Beijing and Moscow deploy AI, QC, automation, and robotics in military and political policing structures to avoid human input, which they fundamentally distrust (NSCAI, 2020; Talmadge, 2016).

In both the Russian and Chinese cases, the authoritarian practices of political policing and control in the physical domain extend to the virtual domain (Imbrie et al, 2020). This forms a system of "algorithmic governance" (Mozur, 2018). For example, in China's "social credit system", AI algorithms and big-data analysis are fed by huge amounts of government-collected data from CCTV cameras, biometrics, IoT devices, 5G networks, smart city infrastructure, and smart identity cards. Even though it lacks a comprehensive nationwide framework, Chinese mass surveillance systems track 20–30 million people who are suspected of different offenses, including political dissidence. Several Chinese municipalities have deployed an algorithm that monitors the use of utilities in Chinese households and reports the data to the authorities for the purpose of predictive policing (Mozur, 2018). Such systems have provided Beijing with an effective, cost-efficient, and covert tool for maximizing a perennial political interest: To increase governmental and party control over society to maintain social and political stability. China's wide use of these tools in massive repression campaigns against the Uighur minority in its Xinjian province is well documented, although party officials and the state-run media usually portray the campaigns as counterterrorism efforts.[2]

Data integrity and control concerns fuel Chinese and Russian tendencies to decouple certain aspects of their critical networks and information environment from the international system. These concerns reportedly influenced China's December 2019 decision to replace all foreign digital hardware and software used in its public sector within three years (Yang & Liu, 2019). Beijing regularly uses WeChat, YY, and TikTok platforms to collect user data and censor content considered undesirable by the party-state (Ruan et al, 2020), as demonstrated during the COVID-19 pandemic. Beijing and Moscow increasingly regard access to data as a zero-sum game and any perceived breaches are met with retaliation, raising the specter of inadvertent escalation. In 2015, China used offensive cyber capabilities against the American company GitHub, which was accused of promoting open source code and content on how to subvert Chinese online censorship. Importantly, the more regime legitimacy in China and Russia depends on their ability to control their information environments, the more these regimes are likely "to consider information control as a core national interest – equivalent to even economic and physical security interests" (Rosenbach & Mansted, 2019).

Authoritarian regimes – much like democratic ones – pursue emerging technologies for geopolitical reasons. Russia and China's investment in emerging technologies for military applications are motivated by strategic interests to avoid technological surprise and narrow the technological gap with the US (Kania, 2020, p. 2; Sukhankin, 2019). Beijing and Moscow are investing in a broad range of military applications for emerging technologies, including autonomous munitions and unmanned aerial, maritime, and ground-based autonomous robots for different military functions as well as integrating AI and autonomy in space, cyber, electronic, and information campaigns. However, supply chain bottlenecks[3] and dependencies – particularly in semiconductor and AI chip manufacturing technologies controlled by American, Taiwanese, South Korean, and Dutch companies (Khan & Flynn, 2020) – are perceived as critical vulnerabilities, further fueling Russian and Chinese technonationalism.

Chinese and Russian efforts in the past five years have demonstrated military capabilities enabled by emerging technologies for offensive and defensive purposes alike. Since 2015, a Russian TV documentary revealed classified Russian plans to build a new nuclear torpedo capable of destroying enemy coastal installations, Russia has tested its new Avangard hypersonic system, and Chinese military parades have revealed new anti-ship missiles capable of sinking aircraft carriers. Since China and Russia already deploy such systems with varying levels of autonomy, and may even use some AI-enabled capabilities, Beijing and Moscow's lack of transparency around the development and fielding of these military technologies, including testing, evaluation, verification, and validation (TEV&E), has been a cause for concern in the West. American analysts have expressed fears that Russia and China could sacrifice safety for speed in the deployment of military AI (Kania, 2020; Scharre, 2019). China and Russia are equally opaque about the operational concepts they are developing for the use of

AI-enabled and autonomous military capabilities (Bendett, 2019) which may, in turn, impact escalation and deterrence dynamics.

Then there is the challenge posed by prolific Sino-Russian arms exports. Russia is the world's second and China the fifth largest arms exporter (Wezeman, et al, 2020). Beijing is supplying 90% of the world's sales of partially autonomous medium-altitude UAVs (Kania, 2020). A distinct pattern of technology diffusion along political regime lines seems to be at work: Between 85% and 90% of the countries that import Chinese and Russian technologies are authoritarian or hybrid regimes and a significant majority imitate Russian and Chinese information control models (Weber, 2019). China's Belt and Road Initiative (BRI) and Russia's Commonwealth of Independent States (CSI) act as emerging technologies diffusion belts for distinctly authoritarian purposes: 25% of Russian and 82% of Chinese technology transfers go to CSI and BRI countries, respectively (Weber, 2019; Rosenberger, 2020). This allows Moscow and Beijing "to shape the global information architecture of platforms, applications, and surveillance systems" and collect data used for security and technological purposes (e.g. training AI) (Rosenberger, 2020). China is the world's largest builder of digital infrastructure – 5G, smart cities, and undersea data cables – which it increasingly leverages for political gains. China reportedly used its contract to build the African Union's headquarters in Addis Ababa, Ethiopia to extract data from the installed computer systems. Russia has used surveillance gear installed in Kyrgyzstan to funnel data back to Moscow, and FaceApp to collect the pictures of millions of unsuspecting users around the world to train facial recognition software (Weber, 2019).

Beijing and Moscow opportunistically use crises in other countries to promote technology transfers and legislative imitation of their models. For example, during the crisis caused by the contested Venezuelan political transition, the government in Caracas rolled out a new smart identity card for its citizens, through which they have access to subsidized food programs, bank loans, health, and other social services. The political and humanitarian crisis provided an opportunity to implement a Chinese-built system through which the Venezuelan government collects ample data on its citizens and, thus, better tracks and represses political dissidence (Berwick, 2018). Similar scenarios unfolded in Ecuador, Angola, Iran, and Pakistan.

There is concern that Beijing and Moscow's efforts are contributing to an expansion of authoritarianism.[4] Democracies are not immune as targets of this phenomenon. Over 65 countries, including 34% of NATO allies and 48% of NATO partners (Soare, 2020, p. 6), were using Chinese AI-enabled smart city, smart policing, and facial recognition technologies in 2019, but the second (50 million) and third largest (5.2 million) numbers are in the US and Germany, respectively. (Feldstein, 2019; Privacy International, 2016). China deploys the highest number of CCTV cameras in the world – 200 million, but the second (50 million) and third largest (5.2 million) numbers are in the US and Germany, respectively (Bischoff, 2019). Kania (2020) fears that to "the extent this trend continues, China will also drive the diffusion of AI-enabled and autonomous weapons systems". Other scholars suggest closer monitoring of Chinese and Russian technology environments is needed to prepare and anticipate waves of advanced and emerging technologies diffusion (Weber, 2019).

Beijing is using TikTok, Facebook, and Twitter to promote pro-Chinese narratives outside China, including in the context of the COVID-19 pandemic. Unlike Russia, whose campaigns are mainly interested in sowing confusion, facilitating online political polarization, and undermining democratic institutions, Beijing is driven by the interest to maintain a positive image for the regime domestically and internationally. China is reportedly paying an army of individuals to post positive messages about the nation, the party, and its policies in an attempt to shape the information environment to reflect the regime in a positive light. In 2019, Beijing launched its own set of ethical principles for AI, though doubts persist whether the party-state will uphold them, especially when they clash with its desire to exercise maximum public control (Kania, 2020). In the framework of the UN Convention on Certain Conventional Weapons (CCW) Group of Governmental Experts (GGE), China supported the call for a ban on LAWS, calling for a prohibition of the use (but not development) of indiscriminate, fully autonomous weapons capable of real-time online self-learning (Campaign to Stop Killer Robots, 2018). Beijing did so despite its ongoing R&D projects and frequent domestic use of indiscriminate, politically-motivated, AI-enabled mass surveillance. Incidentally, the group of countries calling for the ban on the use of LAWs was largely the same group of so-called "digital deciders" (countries like India, South Africa, Mexico, Brazil, Argentina, Colombia, Indonesia, and Singapore) that seem at least partially persuaded by China and Russia's "sovereign control" approach to internet governance (Morgus et al, 2018). Echoing concerns about the dangers of weaponizing digital networks (Burton & Soare, 2019), Rosenberger (2020) argues:

> By developing control over part of the Internet, China could turn connectivity into a geopolitical weapon, insisting that countries submit to Chinese terms and conditions. The threat of disruptions on 5G networks that Chinese companies control could provide similar leverage for geopolitical manipulation in the future.

As such, Chinese and Russian exports of emerging technologies serve a double purpose of feeding domestic sources of economic growth and consolidating international technospheres to maximize their geopolitical influence.

Digital democrats: Forging a democratic way of AI?

The adoption of emerging technologies in democratic regimes is a value-driven process (Imbrie et al, 2020; Rosenberger, 2020), shaped by core aspects of democracy – freedom, rule of law, human rights, equality, and democratic oversight. These inform the democratic pursuit of a human-centered approach to AI (DoD, 2019, pp. 4–6) – namely, emerging technologies are not used to replace individuals but, rather, to enhance human intelligence and well-being (European Commission, 2019). While some authoritarian regimes, including China and Singapore have published principles for the ethical use of AI, democracies have been focused on the ethical use of AI and other emerging technologies that are codified in institutional or national regulations. Both the European Commission and the DoD are running pilot tests of how their ethical AI guidelines and principles fare in the real world – the former focusing on the digital economy and the latter on the military (European Commission, 2019; DoD, 2020). In 2019, the US and the EU cooperated closely in the adoption of the Organisation for Economic Co-operation and Development (OECD) Principles on Artificial Intelligence – a forum in which neither Russia nor China participate. Over the past six years the US, Canada, Australia, and Europe cooperated within the framework of the dedicated group of governmental experts within the UN CCW on the adoption of voluntary norms of responsible state behavior in the area of LAWS.

Governmental and private-sector legal and ethical concerns around the application of emerging technologies in democratic contexts are already shaping political outcomes. Six American cities have banned governmental use of facial recognition technologies. Microsoft reportedly refused to participate in a Californian program due to fears the algorithm training dataset might have been biased (Weber, 2019), and Axon, the largest American manufacturer of police body and car cameras, banned the use of facial recognition technology on its devices (Warzel, 2019). The EU was expected to announce a five-year moratorium on the use of facial recognition in early 2020, but the decision was postponed (Espinoza & Murgia, 2020).

In the context of a new wave of populism and pervasive Russian and Chinese disinformation campaigns against the West, domestic vulnerabilities have been a primary source of concern for democratic regimes. Russia used automated capabilities (social media chat bots and trolls) during the 2016 US elections, Cambridge Analytica used micro-targeting campaigns during the 2016 US elections and the UK Brexit referendum and, more recently, Chinese and Russian disinformation campaigns used automated online content generation during the COVID-19 pandemic. These actions have raised concerns that "democracies remain fundamentally unprepared for strategic competition in the Information Age" (Rosenbach and Mansted, 2019). Western responses to Sino-Russian hybrid and information threats have mainly been defensive and focused on building resilience against adversarial activities. So far, they have failed to recognize the strategic challenge posed by these hostile acts at the levels of "information (the propagation, control, and manipulation of narratives), architecture (the systems and platforms that transmit, order, and collect information), and governance (the laws, norms, and, in some cases, standards for content, data, and technology)" (Rosenberger, 2020). Democracies have not done enough to close vulnerabilities, such as back doors into smart city infrastructure that enable data theft or have used AI-enabled systems, such as smart-policing algorithms which could produce biased results. Democracies have also facilitated surveillance through restrictions on end-to-end encryption , through digital contact tracing apps and through drone surveillance of urban areas to curb the spread of COVID-19. This demonstrates how challenging it is to access the benefits of emerging technologies without compromising the foundations of democracy.

Much of the scale, ease, rapidity, and cheapness of Russian interference in US elections has been inadvertadly facilitated by poorly adapted or non-existent sectoral regulation (in online political advertising and political campaign funding) and by the participation of (unsuspecting) US domestic actors. This has enabled China and Russia to "change a democracy's behavior by influencing its citizens at scale and in real time" without resorting to either conflict or diplomacy. Malign foreign influence is so pervasive that "the weaponization of information technologies threatens to jeopardize democracies' ability to govern and protect their national security and to undermine people's trust in democracy as a system of government" (Rosenbach & Mansted, 2019).

A more proactive democratic effort to deter such disinformation campaigns and defend critical infrastructure, including elections, is clearly needed, but has not been forthcoming for a number of reasons. First, there is no consensus on whether social media and AI-enabled disinformation are undermining democracy and enhancing political polarization (Kundnani, 2020, p. 11). Nevertheless, the way democracies work is changing (Gaub, 2019, p. 34). Over the past decade, the number of democratic regimes and the quality of civil and political liberties has been in decline (Economic Intelligence Unit, 2020, Freedom House, 2020). Even before the advent of deepfakes and synthetic media, the coercive use of civilian and commercial infrastructure by domestic and foreign agents to disseminate disinformation had blurred the lines between legitimate actions for commercial and political gain and illegitimate interference and subversion (Kornbluh & Goodman, 2020). This makes it even more challenging to efficiently apply law enforcement or hard security measures to defend against and deter the threat. The openness of democracies could also limit their ability to use technology to defend themselves (Rosenberger, 2020). The role of emerging digital technologies in creating a virtual relationship between adversarial powers like Russia and China and the citizens of Western democratic states remains under-explored.

Furthermore, necessary adaptations depend in large measure on the active participation of the private sector. This is one of the greatest challenges democracies face in adopting, using, and mitigating the unintended consequences of emerging technologies. While basking in the economic rewards of "surveillance capitalism" (Zuboff, 2019), which scholars warn is dangerously "narrow[ing] the gulf between the application of digital technologies in democracies and their application in autocracies such as China" (Rosenberger, 2020), the private sector has sometimes resisted broader cooperation with governments (DeJonge Schulman, 2017). The track record of private-public partnerships remains mixed and private companies are developing their own ethical guidelines and practices around the use of emerging technologies such as facial recognition. Google employees walked out of Project Maven, a DoD initiative which used AI-enabled video-analysis for target acquisition in the anti-Daesh campaign in Iraq and Syria. Simultaneously, Google enables police tracking by sharing Google GPS phone locations with local law enforcement in murder investigations (Valentino-DeVries, 2019). Even in countering violent extremism, governments have been skeptical of direct regulation of online content; instead, they rely on a complex matrix of limited regulation and platform self-regulation. The EU's 2020 Data Strategy includes provisions for reciprocal public–private data sharing, but it also compels the private sector to share data with European authorities when the absence of access to this data would harm public safety and security.

Uncoordinated national regulation and practices of data, 5G, and other emerging technologies have further compounded the political dilemma faced by democratic regimes and have even introduced tensions in the community of democracies. While the EU is increasingly concerned with regulating external access to its data (i.e. GDPR, 2020 EU data strategy), the US continues to resist any federal regulation of this area. Actors like the EU feel compelled to defend their "data sovereignty" and avoid becoming a "digital colony" of other great powers, notably the US and China (Soare, 2020). The EU and other US allies have questioned Trump administration decisions to decouple from China – including plans to limit technology exports to China, monitor Chinese investments, and freeze technology cooperation with Beijing. The Western – particularly American – concern stems from Chinese uses of stolen intellectual property, data, and technologies for power maximization.

In 2019 Australia's Chinese Defense Universities Tracker found that 92 Chinese academic and business institutions presented a very high risk of funneling technology to Beijing, and at least 15 Chinese civil universities were "implicated in cyberattacks, illegal exports, or espionage" (Joske, 2019). In the context of the COVID-19 pandemic, the UK reported malicious Russian, Iranian, and Chinese cyber-attacks against its universities in an attempt to steal sensitive health technologies (Grierson & Devlin, 2020). In 2018, Washington claimed 90% of its espionage investigations are connected to China (Edelman & Roughead, 2018) and at least 20% of European companies active on the Chinese market claim they have been forced to transfer technology to China (Weinland, 2019). The severity of the Chinese threat, Washington argued in 2018, warranted the establishment of democratic "coalitions of caution" to guard against Chinese technological theft and espionage (Ford, 2018), though many US allies remain skeptical of such proposals.

Not only do democratic governments closely scrutinize Chinese Foreign Direct Investment (FDI) in a number of industrial sectors, but even academic cooperation is closely scrutinized. A Government Accountability Office (2019, p. 2) report "expressed concern that the presence of a [Confucius] institute could constrain [American colleges] campus activities and classroom content". In February 2020, MIT canceled a five-year collaboration with Chinese AI company iFlytech after the Department of Commerce banned the company – along with five other Chinese AI companies – from the US (Knight 2020). European universities reported similar Chinese coercive behavior in the framework of cooperation with Confucius Institutes. Nevertheless, the EU has yet to adopt specific measures – despite academic and R&D collaborations being preferred tools of Chinese engagement with European counterparts (Kratz et al, 2020, p. 7).

The EU's approach to data and the development of emerging technologies has been dubbed a "third AI way" (Thornhill, 2018) whereby Europe is trying to assert its own strategic autonomy, its unique social-democratic European values, as well as mitigate the risks emanating from an intensifying Sino-American strategic competition (Soare, 2020). American unilateral efforts to limit Chinese engagement in Europe and to consolidate market control over parts of the supply chains for emerging technologies (e.g. semiconductors, AI chip manufacturing technology, rare earths, etc.) are increasingly causing transatlantic friction. The same US pressures are exerted on non-European allies and partners, such as Australia and India despite their resistance to Washington's demands (Barnes & Satariano, 2019).

To protect democratic values and the rules-based international order, democracies are equally interested in deploying emerging technologies to preserve their comparative military advantage (Soare, 2020). This advantage has eroded in recent decades due to the rapid proliferation of precision-guided munitions and air-defense systems, as well as the Russian and Chinese A2/AD and hybrid strategies (coercion below the threshold of armed conflict). American and European officials have warned that without sustained investment, which is threatened by COVID-related cuts to defense budgets, both the US and Europe will lose their cutting-edge in military technologies.

Successful *adoption* of emerging technologies is as important as their successful *development*. The US is pursuing a comprehensive digital defense modernization agenda, with AI at its center. This will transform the way the US prepares, trains, and fights by creating a comprehensive "networked nervous system for warfare" (Tucker, 2017) for multi-domain operations. The challenge is not just to deploy AI first, although there is a tendency to regard the technical-military competition through a first-mover advantage lens, but to integrate it successfully on legacy and new military platforms alongside other emerging technologies. The EU is also developing a strategy for the digitalization of its armed forces that involves the use of AI to better connect European armed forces. American and European ecosystems are believed to facilitate and incentivise descentralized innovation, by comparison to the more centralized authoritarian models. American and European efforts have developed in parallel, which is not conducive to increase transatlantic interoperability (Soare, 2020; Imbrie et al, 2020). Based on ongoing projects, democracies seem less keen on joint or collaborative R&D on emerging technologies than they are on establishing AI norms and safety standards. This is true both of NATO's Emerging and Disruptive Technologies Roadmap (December 2019) and the Five Eyes framework.

A more robust multilateral approach to forging a democratic way of AI has been called for, with a broader scope to also include deepening R&D cooperation with allies and partners, enhancing interoperability with allies, accessing a broader pool of data and science on emerging technologies (NSCAI, 2020), and limiting technology transfers to China. Scholars are calling for the US to leverage its extensive strategic partnerships and alliance systems as a strategic asset in the AI competition with China and Russia, either by adapting its alliances to be more "agile" (Imbrie et al, 2020) or by establishing what the Center for New American Security (CNAS) calls a "technology alliance" of like-minded countries. In short, it is not enough for democracies to signal their emerging technologies-based capabilities; they need to be more proactive in shaping the international environment around the norms, standards, and capabilities these technologies help build.

"Who's winning?" Digital democracy vs digital authoritarianism

While there are marked differences, and some similarities, between democracies and authoritarian states, as detailed in the previous section, the questions remain: What type of political context is better at shaping the development of emerging technologies? Do democracies have an inherent advantage, or are authoritarians in a stronger position overall?

In their 2013 book *The New Digital Age: Reshaping the Future of People, Nations and Business*, Eric Schmidt and Jared Cohen of Google wrote:

> modern technology platforms, like Google, Facebook, Amazon and Apple [but equally Alibaba, Tencent and others], are even more powerful than most people realize. (…) Almost nothing short of a biological virus can spread as quickly, efficiently or aggressively as these technology platforms.

While such assessments suggest a democratic innovation advantage (Kroenig, 2020; Rosenbach & Mansted, 2019; Stier, 2017), evidence shows authoritarian uses of these technology platforms can easily undermine this perceived advantage. Political interests – regime survival for authoritarian regimes and national security/democratic integrity for democratic ones – are central in shaping narratives about the development, adoption and use of emerging technologies, especially with respect to AI, but to very different ends. Relatedly, digital emerging technologies present different opportunities and constraints when situated in a particular political context, though both democracies and authoritarian regimes are seeking to shield their nations from unintended and harmful consequences of digital technologies. The digital and information sphere is now the new battlefield of ideological and great power competition (Rosenberger, 2020). Because emerging technologies intermediate so many routine human experiences, they are turning the relationship between human and machine into a new and critical zone for political competition, both domestically and internationally. Manipulation in this zone is more perverse and dangerous to society because it is challenging to detect and because it divides and diffuses responsibility. In this sense, Herrera's (2003, p. 588) words of caution are warranted: "the political implications of digital information technologies are up for grabs – not just because the rules governing their use are undecided, but because the technologies themselves are being shaped by politics".

AI and big data analysis are considered the backbone of better human decision-making in democratic contexts and the effort is geared toward eliminating bias and risks from their use to enable them to enhance human well-being. However, the use of AI in cyber-attacks, online or social-media (dis)information campaigns is a core challenge. Democratic regimes are constrained in the magnitude and type of responses they can put in place for fear of eroding the integrity of democratic institutions, rights, and processes. The violation of human rights and rule of law in the use of emerging technologies by democratic states are exposed by a free media, scrutinized by parliaments and international organizations and corrected by independent courts. By contrast, authoritarian regimes see unfettered opportunity in the use of emerging technologies for political purposes. Though they are challenging to control, these technologies provide authoritarian regimes with exponentially more effective tools to create mass surveillance, political policing and information control, which is helping them be more oppressive and survive longer.

Both authoritarian and democratic regimes leverage international networks for strategic advantage. While democracies leverage international alliances to adopt ethical, safe and responsible norms of use for AI and other emerging technologies, China and Russia are using the BRI and CSI networks, respectively, to establish technospheres of geopolitical influence, to transfer their information control technologies, and to export their domestic political models abroad. Scholars who argue there is an asymmetry between authoritarian and democratic regimes in adopting AI and other emerging technologies (Rosenberger, 2020) are right – democracies have less legal and ethical leeway by nature of their institutional design. However, this asymmetry alone does not automatically translate into a strategic disadvantage, especially given democracies' ability to self-correct. There has never been perfect symmetry between different political regimes when it comes to policy performance or output (Stier, 2017). More research is also needed to better understand the vulnerabilities and limits authoritarian regimes themselves may face in adopting AI and other emerging technologies.

Scholars and practitioners alike fear that authoritarians are more effective and perform better in the adoption of AI and other emerging technologies because of their centralized and legally unencumbered institutional framework. For example, authoritarian regimes, like China, find it relatively straightforward to implement "civil-military fusion" in pursuit of military AI, compared with the legal and institutional adaptations required of democracies to foster public–private partnerships on AI. Similarly, it is technically and politically easier (and self-serving) for authoritarian regimes to decouple and sanitize their information environments from the international system than it is for democratic regimes, who have to constantly redraw the line between individual liberties and societal security in their pursuit of national security, as well as preserve the integrity of democratic institutions and processes.

The challenges democracies face in adopting AI do not represent an enduring disadvantage in comparison to authoritarian regimes (Rosenbach & Mansted, 2019). However, they do serve to delineate between emerging technologies that *empower* (in democracies) or *overpower* (in authoritarian regimes) individuals, societies, and states. Moreover, political narratives based on democratic AI help mitigate "the growing perception that the digital technology being developed in the United States [and other democracies] is no different from that being developed in China" (Rosenberger, 2020). This has led analysts to question who is winning the strategic competition between the West, China, and Russia over emerging technologies (Castro, et al, 2019), who is more efficient in the adoption of these technologies, and whether a "race" for techno-political dominance is even an appropriate framework of analysis for the strategic competition over emerging technologies (Scharre, 2019). AI algorithms and models may not be so technically different in China, the US, Europe or Russia, but their uses and sociopolitical and geopolitical consequences certainly are. Without the association with democratic or authoritarian political interests, concepts like "efficiency" or "winning" the competition over emerging technologies have little meaning.

Conclusion

Mainstream IR theory is limited in helping us navigate the complex relationship between political interests and geopolitical outcomes in the application of emerging technologies. This chapter makes a contribution to the debate on emerging technologies and national and international security by shifting focus from great-power competition dynamics to national and political regime dynamics. In doing so, it offers a comparative analysis of how emerging technologies are shaped by political goals and how, in turn, they shape the ways and means of regimes to pursue perennial interests. The analysis highlights how both democratic and authoritarian regimes are interested in shielding their societies from potential vulnerabilities created by these new technologies, while also pursuing international opportunities created by them or defending against international challenges generated by their proliferation. Nevertheless, applying the lens of authoritarian and democratic political interests reveals more about the use of and narratives around emerging technologies and how seemingly similar applications of technology, in different institutional and political frameworks, lead to important differences in consequences, including from a national and international security perspective. This is not an exhaustive analysis of democratic and authoritarian use of technology, but rather a conversation starter. Undoubtedly, more crosscutting interdisciplinary research, involving psychology, sociology, political sciences, computer sciences, ethics, history and foresight – is needed to help us navigate the complex relationship between society, politics, and international security in the years to come.

Notes

1 I use the definitions of democratic and authoritarian regimes used by Freedom House and the Economist Intelligence Unit. Democratic regimes are defined by pluripartidism; regular, free, and fair elections; institutional checks and balances; an independent judiciary and a free media; the rule of law; the respect for individual and minority civil and political liberties. Authoritarian regimes are states where pluripartidism and democratic institutions do not exist or are significantly curtailed, there is no rule of law or institutional checks and balances, there is no independent judiciary or media, civil and political liberties are regularly infringed upon, including through media censorship and political repression. Countries in between these markers are considered hybrid regimes. See Economist Intelligence Unit, Democracy Index 2019, p. 50–53.

2 Authoritarian regimes are not the only ones conducting mass surveillance. The Patriot Act and special classified court mandates made it possible for the US government to use surveillance technologies to track, identify, and incarcerate individuals suspected of terrorism. However, unlike authoritarian regimes, democratic cases of such violations are generally fostered by crises, are transitional, for the most part, and can and have been exposed and corrected by the role of the independent judiciary and the free media, which are missing in authoritarian regimes.
3 Washington and its European allies experience bottlenecks and dependencies in their military supply chains, as well, particularly in relation to Chinese raw materials or components. See White House. (2018). Assessing and Strengthening the Manufacturing and Defense Industrial Base and Supply Chain Resiliency of the United States: Report to President Donald J. Trump by the Interagency Task Force in Fulfillment of Executive Order 13806, September and Pavel, C.P. and Tzimas, E. (2016). Raw materials in the European defence industry, European Commission Joint Research Center Science for Policy Report.
4 Democracies are leading the market in the sale of surveillance tech – 87% of companies selling such technologies originate in NATO countries (Feldstein, 2019), indicating the need of tighter export controls and democratic oversight. Here again, it is the use of the technology that is important, which differs markedly across democratic and authoritarian regimes.

References

Auslin, M (2018) 'Can the Pentagon win the AI arms race? Why the U.S. is in danger of falling behind', *Foreign Affairs*. October 19. https://www.foreignaffairs.com/articles/united-states/2018-10-19/can-pentagon-win-ai-arms-race

Barnes, J and Satariano, A (2019) 'U.S. Campaign to Ban Huawei overseas stumbles as allies resist', *New York Times*. March 17. https://www.nytimes.com/2019/03/17/us/politics/huawei-ban.html.

Bendett, S (2019) 'Russia's military is writing an armed-robot playbook', *DefenseOne*. November 26. https://www.defenseone.com/ideas/2019/11/russias-military-writing-armed-robot-playbook/161549/

Berwick, A (2018) 'How ZTE helps Venezuela create China-style social control', *Reuters* special report. November 14. https://www.reuters.com/investigates/special-report/venezuela-zte/

Bischoff, P (2019) 'Surveillance camera statistics: which cities have the most CCTV cameras?', *Comparitech*. August 15. https://www.comparitech.com/vpn-privacy/the-worlds-most-surveilled-cities/

Black, J (2013) *War and technology*. Indianapolis: Indiana University Press.

Burton, J and Soare, SR (2019) 'Understanding the strategic implications 249 of the weaponization of artificial intelligence', in Minárik, T et al (eds) *11th international conference on cyber conflict: silent battle*. Tallinn: NATO CCDCOE, pp. 249–267.

Campaign to Stop Killer Robots (2018) *Country views on killer robots*. November 22. https://www.stopkillerrobots.org/wp-content/uploads/2018/11/KRC_CountryViews22Nov2018.pdf

Castro, D et al. (2019) 'Who is winning the AI race: China, the EU or the United States?', *Center for Data Innovation*. August 19. https://www.datainnovation.org/2019/08/who-is-winning-the-ai-race-china-the-eu-or-the-united-states/

DeJonge Schulman, L et al. (2017) 'The Rocky relationship between Washington and Silicon Valley: clearing the path to improved collaboration', *Center for a New American Security Report*, July 19.

Diamond, L (2010) 'Liberation technology', *Journal of Democracy*, 21(3), pp. 69–83.

Economists Intelligence Unit (2020) *Democracy index 2019: a year of democratic setbacks and popular protest*. April. http://www.eiu.com/Handlers/WhitepaperHandler.ashx?fi=Democracy-Index-2019.pdf&mode=wp&campaignid=democracyindex2019

Edelman, E and Roughead, G (2018) *Providing for the common defense: the assessment and recommendations of the national defense strategy commission, US congress mandated commission on the national defense strategy*. November 27. https://www.usip.org/sites/default/files/2018-11/providing-for-the-common-defense.pdf

Espinoza, J and Murgia, M (2020) 'EU backs away from call for blanket ban on facial recognition tech', *Financial Times*. February 11. https://www.ft.com/content/ff798944-4cc6-11ea-95a0-43d18ec715f5

European Commission (2019) *High-level expert group on AI: ethics guidelines for trustworthy AI*. April 8. https://ec.europa.eu/digital-single-market/en/news/ethics-guidelines-trustworthy-ai

European Parliament (2015) 'Parliament to resume investigations into CIA-led operations in EU countries', *Press Release*. June 2. https://www.europarl.europa.eu/news/en/press-room/20150206IPR21212/parliament-to-resume-investigations-into-cia-led-operations-in-eu-countries

Feldstein, S (2019) 'The global expansion of AI surveillance', *Carnegie Endowment for International Peace*. September. https://carnegieendowment.org/files/WP-Feldstein-AISurveillance_final1.pdf

Ford, CA (2018) Coalitions of caution: building a global coalition against Chinese technology-transfer threats, United States department of state. Indianapolis. September 13. https://www.state.gov/remarks-and-releases-bureau-of-international-security-and-nonproliferation/coalitions-of-caution-building-a-global-coalition-against-chinese-technology-transfer-threats/

Freedom House (2020) 'Nations in transit 2020', *Dropping the Democratic Façade*. April. https://freedomhouse.org/sites/default/files/2020-04/05062020_FH_NIT2020_vfinal.pdf?mc_cid=3ef832d93d&mc_eid=b4b026f0d5

Gaub, F (2019) 'Global trends to 2030: challenges and choices for Europe', *European Commission*. April. https://ec.europa.eu/assets/epsc/pages/espas/index.html

Gigova, R (2017) 'Who Vladimir Putin thinks will rule the world', *CNN*. September 2. https://edition.cnn.com/2017/09/01/world/putin-artificial-intelligence-will-rule-world/index.html

Grierson, J and Devlin, H (2020) 'Hostile states trying to steal coronavirus research, says UK agency', *The Guardian*. May 3. https://www.theguardian.com/world/2020/may/03/hostile-states-trying-to-steal-coronavirus-research-says-uk-agency

Hamilton, I (2018) 'Google's ex-CEO Eric Schmidt says the internet will split in two by 2028', *Business Insider*. September 21. https://www.businessinsider.fr/us/eric-schmidt-internet-will-split-in-two-2028-china-2018-9

Herrera, GL (2003) 'Technology and international systems', *Millennium*, 32(3), pp. 559–593.

Ikenberry, JG (2018) 'The end of liberal international order?' *International Affairs*, 94(1), pp. 7–23.

Imbrie, A et al. (2020) *Agile alliances: how the US and its allies can deliver a democratic way of AI*. Center for Security and Emerging Technology. February. https://cset.georgetown.edu/wp-content/uploads/CSET-Agile-Alliances.pdf

Joske, A 'The China defence universities tracker: exploring the military and security links of China's universities', *Australian strategic policy institute policy brief no. 23*. https://s3-ap-southeast-2.amazonaws.com/ad-aspi/2019-11/The%20China%20Defence%20Universities%20Tracker_0.pdf?ozIi2cWm.kXpe7XsEZ44vUMMQBNfnR_x

Kania, EB (2020) 'AI weapons', in *China's military innovation*. Brookings Institution. April. https://www.brookings.edu/research/ai-weapons-in-chinas-military-innovation/

Kendall-Taylor, A, Frantz, E and Wright, J (2020) 'The digital dictators: how technology strengthens autocracy', *Foreign Affairs.* March/April.

Khan, SM and Flynn, C (2020) 'Maintaining China's dependence on democracies for advanced computer chips', *Center for Security and Emerging Technology*. April. https://cset.georgetown.edu/wp-content/uploads/Khan-Flynn—Maintaining-Chinas-Dependence-on-Democracies.pdf

Kliman, D et al. (2020) 'Dangerous synergies: countering Chinese and Russian digital influence operations', *Center for a New American Security*. May 7. https://s3.amazonaws.com/files.cnas.org/documents/CNAS-Report-Dangerous-Synergies-May-2020-DoS-Proof.pdf?mtime=20200506164642

Knight, W (2020) 'MIT cuts ties with a Chinese AI firm Amid human rights concerns', *Wired*. April 21. https://www.wired.com/story/mit-cuts-ties-chinese-ai-firm-human-rights/

Kornbluh, K and Goodman, EP (2020) 'Safeguarding digital democracy: digital innovation and democracy initiative roadmap', *German Marshall fund research papers no. 4*. March. https://www.gmfus.org/publications/safeguarding-democracy-against-disinformation

Kratz, A (2020) 'Chinese FDI in Europe: 2019 update. Special topic: research collaborations', *MERICS*. April 8. https://www.merics.org/en/papers-on-china/chinese-fdi-in-europe-2019

Kroenig, M (2020) *The return of great power rivalry: democracy versus autocracy from the ancient world to the U. S. and China*. London: Oxford University Press.

Kundnani, H (2020) 'The future of democracy in Europe: technology and the evolution of representation', *Chatham house research paper*. March. https://www.chathamhouse.org/sites/default/files/CHHJ7131-Democracy-Technology-RP-INTS-200228.pdf

Miller, C (2020) 'How the CIA used Crypto AG encryption devices to spy on countries for decades', *New York Times*. February 11. https://www.washingtonpost.com/graphics/2020/world/national-security/cia-crypto-encryption-machines-espionage/?itid=ap_gregmiller

Morgenthau, H (1988) *Politics among nations: the struggle for power and peace* (6th ed.). New York: Random House.

Morgus, R et al. (2018) 'The digital deciders: how a group of often overlooked countries could hold the keys to the future of the global internet', *New America*. October. https://d1y8sb8igg2f8e.cloudfront.net/documents/The_Digital_Deciders_2018-10-22_132508_krSpfSY.pdf

Mozur, P (2018) 'Inside China's Dystopian dreams: A.I., Shame and lots of cameras', *New York Times*. July 8. https://www.nytimes.com/2018/07/08/business/china-surveillance-technology.html

Murray, W (2011) *War, strategy and military effectiveness*. Cambridge: Cambridge University Press.

National Security Commission on Artificial Intelligence (2020) *First quarter recommendations*. March. https://drive.google.com/file/d/1wkPh8Gb5drBrKBg6OhGu5oNaTEERbKss/view

New York Times (2000) *Clinton's words on China: trade is the smart thing*. March 9. https://www.nytimes.com/2000/03/09/world/clinton-s-words-on-china-trade-is-the-smart-thing.html

Pavel, CP and Tzimas, E (2016) 'Raw materials in the European defence industry', *European commission joint research center science for policy report*.

Privacy International (2016) *The global surveillance industry report*. July. https://www.privacyinternational.org/sites/default/files/2017-12/global_surveillance_0.pdf

Rosenbach, E and Mansted, K (2019) 'The geopolitics of information, belfer center for science and international affairs', *Harvard Kennedy school research paper*, May 28.

Rosenberger, L (2020) 'Making cyberspace safe for democracy: the new landscape of information competition', *Foreign Affairs*. May/June.

Ruan, L et al. (2020) 'Censored contagion: how information on the coronavirus is managed on chinese social media', *CitizenLab*. Toronto. March 3. https://citizenlab.ca/2020/03/censored-contagion-how-information-on-the-coronavirus-is-managed-on-chinese-social-media/

Sayler, KM (2019) 'Artificial intelligence and national security', *Congressional research service report R45178*. November 21. https://fas.org/sgp/crs/natsec/R45178.pdf

Scharre, P (2019) 'Killer apps: the real danger of an AI arms race', *Foreign Affairs*, May/June.

Schneider-Petsinger, M et al. (2019) 'US–China strategic competition: the quest for global technological leadership', *Chatham house research paper*. November. https://www.chathamhouse.org/sites/default/files/publications/research/CHHJ7480-US-China-Competition-RP-WEB.pdf

Shahbaz, A (2018) *Freedom on the net 2018: the rise of digital authoritarianism*. Freedom House. https://freedomhouse.org/report/freedom-net/2018/rise-digital-authoritarianism

Skolnikoff, EB (2002) 'Will science and technology undermine the international political system?', *International Relations of the Asia-Pacific*, 2(1), pp. 29–46.

Soare, SR (2020) 'Digital divide? Transatlantic defence cooperation on AI', *European Union Institute for Security Studies*. March 5. https://www.iss.europa.eu/content/digital-divide-transatlantic-defence-cooperation-ai

Statista. (2019) *Amount of text messages in China from June 2017 to June 2018*. September 23. https://www.statista.com/statistics/278205/china-amount-of-text-messages/, accessed February 12, 2020.

Stier, S (2017) 'Internet diffusion and regime type: temporal patterns in technology adoption', *Telecommunications Policy*, 41(1), pp. 25–34.

Sukhankin, S (2019) 'The three "Faces" of Russia's AI strategy', *Eurasia Daily Monitor*. 16, p. 154. November 5.

Talmadge, C (2016) 'Different threats, different militaries: explaining organizational practices in authoritarian armies', *Security Studies*, 25(1), pp. 111–141.

Thornhill, T (2018) 'There is a "third way" for Europe to navigate the digital world', *Financial Times*. November 19. https://www.ft.com/content/9da4156c-ebd4-11e8-89c8-d36339d835c0

Tucker, P (2017) 'The future the US military is constructing: a giant, armed nervous system', *DefenseOne*. September 26. https://www.defenseone.com/technology/2017/09/future-us-military-constructing-giant-armed-nervous-system/141303/

United States Department of Defence (2019) *Summary of the 2018 department of defence artificial intelligence strategy: harnessing AI to advance our security and prosperity*. February 19. https://media.defense.gov/2019/Feb/12/2002088963/-1/-1/1/SUMMARY-OF-DOD-AI-STRATEGY.PDF

United States Department of Defence (2020) 'DoD adopts ethical principles for artificial intelligence', *Press Release*. February 24. https://www.defense.gov/Newsroom/Releases/Release/Article/2091996/dod-adopts-ethical-principles-for-artificial-intelligence/

United States Governmental Accountability Office (2019) *Observations on confucius institutes in the United States and U.S. universities in China*. March 28. https://www.gao.gov/assets/700/697156.pdf

Valentino-DeVries, J (2019) 'Tracing phones, Google is a Dragnet for the police', *New York Times*. April 13. https://www.nytimes.com/interactive/2019/04/13/us/google-location-tracking-police.html

Warzel, C (2019) 'A major police body cam company just banned facial recognition', *New York Times*. June 27. https://www.nytimes.com/2019/06/27/opinion/police-cam-facial-recognition.html?searchResultPosition=3

Waltz, KN (1979) *Theory of international politics*. Long Grove: Waveland Press.

Watts, S et al. (2020) *Alternative worldviews: understanding potential trajectories of great-power ideological competition*. RAND Corporation. March 12. https://www.rand.org/pubs/research_reports/RR2982.html

Weber, V (2019) 'The worldwide web of Chinese and Russian information controls', *Center for technology and global affairs, University of Oxford*, working paper no. 11. September. https://www.ctga.ox.ac.uk/files/theworldwidewebofchineseandrussianinformationcontrolspdf

Weinland, D (2019) 'European companies forced to hand tech to China', *Financial Times*. May 20. https://www.ft.com/content/f2f4dca0-7abc-11e9-81d2-f785092ab560

Weiss, JC (2020) 'A world safe for autocracy? China's rise and the future of global politics', *Foreign Affairs*. June/July.

Wendt, A (1995) 'Constructing international politics', *International Security*, 20(1), pp. 71–81.

Wezeman, PD et al. (2020) *Trends in international arms transfers, 2019*. SIPRI. March. https://www.sipri.org/sites/default/files/2020-03/fs_2003_at_2019_0.pdf

White House (2018) *Assessing and strengthening the manufacturing and defense industrial base and supply chain resiliency of the United States: report to President Donald J. Trump by the interagency task force in fulfilment of executive order 13806*. September.

Wright, N (2018) 'How artificial intelligence will reshape the global order: the coming competition between digital authoritarianism and liberal democracy', *Foreign Affairs*. July 10.

Yang, Y and Liu, N (2019) 'Beijing orders state offices to replace foreign PCs and software', *Financial Times*. December 8. https://www.ft.com/content/b55fc6ee-1787-11ea-8d73-6303645ac406

Zuboff, S (2019) *The age of surveillance capitalism: the fight for a human future at the new frontier of power*. New York: PublicAffairs.

7 Inequitable Internet

Reclaiming digital sovereignty through the blockchain

Richard Wilson and Andrew M. Colarik

Introduction

Science fiction writer William Gibson is widely credited for saying, "The future is already here, it's just not very evenly distributed" (Fresh Air, 1993). The words, although spoken decades ago, seem particularly poignant today. The advent of the Internet was nothing short of revolutionary. The ability to share and access information bypassing the bounds of time and space fundamentally changed society on the whole. Economies were particularly disrupted as entirely new business models emerged leading to productivity gains unseen since the first Industrial Revolution. Even as the hockey-stick trend of growth continued, a parallel and negative trend emerged alongside; growing inequity. Today, Technology's Frightful Five, (Manjoo, 2017, p. B1) a moniker applied to Amazon, Apple, Alphabet, Facebook, and Microsoft, is worth a staggering three trillion dollars. The success of these companies, once valued for their innovation and disruption, are increasingly an illustration of the trend of consolidation among an exclusive few.

The rise of monopolies in the increasingly important digital space carries with it the conventional dangers associated with traditional monopolies, such as reduced innovation and artificially inflated prices. But the fact is that so much of the content and services society relies on is accessed through the platforms provided by these powerful corporations that their monopolistic tendencies have increased societal and security risks. Even as the disparity between content creator and platform provider grows, a new and profitable asset class, data, the question of sovereignty and security persist. The process by which data is gathered, stored, and distributed is convoluted and opaque to the majority of users interacting with the collecting institutions. These same organizations continue to demonstrate an inability to properly safeguard users' digital identities which erodes social trust in digital technologies. A new technology, the blockchain, has the potential to assist in mitigating these issues. If the proper stakeholders are involved in the development and governance of this technology, the social and security benefits may be prodigious.

With the above in mind, this chapter first offers a synopsis of the consolidation of market power among a few technology corporations and some of the implications this presents; a discussion of the opaque, one-sided nature of the data economy; and a supported view of the fractured and increasingly vulnerable ecosystem of digital identity management. The authors contend that these issues are interrelated and mutually reinforcing, which pose both a social and security risk to fundamental digital interactions. A preferred outcome to the issues is then proposed and the blockchain is offered as a potential mitigating technology. The benefits of public interest organizations participating in governance consortiums early are then discussed, and key measures are proposed to address the hurdles to the widespread adoption of the technology.

To establish a case for growing disparity in the digital economic space, it is necessary to delineate the concentration of wealth and power and demonstrate the "tendency towards the creation of natural monopolies on the Internet" (Esteve, 2016). The headlines of the business section hint at monopolistic practices regularly. Facebook's acquisitions, WhatsApp and Instagram – two start-up companies operating in the social media space – illustrate the trend of the disruptive competitors being swallowed by the digital giants (Allen, 2017). But these examples are more anecdotal than complete representations of the magnitude of consolidation in the digital space. It is necessary to first understand what these companies are truly selling to understand how far the concentration has progressed. Consider that Google has an 88% market share of the highly lucrative market of search advertising. Facebook owns 77% of social traffic on mobile devices, while Amazon controls three-quarters of the e-book market (Taplin, 2017). These alarming figures illuminate the power and control these companies have acquired in highly coveted and profitable revenue streams, and they are increasingly representative of the rule rather than the exception. The result is a landscape of a very few companies that "create, apply, and optimize digital technology to control massive consumer and business markets" (Andriole, 2017). The vertical and lateral integration business practices on display lead to a system in which Amazon controls retail but also cloud-computing; where Google controls mobile search and online payments. The monopolistic practices of these technology firms carry with them the traditional fears, such as reduced innovation and price-fixing. But when companies control vast swaths of the digital landscape, they control all social and economic interactions in that space in a distorted power relationship with the user.

Value creation vs. value capture

An example of the risks involved when disproportionate power relationships arise can be seen in the tumultuous move of the media industry into the digital realm. As the Internet matured and media became a digital product, traditional marketplaces began to erode into the digital economic space. Technology companies quickly leveraged their user bases to form and operate transaction model platforms. YouTube connects video bloggers with an audience base and the App Store connects Apple device users with software application programmers. In each case the creators of the content are matched with the potential customers and the platform extracts a percentage of the profit (Allen, 2015). Benefitting from creating a marketplace is a perfectly reasonable business model; the problem the media industry faces is one of the disproportionality of returns between creators and content providers. Consider the current value appropriated by content creators that these transaction platforms rely on. An app developer often cedes 30% of the sale revenue to the App Store, a musician is paid an average of 0.005 cents per stream, and a YouTube star makes $150.00 per one million views (Conte, 2017). These figures point to a disparity in the power relationship between an industry which must utilize the digital space to remain viable and the powerful companies that control market entry and continued sustainability.

The disparity has arisen as a contention between the forces of value creation versus value capture. Value creation is the work that is put toward offering a service, resource, capability, or product that is higher than the cost of production. Value capture is the ability to realize profits from the sale of a product or service (Allen, 2017). The digital economy is becoming one in which the ability to realize profits from value creation is unevenly distributed to the few that excel in value capture. The result is a reallocation of billions in revenue from content creators to the powerful monopoly platforms on which they must now reside (Taplin, 2017), as well as leveraging the continued domination of user attention as an advertising revenue stream and control over another revenue source, user data.

Data is the new oil

Every time a person interacts with the digital space through an intermediary, data is created, collected, and stored. Every Google search, every Facebook post and every Amazon purchase generates data. The sheer scope of data that is created each day is staggering, already difficult to comprehend, and is increasing exponentially year by year. According to the International Data Corporation, the global datasphere will expand from 33 zettabytes in 2018 to 175 by 2025 (Reinsel, Gantz, and Rydning, 2018). Additionally, with new users joining every day, the Internet of things is a rapidly (IoT) growing space. Houses, cars, and even jewelry are coming online and every new device is filled with sensors that collect data that is sent to be processed and stored on the Internet. More and more of people's environment and interactions are moving into the digital sphere to be "mediated by digital services" (Kosinski, Sitwell, and Graepei, 2012). But here again, the uneven power relationship of users with digital intermediaries is evident in the way user data is gathered, analyzed, stored, and distributed.

An entire economy has arisen around, "collecting, aggregating, analyzing, and monetizing personal data" (World Economic Forum, 2011). The scope of value in this new economic realm is so vast experts have commonly compared it to the oil industry. In this context it is surprising how little the questions around data ownership and the rights of users arise. When questioning who has sovereignty of user data, it seems natural to believe that users would have at least some control over how their data are managed. The General Data Protection Regulation (GDPR), otherwise known as the Right to be Forgotten, was derived out of the European Union's case against Google and is but one of the first steps needed in challenging the prevailing control and ownership of data. In the current digital landscape, however, this is rarely the case unless a person resolves to opt-out of the increasingly essential services that digital intermediaries administer. While the GDPR is a first step, the precedent has already been set and is now the norm for users to exchange their rights to control their personal data for the ability to access digital services (Foer, 2017). For example, every time a user searches Google their search is aggregated into metadata owned by Google in exchange for benefitting from Google's search algorithms. Also, the data from social media interactions on platforms, such as Facebook, are exchanged for access to the platform. The economy of data has evolved into an opaque environment in which many users are unaware of how much data they are trading, how that data is being used, and who in fact controls and stores it (World Bank, 2016). The exchange of data for services also creates a byproduct; significant privacy and security risks to the user.

In the digital space it seems that privacy is in exceedingly short supply. Polling data from the US has shown that a majority of respondents are concerned about digital privacy (Rainie and Duggan, 2016). Proponents of the current framework attest that the privacy concerns accompanying data collection are conflated and benign. But research has shown that the process for identifying a private user from metadata is startlingly feasible and, in many cases, a simple algorithmic process (Kosinski, Sitwell, and Graepei, 2012). In this way, data that a user considered to be private, or at least not directly linked to their personal identity, may be stored and sold by an entity outside of their control and later linked to their identity. In many cases, the pretense of anonymity is not even pursued. In China, it is estimated that 70% of users will have personally identifiable information leaked from sources they viewed as protected (Han, 2017). When companies and institutions maintain a wealth of user data, the consumer becomes transparent and opportunities to discriminate, through prices or otherwise, are available (Martin, 2015; Schudy and Utikal, 2015). The risks to privacy are significant, and they will continue to grow as the business of aggregating and selling data matures. These concerns arise within the realm of controlled and deliberate distribution of data. The problems compound when sensitive data is maliciously seized by unknown actors.

Vulnerable digital identities

To transact and interact with institutions in the digital space, users must possess and share a digital identity. Internet storefronts, social media platforms, government institutions, etc. must have a means of authenticating users to process transactions or interactions within a network. That identity is then stored with the interaction data which can range from cell phone call records to financial information to employment background checks. This information can range from innocuous to vitally important, and institutions have an obligation to maintain the security of private records. However, this is increasingly proving a challenging task. In 2017, the consumer credit reporting agency, Equifax, reported a breach in their security systems that resulted in the theft of 143 million files (Rosenzweig, 2017). These files contained extremely sensitive financial information for millions of users. In 2015, the Office of Personnel Management of the US government experienced a breach of 21.5 million records. Among the lost records were fingerprint scans, background checks, and Social Security Numbers (Naylor, 2016). When these two instances are combined with the mounting corporate breaches of customers' personal and financial information, the unforeseen consequences are staggering. In all of these cases the people injured were not in control of their digital identity and records.

Consumers now have a fragmented and decentralized digital identity that is maintained with every institution they interact with (Chester, 2017). This is in stark contrast to a centralized system such as a consumer maintaining their personal driver's license and using it for identification on a case by case basis. The digital space is more akin to a system in which each institution generates its own driver's license and maintains a functioning copy indefinitely. In this environment a person does not necessarily own their identity. A measure of sovereignty is held by every organization the user interacts with in a range of security settings. For an individual to cede such precious control creates a poorly considered trust environment between the user and the institution, and it is becoming increasingly apparent that the consumer is shouldering more risk than the institutions charged with protecting their identities. As power and wealth in the digital space continue to condense, these security risks will continue to intensify.

The goal

As presented above, the current path would appear untenable for the whole of society. Monopolistic practices; opaque, one-sided data economies; and a fragmented, vulnerable identity ecosystem are all interrelated problems. Together they pose a security risk that is so vast in scope and potential that a course correction is absolutely necessary if security is to be a societal focus and a true balance of power between producer and consumer is to be restored. So what should be the goal in considering a change of course?

The consolidation of wealth and power in the digital economy may be summarized as an issue of the inability for consumers to opt-out of the essential services and the disproportionate ability of powerful companies to capture value from products and services. An attainable goal in this space would be simply the creation of platforms to challenge the incumbent corporations monopolizing their fields (Catalini and Ganz, 2016). The mere availability of choice for consumers and a reduced barrier to entry for content creators addresses the issue in a manner that can disrupt market power and force corporations to better serve their user base.

The problems of the metadata economy and the current authentication regime are similar. In the case of data, users are unable to practically control and track what information is collected and trafficked from their digital interactions. To interact with any service, users must authenticate their identity by providing sensitive information to be stored with a myriad of institutions indefinitely. In these arenas, the way data and identity ownership are conceived must be completely reversed in which data belong to the user and the services are treated "as guests with delegated and finite permissions" (Zyskind, Nathan, and Pentland, 2015). In this way, users regain the ability to control their data and provide consent to services using it. This approach allows the benefits derived from sharing data to be retained while lowering the individual's risk and increasing data usage transparency.

The barriers

The goals seem achievable: Greater ability to participate in markets and consumers retaining greater control of their data and identities. Why then, do the problems persist? Perhaps powerful lobby interests and a dearth of viable alternative systems are possible challenges? They do not, however, fully explain why regulation has been so ineffectual or why innovative new models of data protection have not disrupted the status quo. In many countries any proposed regulation will face an arduous march toward enactment. Governments are faced with the burden of creating a bureaucracy to enforce regulation while corporations face challenges in compliance. Regulating Internet marketplaces and data flows is especially limited as they inherently are not bound to any single jurisdiction (Cuomo et al, 2017). Effective regulation must then achieve consensus among many stakeholders. For example, the European Union has been very assertive in their push to regulate monopolies and data protection while the US has been far more willing to let industry self-regulate (Goldfarb and Tucker, 2010). The $2.7 billion dollar fine levied against Google and the Right to be Forgotten exemplifies the differing regulatory climates leading to a patchwork system that can be difficult to navigate.

Alternatively to regulation, when problems arise in a socioeconomic system, the market is sometimes able to self-correct with innovation in business models and technology. Unfortunately, the market is proving unable to match the task. The incumbent corporations are integrating both vertically and laterally, leveraging their market power to absorb new entrants. The data economy precedent is set and current market players are content with the status quo. The multitude of identity management systems that have been proposed have not achieved the critical mass required to be a centralized, secure system (Lazarovich, 2015). Businesses have simply not been able to produce viable solutions in these spaces. Fortunately, a technology is emerging that, although not a panacea, has the potential to significantly alter the current system and is presented in the next section.

The blockchain

In 2008 a pseudonymous group named Satoshi Nakamoto released a white paper introducing the blockchain, the technological underpinnings of the Bitcoin cryptocurrency. A blockchain is essentially a distributed ledger that records interactions on a network. A full copy of the ledger is maintained on a multitude of network nodes, and, periodically, all nodes participate to achieve a consensus that the ledger is correct; the ledger version is then time-stamped and appended to the chain of all previous versions (Hoser, 2016). Essentially, the ledger records a chain of sequential transactions in a non-reputable manner and then distributes the updated ledger across a network. The distributed element of blockchain provides some unique characteristics not offered by legacy client-server databases. Because the ledger is distributed rather than centralized, all network participants can view transactions in their entirety giving the ledger a transparency unavailable in centralized systems. User authentication and authorization to transact are handled with public key infrastructure (PKI) cryptography. Because the ledger is validated by all users through consensus algorithm processing and each ledger chained to the previous with a hashed timestamp, the ledger is effectively tamper-proof, immutable, and easily auditable (Arun and Carmichael, 2017). The transparency and immutability make the blockchain a particularly compelling technology in a myriad of use cases. Because of this, although initially conceived as an alternative to third-party institutions, research has shown that institutions stand to gain from employing blockchains in closed, "permissioned" networks as well (Underwood, 2016).

At first it can be difficult to understand how a ledger recording the transactions of digital coins could be a technology able to disrupt the digital economy and return data sovereignty to Internet users. It must be remembered that Bitcoin is merely structured data in the same way a Google search, a smartphone application, and a password are all simply blocks of data. Therefore, any data that can be grouped for processing can be envisioned as a coin or some other moniker. On this foundation, it becomes apparent that the benefits of the blockchain – transparency, integrity, and immutability – can be applied to almost any form of data (Arun and Carmichael, 2017; Catalini and Ganz, 2016). What can be done with coins can be done to any rational grouping of data or information.

Further work in this space has also been done to introduce smart contracts into blockchain technology. These are simple codified rules that can self-execute within the network (i.e. when x conditions are met, execute action y). This again spawns the possibility that the ledger technology can automate and self-execute transactions within a codified rule set (Snow et al, 2014). The concept that all forms of data may reside on a distributed ledger and that ledger can automate actions within codified conditions open the technology to uses far beyond transacting cryptocurrencies. There are many models and frameworks of the initial research of blockchain technology that can be deployed as a digital marketplace, a data transaction platform and, more relevant to this paper, as an identity management tool. The most relevant to this discussion are presented in the following sections and begin to outline a possible solution to restoring digital sovereignty.

Peer to peer business models

Blockchain, from its inception, was developed as a transaction platform. Bitcoin, despite its history, has proven that trust in value transactions can be maintained on a peer to peer network outside of the control of a central authority. It is a short leap then, as previously discussed, to imagine other forms of value being transacted on a blockchain network. Peer to peer marketplaces such as journalism, music, applications, etc. can arise with very low barriers to entry. In fact, the ability to cheaply transact could spawn entirely new business models in which micropayments dominate the manner in which media and services are consumed (Tang, 2018). Freelance journalists could receive a payment directly from their readers while musicians could distribute their content directly to fans at any pricing model they see fit. Content creators can also benefit from the transparency of the blockchain (e.g. a team of academic researchers could track the use and distribution of a white paper) (McConahey and Holtzman, 2015). An auditable trail of intellectual property (IP) use stands to benefit any content creator and could reduce IP theft.

These possible developments point toward a future in which the modes of value capture are not completely dominated by a few powerful tech companies. The ability of marketplaces to arise on a peer to peer network with low transaction costs has the real possibility to challenge the current makeup of the digital economy (Mainelli, 2017). This is not to say that technology companies will be completely eroded, but a compelling alternative has the potential to disrupt the current market domination. Blockchain, employed as a marketplace, has the power to forestall antitrust regulation and spur innovation in the digital economy.

User-controlled data

The blockchain has the potential to reallocate control of user-generated data from the collecting corporations back to users themselves. The foundations of this concept are similar to those governing the transactions of media discussed above. Because blockchains are governed by code, the rules, both technical and legal, can be programmed into the ledger itself. Once the code is written and understood by transacting parties, all members of the network may be reasonably assured that their data will be used in accordance with the governing principles of the network (Zyskind, Nathan, and Pentland, 2015). This, in theory, leads to early adoption and increased usage and deployment.

On the blockchain, permission to use data would be transacted and a pair of keys, guest and owner, would be generated. A hash pointer would allow the guest to access the specified data in exchange for a service or alternate form of value (Lazarovich, 2015). This model allows the user to control their data through the ability to consent and revoke use. The user is also able to benefit from the transparency inherent in blockchains to track how their data is being used. Research has shown that when users are in control of their data, they become more willing to share it (Arun and Carmichael, 2017). This model allows for user control while retaining the benefits that corporations and governments derive from the use of metadata.

Identity sovereignty

Similar to metadata sharing, the blockchain could potentially return sovereignty of digital identities back to individuals. Where data may or may not be linked to a specific user, digital identities exist purely to identify users in the digital space. Current uses of identities are used to authenticate users by means of passwords, biometric scans, etc. and to authorize users or control their interaction capability (Baars, 2016). Identity management can be built into a blockchain network, in much the same manner as data, through transacting permissions. In the case of identity management the data store could contain, rather than raw data, certified documents that verify identity and personal attributes. The obvious uses here would be documents such as passports and driver's licenses but the opportunities are much greater than basic authentication documents. Tax and financial records, student report history, medical records, and more could be placed in a user-controlled identity store following their certification by trusted third parties (Mainelli, 2017). Once the identity store is in place, the transaction ledger operates in much the same fashion as a raw data ledger. The user transacts permission to access the data store to authenticate and authorize interactions with the institution. Once an organization has authenticated an individual and determined their authorization level, an identity token can be generated. The identity token, based on PKI cryptography, would be used to control all interactions with the institution eliminating the need for a corporation to store private records linked to individuals (Pratini, 2017). This model would allow users to share only the personal information required while retaining the ability to revoke access to records. For example, a user could share healthcare records with their care provider and completely revoke access when they change doctors. This model of identity management is more centralized and reduces the fragmented redundancy of the current system while maintaining access rights in a distributed manner.

Policy Discussion

This paper has argued thus far that the blockchain is one viable solution to the unequal power relationships that have emerged from the deployment of digital technologies. The authors would be remiss if they did not mitigate the exuberance that blockchain technology offers with a larger policy discussion that clarifies some of the major issues moving forward.

Regulation vs. governance

The preceding examples show that, even in the conceptual stages, the blockchain holds the potential to fundamentally disrupt the trend of power consolidation among technology corporations. It also introduces transparency and control into the currently opaque and vulnerable regimes of data sharing and identity management. Excitement in the business community has reached a fever pitch spawning 130 start-up blockchain companies and over $1.5 billion in investment (Michalik, 2017). Public response to the technology has generally been characterized in two fashions. Markets such as the European Union and the US have adopted a wait and see approach (Parker, 2017). This approach is far from passive but also not explicit, characterized instead by regulators gathering knowledge to avoid premature legislation. On the other end of the spectrum, smaller countries such as the UAE, Ukraine, Estonia, and Sweden are moving substantial government records and processes onto the blockchain in a bid to embrace and sponsor the fledgling technology (Finck, 2018). This approach does much to bolster blockchain companies but may be difficult to scale and potentially premature to deploy in large-scale jurisdictions. Estonia's E-Residency program currently serves approximately 30,000 citizens (Prisco, 2015), a far cry from the hundreds of millions in the US. Recognizing these limitations in the current regulatory approaches, a middle ground appears increasingly necessary to address large markets.

A strong case can be made that public interest in regard to blockchain deployment would be better served with a shift in focus from regulation towards governance. Regulation, a means of top-down control, carries with it the pitfalls of stifling innovation, introducing costs associated with compliance and enforcement, and being too slow to evolve with the technology and market (Byrne, 2016). Governance, however, is a means for stakeholders to collaboratively guide the development of a technology (Tapscott and Tapscott, 2016). Governance consortiums are being convened, but participation is invariably industry-dominated and the focus is on the development of common protocols rather than the fostering of public good (Gabinson, 2016). There is an opportunity here for governments and public interest groups to sponsor consortiums that place issues, such as financial inclusion, data protection, and digital identity, front and center. Conversations between public and private sectors early could have the added benefit of forestalling the need for regulations enacted when the technology matures. Blockchains can be built around agreed governance and regulation, essentially self-automating compliance through code. Regulations rely on extrinsic compliance through consequences, but when rules are breached on a blockchain, the transaction simply fails to process (Yeoh, 2017). For these reasons, it is imperative that the public sector join the governance conversation early to foster the public good they are charged to protect.

Next steps

Blockchain holds promise to fundamentally alter societies' interactions with the digital space. The technology, however, is far from mature. User interfaces must be developed, Interledger protocols are in their infancy, and privacy remains a concern. The most glaring issue is bringing the technology to scale. Every day new users and devices come online with associated data flows and digital identities. Blockchains, if deployed as a central platform of markets, data sharing, and single sign-on identity management must process massive quantities of transactions, safely and securely. To date, publicly funded research has not been concordant with need. Most public funding is awarded to private industry as innovation grants in specific use cases (Cheng et al, 2017; Higgins, 2017). This stands in stark contrast to the development of the TCP/IP framework for the Internet. In that case, much of the funding and innovation into the plumbing of the Internet was conducted through general research government funding. The technology was then deployed and the private sector innovated atop the infrastructure. An increase of general research funding may bring about innovations in blockchain foundations to address problems facing scalability (Walport, 2016). This may even lead to a widespread public blockchain upon which private entities could establish community protocols associated with specific use cases.

The second hurdle to widespread adoption will inevitably be the considerable cost of implementing blockchain technology. The financial services industry has conducted exhaustive cost-benefit analyses concerning blockchain deployment and consensus has generally returned that the cost savings are considerable (Cocco, Pinna, and Marchesi, 2017). The spheres of identity management, data sharing, and market inclusion stand to benefit from similar analyses. To be sure, costs do arise when technology companies dominate their markets, or when large-scale personal information breaches occur. These costs must be quantitatively weighed against the cost of replacing legacy IT systems to provide a clearer picture to decision-makers. Both the public and private sectors would benefit from increased clarity on the costs and benefits of widespread deployment in the aforementioned spheres.

Conclusion

Inequality in both power dynamics and economic opportunity are growing in the digital space. The most formidable technology companies are consolidating their power and controlling more and more of fundamental societal interactions. The consolidation trend has contributed to a disparate ability among the powerful to realize profit from products and services, an opaque data economy, and a fractured and increasingly vulnerable identity management system. The problems are interrelated and growing in scale. The trend cannot continue with significant harm done; societies must enact measures that foster market inclusion and restore sovereignty of data and identities back to Internet users. Thus far, regulation and market course corrections have proven unmatched to the task, but a new technology, the blockchain, shows promise. The transparency and immutability offered by the blockchain can mitigate current trends, reduce consumer risk, and restore a measure of public trust. Although blockchain technology is in its infancy, the public sector must be involved in governance discussions early. Government-sponsored consortiums can focus the conversation on public good, and blockchains can be developed with consumer protections built into the code. To spur widespread adoption of the technology, governments should divert funding from specific use case innovation grants to general research into the foundations of the technology itself, focusing on scalability and security. Cost-benefit analysis research must also branch from solely financial services to the realms of data sharing, market inclusion, and identity management. Such research would provide policy makers with vital information to determine where and when deployment is appropriate. Public sector research and funding can bring about a future of growth and security so desperately needed to restore public trust and return equity to the Internet.

References

Allen, JP (2017) *Technology and inequality: concentrated wealth in a digital world*. Cham: Palgrave Macmillan.

Allen, K (2015) 'Big tech's big problem – its role in rising inequality', *The Guardian*. August 2. https://www.theguardian.com/business/economics-blog/2015/aug/02/big-techs-big-problem-rising-inequality, accessed September 1, 2019.

Andriole, S (2017) 'There will be 30 technology companies in 2030, 10 in 2050, and then there will be none', *Forbes*. May 25. https://www.forbes.com/sites/steveandriole/2017/05/25/there-will-be-20-technology-companies-in-2030-10-in-2050-and-then-there-will-be-none/#4af13f77132b, accessed September 1, 2019.

Arun, J and Carmichael, A (2017) 'Trust me: digital identity on blockchain', *IBM Institute for Business Value*. January. https://www-01.ibm.com/common/ssi/cgi-bin/ssialias?htmlfid=GBE03823USEN&, accessed September 1, 2019.

Baars, DS (2016) *Towards self-sovereign identity using blockchain technology*. Master's thesis, University of Twente.

Byrne, M (2016) 'Blockchain: from "what" and "why" to regulating "how"', *Lawyer (Online Edition)*. May 20, p. 5. https://www.thelawyer.com/issues/online-may-2016/blockchain-from-what-and-why-to-regulating-how/, accessed September 1, 2019.

Catalini, C and Ganz, J (2016) *Some simple economics of the blockchain*. NBER Working Paper No. 22952. The National Bureau of Economic Research. December. doi:10.3386/w22952

Cheng, S, Daub, M, Domeyer, A and Lundqvist, M (2017) 'Using blockchain to improve data management in the public sector', *McKinsey Digital*. February. https://www.mckinsey.com/business-functions/digital-mckinsey/our-insights/using-blockchain-to-improve-data-management-in-the-public-sector, accessed September 1, 2019.

Chester, J (2017) 'How the blockchain will secure your online identity', *Forbes*. May 3. https://www.forbes.com/sites/jonathanchester/2017/03/03/how-the-blockchain-will-secure-your-online-identity/#53b11ca65523, accessed September 1, 2019.

Cocco, L, Pinna, A and Marchesi, M (2017) 'Banking on blockchain: costs savings thanks to the blockchain technology', *Future Internet*, 9(25), 1–20. doi:10.3390/fi9030025

Conte, J (2017) *How artists can (finally) get paid in the digital age*. August 23. https://www.ted.com/talks/jack_conte_how:artists_can_finally_get_paid_in_the_digital_age, accessed September 1, 2019.

Cuomo, J, Nash, R, Pureswaran, V, Thurlow, A and Zaharchuck, D (2017) *Building trust in government*. IBM. March 17. https://www.ibm.com/downloads/cas/WJNPLNGZ, accessed September 1, 2019.

Esteve, F (2016) 'The concentration of wealth in the digital economy', *Technology and Inequality*. May 3. http://lab.cccb.org/en/technology-and-inequality-the-concentration-of-wealth-in-the-digital-economy/, accessed September 1, 2019.

Finck, M (2018) 'Blockchains: regulating the unknown', *German Law Journal*, 19(4), pp. 665–692.

Foer, F (2017) 'How silicon valley is erasing your individuality', *Washington Post*. September 8. https://www.washingtonpost.com/outlook/how-silicon-valley-is-erasing-your-individuality/2017/09/08/a100010a-937c-11e7-aace-04b862b2b3f3_story.html?utm_term=.aa99307cff12, accessed September 1, 2019.

Fresh Air (1993) 'Science fiction writer WILLIAM GIBSON', *Terry Gross National Public Radio*. August 31.

Gabinson, G (2016) 'Policy considerations for the blockchain public and private applications', 19 SMU. *Science Technology & Law Review*, 327.

Goldfarb, A and Tucker, CE (2010) *Privacy regulation and online advertising*. http://dx.doi.org/10.2139/ssrn.1600259, accessed September 1, 2019.

Han, D (2017) 'The market value of who we are: the flow of personal data and its regulation in China', *Media and Communication*, 5(2), pp. 21–30.

Higgins, S (2017) 'US government awards $2.25 million to blockchain research projects', *Coindesk*. May 12. https://www.coindesk.com/us-government-awards-2-25-million-blockchain-research-projects/, accessed September 1, 2019.

Hoser, T (2016) 'Blockchain basics, commercial impacts and governance challenges', *Governance Directions*, 68(10), 608–612.

Kosinski, M, Sitwell, D and Graepei, T (2012) 'Private traits and attributes are predictable from digital records of human behavior', *Proceedings of the National Academy of Sciences of the United States of America*, 110(15), pp. 5802–5805.

Lazarovich, A (2015) *Invisible Ink: blockchain for data privacy*. Massachusetts Institute of Technology. June. https://dspace.mit.edu/bitstream/handle/1721.1/98626/920475053-MIT.pdf?sequence=1, accessed September 1, 2019.

Mainelli, M (2017) 'Blockchain will help us prove our identities in a digital world', *Harvard Business Review*. March 16. https://hbr.org/2017/03/blockchain-will-help-us-prove-our-identities-in-a-digital-world, accessed September 1, 2019.

Manjoo, F (2017) 'Tech's frightful five: they've got us', *The New York Times*. May 10, p. B1.

Martin, K (2015) 'Ethical issues in the big data industry', *MIS Quarterly Executive*, 14(2), pp. 67–85. http://www.misqe.org/ojs2/index.php/misqe/article/viewFile/588/394, accessed September 1, 2019.

McConahey, T and Holtzman, D (2015) 'Towards an ownership layer for the internet', *ascribe GmbH*. June 24. https://assets.ctfassets.net/sdlntm3tthp6/resource-asset-r391/d110e1250fe31959150659144c424feb/5d5f7fde-646f-4b1c-8fe8-e939080348a0.pdf, accessed September 1, 2019.

Michalik, V (2017) *Frost & Sullivan and outlier ventures identify the 2017 global blockchain startup map*. March 27. https://ww2.frost.com/news/press-releases/frost-sullivan-identifies-2017-global-blockchain-startup-map/, accessed September 1, 2019.

Naylor, B (2016) *One year after OPM data breach, what has the government learned?* National Public Radio, Inc. June 6. http://www.npr.org/sections/alltechconsidered/2016/06/06/480968999/one-year-after-opm-data-breach-what-has-the-government-learned, accessed September 1, 2019.

Parker, L (2017) 'European commission "actively monitoring" blockchain developments', *Brave New Coin*. February 17. https://bravenewcoin.com/insights/european-commission-actively-monitoring-blockchain-developments, accessed September 1, 2019.

Pratini, N (2017) 'Identity, privacy, and the blockchain: what do identity and privacy mean in the digital world, and how might blockchain technology play a role?', *Insights on the Future of Finance from Plaid*. May 4. https://fin.plaid.com/articles/identity-privacy-and-the-blockchain, accessed September 1, 2019.

Prisco, G (2015) 'Estonian government partners with bitnation to offer blockchain notarization services to e-residents', *Bitcoin Magazine*. November 30. https://bitcoinmagazine.com/articles/estonian-government-partners-with-bitnation-to-offer-blockchain-notarization-services-to-e-residents-1448915243/, accessed September 1, 2019.

Rainie, L and Duggan, M (2016) Privacy and information sharing. Pew Research Center. January 14. http://www.pewinternet.org/2016/01/14/privacy-and-information-sharing/, accessed September 1, 2019.

Reinsel, D, Gantz, J and Rydning, J (2018) 'The digitization of the world: from edge to core', *IDC White Paper – #US44413318*. https://www.seagate.com/www-content/our-story/trends/files/idc-seagate-dataage-whitepaper.pdf, accessed September 1, 2019.

Rosenzweig, P (2017) *The equifax hack--bad for them, worse for us*. Scientific American. September 12. https://blogs.scientificamerican.com/observations/the-equifax-hack-bad-for-them-worse-for-us/, accessed September 1, 2019.

Schudy, S and Utikal, V (2015) '"You must not know about me" – on the willingness to share personal data', *Journal of Economic Behavior & Organization*, 141(1), 1–11.

Snow, P, Deery, B, Lu, J, Johnston, D, Kirby, P, Sprague, AY and Byington, D (2014) 'Business processes secured by immutable audit trails on the blockchain', *Brave New Coin*. November 16. https://bravenewcoin.com/insights/business-processes-secured-by-immutable-audit-trails-on-the-blockchain, accessed September 1, 2019.

Tang, G (2018) *Peer-to-peer decentralized marketplace based on blockchain technology. Version 5.0*. https://lookrev.com/doc/lookrev-whitepaper.pdf, accessed September 1, 2019.

Taplin, J (2017) 'Is it time to break up google?', *The New York Times*. April 22. https://www.nytimes.com/2017/04/22/opinion/sunday/is-it-time-to-break-up-google.html, accessed September 1, 2019.

Tapscott, D and Tapscott, A (2016) *Blockchain revolution: how the technology behind bitcoin is changing money, business, and the world*. New York: Penguin.

Underwood, S (2016) 'Blockchain beyond bitcoin', *Communications of the ACM*, 59(11), pp. 15–17. doi:10.1145/2994581

Walport, M (2016) *Distributed ledger technology: beyond block chain*. London: Government Office for Science.

World Bank (2016) *World development report 2016: digital dividends*. Washington, DC. doi:10.1596/978-1-4648-0671-1

World Economic Forum (2011) *Personal data: the emergence of a new asset class*. January. http://www3.weforum.org/docs/WEF_ITTC_PersonalDataNewAsset_Report_2011.pdf?utm_source=datafloq&utm_medium=ref&utm_campaign=datafloq , accessed September 1, 2019.

Yeoh, P (2017) 'Regulatory issues in blockchain technology', *Journal of Financial Regulation and Compliance*, 25(2), pp. 196–208. doi:10.1108/JFRC-08-2016-0068

Zyskind, G, Nathan, O and Pentland, A (2015) *Enigma: decentralized computation platform with guaranteed privacy*. https://www.enigma.co/enigma_full.pdf, accessed September 1, 2019.

8 The evolution of the Russian way of informatsionnaya voyna

Sean Ainsworth

Introduction

In recent years, numerous policy makers, military officials, and scholars have warned of the threat posed by a "new Russian way of war" (Thomas, 2016; Seely, 2017) which blends a range of means, including both kinetic and non-kinetic, military and non-military, as well as overt and covert. This "new way of war" has been employed in efforts to assert regional hegemony and dominance over what Russia regards as the "near abroad" of former Soviet republics and to project Russian power to defend Russian interests further afield, such as in Syria.

One of the core components of the "new way of war" is Russia's exploitation of emerging technologies, employing information-technology systems and networks for coercive purposes. In recent years, Russia has grown increasingly bold, engaging in cyber and information warfare operations, targeting the democratic processes and systems of several Western states, and bringing the specter of Russian cyber and information warfare to the fore of Western strategic thought (Greenberg, 2017; Office of the Director of National Intelligence, 2017; Saeed, 2017).

However, this is not an entirely new phenomenon or way of war. Russia's cyber and information warfare operations have been informed and shaped by its experiences in post–Soviet conflicts in addition to Moscow's strategic understanding of the West's actions, motivations, and Russia's newfound place in the post–Cold War international system. These experiences have driven Russia to adapt Soviet-era military doctrine and geopolitical strategies to take advantage of the virtual strategic environment created by the information revolution and emerging technologies.

This chapter first explores the historical foundations and precedent of Russia's strategic thought and doctrine concerning information warfare. It then examines the factors that have motivated Russia's adaptation of these earlier theories to the new fifth domain of war – cyberspace. This includes a consideration of Russia's experiences during the wars in Chechnya and concern over Western influence and interference within the near abroad of former Soviet satellites. The chapter analyzes how these adaptations have been actively employed by Russia in the pursuit of its interests in the near abroad, including Estonia, Lithuania, Georgia, and

Ukraine. Finally, the chapter discusses the potential future capabilities of artificial intelligence (AI)-enabled information warfare.

Russian doctrine and strategic thought

The development of military doctrine and strategy is shaped by a state's unique historical, cultural, and political background. As a result, there are substantial differences between Western military conceptualizations of the role of emerging technologies and means or domains of conflict, such as cyberspace, and how they are understood within Russian military doctrine and strategic thought.

Russia's strategic military theorists, for example, tend to use terms such as "cyberwarfare" only when translating and discussing Western strategic thought and doctrine (Giles and Hagestad, 2013). Similarly, Russian strategic thought refers to the "information space" (*informatsionnye prostranstvo*) rather than "cyberspace", the term commonly used in the West. While seemingly similar, these concepts are, in fact, substantially different from one another with implications for the ability of states to understand and predict Russian strategy and military actions. Western definitions of cyberspace tend to focus on the hardware and infrastructure; both the US military and North Atlantic Treaty Organization (NATO) define cyberspace as a global domain of interconnected technology and communications infrastructure, including telecommunications networks and computer systems (US Joint Chiefs of Staff, 2018, p. GL-4; NATO Standardization Office, 2019). The focus on the infrastructural hardware establishes the boundaries of Western approaches to cyberwarfare as offensive cyber operations designed to deny, degrade, disrupt, or destroy an adversary's cyberspace assets and capabilities. The Russian information space concept, by contrast, extends this definition to also include human cognitive domains and social consciousness elements (Ministry of Defence of the Russian Federation, 2011, p. 5). This is a significant definitional difference, affecting Russian and Western strategic thought and understanding concerning the role of offensive cyber operations. For example, Western militaries tend to delineate between cyberwarfare, information operations, and psychological warfare as separate, though closely related, and often interdependent tools in the toolkit. Russia instead views all three as falling within a broad overall concept of "information war" (*informatsionnaya voyna*), which is a continuous ongoing confrontation not necessarily limited to wartime (Giles, 2016, p. 4). Within this concept, Russian military theorists do recognize a division between "information-technical" and "information-psychological" means, which roughly align with Western conceptualizations of "cyberwarfare" and "information warfare" respectively (ibid, p. 9).

Russia's more comprehensive information warfare concept is arguably a continuation of long-running Russian strategic thought and historical military doctrines that have been adapted to the cyber domain. These adaptations have been driven by both internal and external factors and strategic threats or possibilities that have emerged alongside the technologies of the information revolution. In the mid-1990s, as Russia was adjusting to its substantially weakened military position following the dissolution of the Soviet Union, Russian military theorists began to emphasize the growing role of non-kinetic means of projecting power, most importantly those relating to the ongoing information revolution (Blank, 2013, p. 34; Gerasimov, 2016). The value of these methods was that they depended on means other than the use of raw military power or kinetic force during a period when the Russian military was declining, and it was preparing for, and undertaking, substantial modernization and reform. Indeed, the Russian military has long emphasized the importance of the informational domain of conflict in an effort to overcome its own economic or military weaknesses. Russian information warfare stems from the adaptation and application of several historical Russian strategies to the cyber domain, most notably the Soviet-era military doctrine of *maskirovka* (typically translated as camouflage or deception). Emphasizing the strategic importance of operational concealment and deception, *maskirovka* comprised "a set of processes … designed to mislead, confuse, and interfere with anyone accurately assessing [the Soviet Union's] plans, objectives, strengths, and weaknesses" (Shea, 2002, p. 63).

Maskirovka is therefore closely aligned with Soviet intelligence services employment of "active measures" to coerce and subvert during the Cold War, in addition to the Soviet military strategy of "reflexive control". Reflexive control forms a critical component of Russian information warfare. The author of the theory and a key figure in its development under the Soviet military, Vladimir Lefebvre, defines reflexive control as "conveying to a partner or an opponent specially prepared information to incline him to voluntarily make the predetermined decision" (cited in Chotikul, 1986, p. 5). In essence, reflexive control is the creation of a reality that leads the target to make a decision of their own "free will" that benefits the controller, akin to a chess player baiting their opponent into exposing their queen. Reflexive control formed "an integral, valuable, and potentially very lethal part of the Soviet decision making process" (Chotikul, 1986, p. 90), though some Western analysts expressed doubts regarding its efficacy due to the "impossibility of reducing thought processes and psychological functioning to quantitative, exact objects of control" (ibid, p. 96).

As this chapter demonstrates, the technological developments of the information revolution and the emergence of a "global village" connected through cyberspace has created opportunities for the identification and exploitation of thought processes and psychological functioning as specific "objects" of control. The tracking and data harvesting conducted by social media websites and other online services has enabled information warfare operations to target audiences with unprecedented specificity: Audiences are now able to be targeted with precision based on specific characteristics including political leanings, geographic location, age, gender, and occupation. Similarly, by exploiting the algorithms that select which content is recommended or displayed to a user, information warfare operations can attach themselves to, or blend into, popular topics and categories of content. User engagement with the inserted material leads the algorithm to recommend more material from the same source or of a similar nature. This is a strategy that has been employed by Russian state broadcaster Russia Today (RT) on

YouTube, achieving user engagement through the upload of non-political "click-bait" videos in the hope that the algorithm recommends additional content with a political focus (EU East Stratcom Task Force, 2017). As such, users' choices are structured for them without their knowledge.

Reflexive control in the informatsionnye prostranstvo

Despite the historical precedent and interest of the Russian state in the mid-1990s to innovate in the information domain owing to its declining material hard power, it was not until the early 2000s that these methods came to the fore of Russian strategic thought. The primary driving factors for this adjustment were the political and military failings Russia experienced during the First Chechen War, in addition to Russia's strategic understanding of the "Color Revolutions", which it saw as a series of connected pro-Western democratic protest movements throughout the near abroad (Cordesman, 2014).

The Chechen Wars

The First Chechen War (1994–1996) exposed substantial failings in Russia's military doctrine, organizational structure, and strategy. It also demonstrated the potential power of the emerging information domain because of the technological developments of the information revolution. In December 1994, tensions between Moscow and the breakaway republic of Chechnya escalated, leading to a brutal 20-month long conflict which resulted in a Russian defeat despite the Russian military's overwhelming advantage in the numbers of available manpower, armored vehicles, firepower, and air support. A substantial contributory factor in Russia's defeat was undoubtedly its reliance on Cold War–era military doctrine designed to fight a peer–competitor, with a substantial focus on mass troop and armored vehicle movements, coupled with the use of devastating supportive fire and aerial bombardment (Arquilla and Karasik, 1999). This approach was ill-suited to combatting the networked and decentralized Chechen separatists, who typically fought in highly mobile bands of 12–20 fighters capable of rapidly swarming and overwhelming Russian forces.

The exploitation of the information space in the conflict was, arguably, more consequential for a Chechen victory than kinetic military operations. Facing an overwhelmingly militarily superior opponent, the Chechen separatists, many of whom were former members of the Soviet military, recognized the military value of the informational-psychological domain of the conflict. The separatists concentrated their information warfare efforts along two lines: First, engendering sympathy and support from the international community, and second, fomenting disillusionment and harming the morale of Russian troops and the Russian public. The separatists created websites and online communities targeting the Chechen diaspora and international non-governmental organizations (NGOs). This provided a direct means of distributing the Chechen narrative, distributing images and video of civilian casualties caused by Russian bombardments as part of a wider effort to portray Chechens as helpless victims of Russian cruelty. These efforts rallied international support for the legitimacy of Chechen separatism and condemnation of Russia (Fayutkin, 2006). Whereas Chechen separatists welcomed journalists and cooperated with media filming and photography, the Russian military's lack of cooperation with, and hostility toward, the media prevented any potential counter-narrative from emerging.

The combined effects of a lack of effective military and political leadership, the brutality of the conflict, and Chechen information warfare proved devastating for the morale of Russian troops (Arquilla and Karasik, 1999, p. 221). Chechen radio broadcasts would address Russian officers by name, listing the names and location of their wives and children, and claiming that they were targets of Chechen "hit-squads" (Arquilla and Karasik, 1999, p. 217). The breakdown in the morale of Russian forces proved of substantial benefit during the Third Battle of Grozny, when Chechen separatists, who were outnumbered approximately eight to one, were able to retake the Chechen capital and force the surrender of approximately 7,000 Russian troops stationed in the city. Information warfare targeting the wider Russian public included "pirate" television and radio broadcasts, in addition to the use of radio-jamming equipment to prevent Russian broadcasts within Chechnya (Arquilla and Karasik, 1999). Taking advantage of public suspicion that the war was a political distraction by the beleaguered Yeltsin government, the separatists were able to substantially undermine the Russian public's support for the war effort. These efforts created the political impetus for a negotiated solution to the conflict, which was only hastened by the success of separatist forces during the Third Battle of Grozny.

Official Russian assessments of the conflict focused on the advantage the separatists gained through their employment of information warfare. Sergei Stepashin, then Director of the Federal Security Service until his resignation following a Chechen terrorist attack in June 1995, later noted that "the information war was lost" (Arquilla and Karasik, 1999, p. 217). Similarly, Dan Fayutkin, the head of the Israeli Defense Force's Doctrine Section, suggests that the conflict "teaches us the importance of information warfare in the realization of political and military goals" (Fayutkin, 2006, p. 55) and that "it was consistent, well-thought-out, and potent propaganda" (ibid) that ensured Chechen victory. Indeed, the war was regarded as a crisis point for the Russian military's public communications systems and state media organizations (Sieca-Kozlowski, 2009, p. 304). In this way, the First Chechen War not only provided the impetus for much-needed Russian military reform; it also served as an early lesson regarding the value and necessity of harnessing the new strategic informational environment created by cyberspace and the growing role that information dominance would play in future conflicts.

Russian military and security services, bruised by the failure to counter Chechen use of information warfare, adapted relatively quickly in the three years interim before the start of the Second Chechen War (1999–2009). Russia recognized the effectiveness of the separatists' exploitation of Russian mass media during the First Chechen War for information warfare purposes, including strategic messaging, damaging troop morale, and weakening public support. As a result, Russia established new censorship regimes and the Russian Information Center (*Rosinformtsentr*). Managed by the Russian Army, the Center was ostensibly intended to filter the flow of information from the conflict zone in Chechnya to prevent the spread of disinformation and Chechen propaganda. In practice, the Center controlled and restricted Russian and foreign media access to information regarding the conflict that could prove damaging to the Russian government. The Center also distributed foreign press articles and material that were supportive of the Russian government's narrative to the Russian press for domestic distribution (Sieca-Kozlowski, 2009, p. 305). Announcing the creation of the Center, Vladimir Putin, then-Prime Minister of Russia, remarked that Russia had "surrendered this terrain some time ago and now we are entering the game again" (Dixon, 1999). These reforms proved decisive in securing, and maintaining, public support for the war effort by restricting Chechen messaging efforts, limiting media reporting to the official government narrative, and insulating the Russian information space from the escalating levels of violence and rising casualties during the prolonged insurgency (Pain, 2000; Thomas, 2003).

Lessons learned from the Chechen wars were incorporated into Russia's 2000 *Information Security Doctrine*. The doctrine outlines a range of threats to Russia's information space and necessary future policy approaches. One of the major internal threats identified by the doctrine was the insufficient levels of coordination between various governmental bodies "in shaping and carrying out a unified state policy in the realm of national information security" (Russian Federation, 2000, p. 7). Other internal threats included insufficient levels of control over the Russian information space coupled with insufficiencies in government messaging efforts (Russian Federation, 2000, p. 8). In line with these identified threats and the 2000 *Russian Military Doctrine*, the *Information Security Doctrine* outlined several urgent measures, including "protecting society against distorted and untrustworthy information" (Russian Federation, 2000, p. 27) and "counteracting information war threats in a comprehensive way" (ibid, p. 29).

The Color Revolutions

These identified threats and countermeasures rapidly grew in importance following the Color Revolutions of the early 2000s, which arguably proved just as significant for the development of Russian strategic thought as the conflict in Chechnya. Presaged by the 2000 "Bulldozer Revolution" and overthrow of Slobodan Milošević in Serbia, the Color Revolutions were a series of pro-democratic youth and civic protests movements that took place throughout the near abroad of former Soviet satellites. The revolutions were able to achieve relatively rapid pro-Western regime change, beginning with the 2003 Rose Revolution in Georgia and extending to include the 2004 Orange Revolution in Ukraine and Kyrgyzstan's 2005 Tulip Revolution. Youth activists critical of the current regimes in these countries took advantage of the spread of new technologies, including internet access and text messaging, to communicate with one another and to circumvent government surveillance and censorship efforts. Internet access and text messaging proved vital for communicating with other activist movements in neighboring countries in addition to international democracy promotion and human rights NGOs (Stent, 2014, pp. 100–102). Using the newly accessible technologies of the information revolution, activists were able to bypass the traditional barriers and restrictions of the existing information space and media environment to provide an alternative critical viewpoint and to attract the attention and support of foreign NGOs and governments (Stent, 2014).

In line with the 2000 *Information Security* and *Russian Military Doctrines*, a consensus view was reached within Russia that these revolutions formed a component of Western information warfare intended to undermine Russia's strategic position within the near abroad (Giles, 2019). Russia's foremost military theorist General Makhmut Gareev, for example, extended Russian suspicion of Western interference and malign machinations to include the collapse of the Soviet Union and Yugoslavia. Gareev described such threats as "assuming not so much military forms as direct or indirect forms of political, diplomatic, economic, and informational pressure, subversive activities, and interference in internal affairs" (Jonsson and Seely, 2015, p. 8). Similarly, in May 2005, Sergei Markov, a Russian political scientist and politician, accused Ukrainian protestors of having been paid $10 a day to protest by the US Central Intelligence Agency (Stent, 2014, p. 115).

This siege mentality and view of the Color Revolutions as a strategic threat to Russian national interests was wholly consistent with the 2000 *Information Security* and *Russian Military Doctrines*. It also aligns with the prevailing view concerning the collapse of the Soviet Union amongst the *siloviki* (former members of Soviet and Russian security services) members of the Russian leadership. Foremost among these *siloviki*, President Vladimir Putin in 2005 lamented the collapse of the Soviet Union as the 20th century's "greatest geopolitical catastrophe" (BBC News, 2005). In the Russian view, the regime changes achieved by the Color Revolutions were heightened by the strategic threat posed by the 1999 and 2004 rounds of NATO enlargement, as the new leadership in Georgia and Ukraine favored pursuing NATO membership (Oliker et al, 2015).

In the Russian leadership's view, the Color Revolutions were a component of Western information warfare intended to isolate and undermine Russia's strategic interests within the region. The Russian leadership therefore adopted a genuine fear that the West would soon extend this information warfare to Russia itself (Duncan, 2013). This siege mentality and growing suspicion is evident in Russia's 2014 *Military Doctrine*, which highlights the enlargement of NATO and stationing of military assets within NATO member-states that border Russia as the principal security threat facing Russia. Other identified threats include the use of information warfare to subvert the sovereignty, political independence, and territorial integrity of states to destabilize individual states and regions, in addition to the use of regime change to establish regimes in contiguous neighboring states with policies that threaten Russian interests (Russian Federation, 2014). Indeed, the oft-cited "Gerasimov Doctrine", stemming from a speech given by the Russian Army's Chief of the General Staff, Valery Gerasimov, was intended as a description not of a new Russian way of war but the supposed information warfare targeting Russia carried out by the US and other Western states (Bartles, 2016; Galeotti, 2018).

The exploitation of the information revolution and new technologies

It was therefore defensive concerns that initially motivated Russia's adoption of an aggressive offensive information warfare strategy. Defeat in the First Chechen War, and ongoing instability in the Caucasus, posed a threat to Russia's territorial integrity, while Western influence and potential machinations targeting the near abroad threatened Russia's dominion over its historical sphere of influence. Russia's declining military hard power and the asymmetry of the post–Cold War balance of power heightened Moscow's perception of its vulnerability to that of an existential threat (Giles, 2016, pp. 36–41).

Viewing wholly defensive measures as too risky in the face of these combined threats, Russia adopted a strategy entailing the opportunistic employment of offensive information warfare. Indeed, in 2007 then-Defense Minister Sergei Ivanov remarked that "information itself [has turned] into a certain kind of weapon… that allows us to carry out would-be military actions in practically [a] theater of war and most importantly, without using military power" (Blank, 2013, p. 34). The foundational element of Russia's information warfare strategy is the employment of disinformation and misinformation-based propaganda.

Traditional propaganda methods, such as those pursued by both superpowers during the Cold War, are typically intended to persuade the reader or listener that the propagandists' objectives, system of government, or ideology are superior and more worthy of support. As such, traditional propaganda tends to emphasize factors such as "trust, credibility, actions, legitimacy, and reputations [which] are critical to success" (Defense Science Board, 2008, p. 39). By contrast, Russian information warfare strategies rely on what a 2016 RAND report described as the "firehose of falsehood" model (Paul and Matthews, 2016). Unlike traditional propaganda methods, the firehose of falsehood eschews any commitment to consistency or even an objective reality. Counterintuitively, the strategy actively incorporates these inconsistent and incredulous aspects into its strategic framework, exploiting inherent cognitive biases through the dissemination of rapid, continuous, and repetitive messaging across multiple channels of communication.

Contemporary Russian information warfare is not intended to portray Russia as superior to the West, but instead to "confuse, befuddle, and distract" (Lucas and Nimmo, 2015) while exploiting divisive social, political, and cultural issues to foster political polarization and division within targeted states. In recent years, many of the targeted states have included Western democracies (US Department of Homeland Security, 2016; Greenberg, 2017; Office of the Director of National Intelligence, 2017). However, of far more strategic importance for Russia are its information warfare campaigns targeting states within the near abroad.

Russian information warfare campaigns targeting the near abroad take the form of opportunistic reflexive control, with a specific focus on the fomenting of political and ethnic tensions. The legacy of Soviet-era "russification" policies has resulted in substantial ethnic Russian minority populations throughout the near abroad. Official Russian policy refers to these diasporas as "compatriots", and a substantial number of Russian-language state media outlets are marketed toward the diasporic populations throughout the near abroad. As discussed above, rather than attempting to engineer political outcomes wholesale, Russia employs these media outlets to instigate and exacerbate divisions between ethnic Russian diasporas and the majority populations of targeted states through the "firehose of falsehoods" propaganda model. Russia's information warfare therefore follows an opportunistic strategy, using political and ethnic divisions to create seemingly "organic" political crises that can then be quickly escalated or subdued in accordance with Russia's interests at the time.

Russia's information warfare in Estonia, Lithuania, Georgia, and Ukraine

In 2007 a political dispute ensued between Russia and Estonia in response to the relocation of a Soviet-era war memorial. By leveraging its information warfare capabilities Russia was able to escalate the dispute into the worst civil unrest Estonia had experienced since the Soviet occupation. Estonian security services suspected the resultant riots were actively orchestrated by Russian intelligence and Special Forces (Cavegn, 2017). Russian state media outlets targeting the ethnic Russian diaspora in Estonia were broadcasting hyper-emotional coverage of the relocation, including false reports of police brutality and portrayals of ethnic Russians as facing a threat from Estonian fascists, to foster outrage amongst the ethnic Russian diaspora. Estonia also found itself the target of a substantial cyber-attack described by then-Estonian Defense Minister Jaak Aavisoo as affecting "the majority of the Estonian population", with "all major commercial banks, telcos, media outlets, and name servers… [feeling] the impact" (Davis, 2007). The attacks, in effect, placed Estonia under a "cyber siege" which lasted three weeks and was only halted by the Estonian government blocking all foreign internet traffic. The technological advances that had until that point been regarded as Estonia's comparative advantage were now a potential source of vulnerability. Several organizations and individuals have since claimed credit for the cyber-attacks, most notably *Nashi*, a Kremlin-linked youth movement. *Nashi* have previously been accused of and implicated in similar activities within Russia against adversaries of the Kremlin, including cyber-attacks targeting media organizations critical of Putin and the Kremlin. While it is possible that the cyber-attacks were carried out by patriotic hacker militias or criminal organizations, the suggestion that the attacks were not coordinated or at least tacitly sanctioned by the Kremlin stretches credulity. Even without any coordination, Russian state media coverage's purposefully emotional misinformation would have most likely served as the catalyst for any decision to attack Estonia by patriotic hacker militias.

By contrast, a similar legislative initiative by Lithuania in 2008 banned the public display of both Nazi and Soviet symbols, including images of Nazi and Soviet leaders, flags, and insignia. This outraged many ethnic Russians, who viewed the law as equating the Soviet Union with Nazi Germany. In response, hundreds of Lithuanian websites were defaced by outraged "patriotic hacker" groups that called for an organized hacking campaign similar to the one that targeted Estonia, with the intention of including other Baltic states as well as Ukraine as targets (Danchev, 2008). However, in this instance, the patriotic hackers were either unwilling or incapable of escalating their attacks, with no organized campaign materializing. The inability to escalate the attacks from low-level "cyber-vandalism", despite enthusiasm from at least some patriotic hacker groups, would suggest that there was little widespread support to do so despite the similarities with Estonia the previous year. Alternatively, an escalation of the dispute may have been viewed as undesirable for Russia's interests. Indeed, there was little to no focus on the legislation by Russia's state media, much less the overly emotional disinformation that occurred in Estonia the previous year. One potential explanation for this lack of media coverage is the disparity in potential audience size: Ethnic Russians comprise 24% of Estonia's population, but total only approximately five percent of Lithuania's population, thereby limiting any opportunity to sow ethnic and political divisions.

Similar methods can be seen in the 2008 August War between Georgia and Russia, provoked by Georgia's aspirations for NATO membership (Kishkovsky, 2008). Russian information warfare portrayed Georgian responses to escalating attacks by South Ossetian separatists as acts of Georgian aggression against Russian compatriots. Russian state media organizations stationed journalists in South Ossetia days in advance of the conflict. When the war began, Russia's state news channels immediately displayed detailed graphics of the ongoing military operations as well as coordinated talking points accusing Georgia of genocide and ethnic cleansing (Whitmore, 2008).

The Russian military's counter-offensive became the first instance of the use of cyber operations in a combined operation with conventional military forces (Hollis, 2011). Georgian government networks were targeted by both cyber and kinetic attacks to disrupt government and military communications capabilities. Civilian communication networks near military areas of operation were also targeted as part of the cyber operation in order to foster panic and confusion amongst the civilian population (Haddick, 2011). These attacks disrupted the Georgian government's capability to communicate with the international community, reducing its ability to counter Russian narratives of the conflict. In an example of reflexive control through information warfare, the numerous conflicting narratives regarding the onset and conduct of the conflict led to widespread confusion within the international community and media (Fawn and Nalbandov, 2012). As a result, Russian information warfare was able to turn a traditional strength of NATO, its large membership, into a weakness. With so many members, the conflicting narratives generated by Russian information warfare stymied any potential for a cohesive international response. This left Western leaders reliant on urging restraint and attempting to negotiate ceasefires, thereby "ultimately tolerating the Russian fait accompli" (Socor, 2008).

Russia refined this synergistic combination of reflexive control and information warfare further still by the 2014 Euromaidan Revolution (also known as the Ukrainian Revolution). As in Georgia, Russia's information was founded in the prevailing Russian view of the Euromaidan protests as a continuation of Western-orchestrated Color Revolutions and a fear of Ukraine's aspirations toward NATO and EU membership. Russian state media presented the Euromaidan Revolution as a fascist coup orchestrated by the CIA, or Ukrainian "Nazis", that threatened the safety of Ukraine's ethnic Russian population and the Black Sea Fleet (Chalupa, 2014). Similar narratives were employed by Russian-financed "web brigades" on social media websites (Sindelar, 2014). These web brigade efforts included social media influence campaigns such as "Polite People", which "promoted the invasion of Crimea with pictures of Russian troops posing alongside girls, the elderly, and cats" (Seddon, 2014). At the same time, Russian government officials provided contradictory narratives and denials of objective fact (*maskirovka*). This brazen disregard for fact gave rise to the "little green men" phenomenon, as Russia continued to resolutely deny any military presence in Ukraine, even as it secured the annexation of the Crimean Peninsula. Western media organizations were unsure how to respond to Russia's strident denials in the face of objective reality, providing a window of opportunity for Russian state media organizations to push the Kremlin's narratives.

Russia attempted to buttress its information warfare portrayal of Euromaidan protesters as fascists by actively manipulating the published results of the 2014 Ukrainian presidential election. Malware was inserted by "CyberBerkut", a group the UK National Cyber Security Center asserts is a Russian military intelligence operation (GCHQ National Cyber Security Centre, 2018). CyberBerkut inserted malware onto the servers of Ukraine's electoral commission programmed to replace the official results page with an identical one that displayed the anti-Russian far-right candidate Dmytro Yarosh as the winner of the election. The malware was programmed to activate after the polls had closed but was discovered and removed by a Ukrainian cybersecurity company minutes beforehand. Despite this, Channel One Russia, a state media outlet, nonetheless reported Dmytro Yarosh as the winner, displaying a graphic of the false results page and citing the electoral commission's website, despite the manipulation and publication of false results having been prevented (Kramer and Higgins, 2017). Russia's annexation of Crimea involved simultaneous cyber-attacks targeting Ukraine's telecommunications infrastructure to disrupt the availability or flow of information to and from the peninsula. Russia's efforts to ensure regional information dominance and control, while confusing and delaying the international community's response, provides an excellent example of Russia's blending of reflexive control theory with contemporary information warfare. Because of these measures, Russia was able to secure the annexation of the peninsula through *fait accompli* in largely the same manner as it secured the de facto sovereignty of Georgia's breakaway regions, South Ossetia and Abkhazia, in the 2008 August War.

Future capabilities

The Kremlin views the development of AI technologies as a matter of strategic importance, encapsulated by President Vladimir Putin's claim that "whoever leads in AI will rule the world" (RT, 2017). As such, the Russian military has grown increasingly interested in the potential military application of AI technologies. This manifested in 2019 when the Kremlin outlined an ambitious national AI strategy primarily focused on supporting domestic research and development of AI and related technologies while preventing, or at least limiting, any dependence on foreign technologies (Office of the President of the Russian Federation, 2019). To date, most of the official discussions and demonstrations of military AI capabilities have focused on potential battlefield applications, such as AI-enabled combat and sensor systems. However, AI technologies may prove most effective, at least in the short-term future, in augmenting and enhancing Russia's information warfare strategies.

Russia's information warfare remains reliant on a significant amount of human labor, much of which is drawn from "troll farm" web-brigades-for-hire, where workers are expected to rapidly produce content, including managing multiple different social media accounts and making hundreds of posts or comments a day (Seddon, 2014). Some of these activities can be automated using simple social media bot software. However, such bot accounts are relatively simple for social media websites to detect and deactivate. Future bot accounts harnessing these AI technologies may not be so easy to counter, employing AI-generated synthesized portraits and photography to appear as a real person. The addition of text generation capable of mimicking human communication and behavioral patterns could theoretically enable such bot accounts to pose as real human beings to the extent that they are able to respond in real time and engage in conversations with other social media users.

Moreover, emerging AI technologies harnessing neural networks can generate increasingly sophisticated, difficult to detect, and hyper-realistic synthesized photo, video, and audio. The most notable example of this being "deepfake" artificial videos, wherein an individual may be depicted performing actions or taking part in an event that never took place. Similarly, ASI Data Science, an AI development company, developed an algorithm capable of producing a convincing recording of US President Donald Trump declaring a nuclear war against Russia using just two hours of source audio processed over five days (Chertoff and Rasmussen, 2019).

Conclusion

The development of Russia's cyber and information operations provides a startling example of the potential threats that can be created by new and emerging technologies. These threats can be particularly pernicious when emerging technologies are combined with unforeseen and innovative strategies designed to exploit fundamental weaknesses in their design and even human nature itself, as in Russia's "firehose of falsehood" model. Despite the unconventional approach adopted by Russian information warfare, these "new tools" are ultimately founded in a long tradition of Russian military strategic thought, including *maskirovka* and reflexive control theory. Their development has, however, been shaped by dominant Russian strategic thought and narratives regarding supposed Western machinations intended to isolate Russia and the opportunistic exploitation of vulnerabilities created by new technological developments.

Russia's successful use of an evolving information warfare strategy throughout the near abroad stands in stark contrast to the failings experienced during the First Chechen War and likely influenced Russia's decision to engage in information warfare campaigns targeting Western democracies themselves. Through the exploitation of state media organizations and social media, Russia's firehose of falsehood is able to flood the information space of the majority of "connected" states around the globe, fomenting political divisions and creating crises and other opportunities for Russia to exploit. Without the development of countermeasures, the West may soon find itself falling prey to Russia's information warfare-enabled reflexive control strategies. Countermeasures must be both information-technical, such as enhanced detection methods for social media bot accounts, and information-psychological, such as enhanced education through the promotion of critical thinking in addition to digital and media literacy programs.

References

Arquilla, J and Karasik, T (1999) 'Chechnya: a glimpse of future conflict?', *Studies in Conflict and Terrorism*, 22(3), pp. 207–229. doi:10.1080/105761099265739.

Bartles, CK (2016) 'Getting gerasimov right', *Military Review*. US Army CGSC, 96(1), pp. 30–38.

BBC News (2005) 'Putin deplores collapse of USSR', *BBC News*. http://news.bbc.co.uk/2/hi/4480745.stm, accessed May 6, 2020.

Blank, S (2013) 'Russian information warfare as domestic counterinsurgency', *American Foreign Policy Interests*. Routledge, 35(1), pp. 31–44. doi:10.1080/10803920.2013.757946.

Cavegn, D (2017) 'Ansip, laaneots: Russian agents present during Bronze night riots', *Eesti Rahvusringhääling*. https://news.err.ee/592127/ansip-laaneots-russian-agents-present-during-bronze-night-riots, accessed May 6, 2020.

Chalupa, A (2014) 'Putin's fabricated claim of a fascist threat in Ukraine', *Forbes*. https://www.forbes.com/sites/realspin/2014/04/04/putins-fabricated-claim-of-a-fascist-threat-in-ukraine, accessed May 6, 2020.

Chertoff, M and Rasmussen, AF (2019) 'The unhackable election: what it takes to defend democracy', *Foreign Affairs*, 98, pp. 156–164.

Chotikul, D (1986) *The Soviet theory of reflexive control*. Monterey, CA: Naval Postgraduate School.

Cordesman, AH (2014) *Russia and the "Color Revolution"*. Center for Strategic & International Studies. https://www.csis.org/analysis/russia-and-color-revolution, accessed May 6, 2020.

Danchev, D (2008) '300 Lithuanian sites hacked by Russian hackers', *ZDNet*. http://www.zdnet.com/article/300-lithuanian-sites-hacked-by-russian-hackers/, accessed May 6, 2020.

Davis, J (2007) 'Hackers take down the most wired country in Europe', *WIRED*. https://www.wired.com/2007/08/ff-estonia/, accessed May 6, 2020.

Defense Science Board (2008) *Report of the defense science board task force on strategic communication*. Washington, DC. https://apps.dtic.mil/docs/citations/ADA476331, accessed May 6, 2020.

Dixon, R (1999) 'Chechens use net in publicity war with Russia', *Los Angeles Times*. https://www.latimes.com/archives/la-xpm-1999-oct-08-mn-20034-story.html, accessed May 6, 2020.

Duncan, PJS (2013) 'Russia, the West and the 2007–2008 electoral cycle: did the Kremlin really fear a "Coloured Revolution"?', *Europe–Asia Studies*, 65(1), pp. 1–25. doi:10.1080/09668136.2012.698049.

EU East Stratcom Task Force (2017) 'RT goes undercover as in the now', *EUvsDisinfo*. https://euvsdisinfo.eu/rt-goes-undercover-as-in-the-now/, accessed May 6, 2020.

Fawn, R and Nalbandov, R (2012) 'The difficulties of knowing the start of war in the information age: Russia, Georgia and the War over South Ossetia', *European Security*, 21(1), pp. 57–89. doi: 10.1080/09662839.2012.656601

Fayutkin, D (2006) 'Russian-Chechen information warfare 1994–2006', *RUSI Journal*, 151(5), pp. 52–55. doi:10.1080/03071840608522874.

Galeotti, M (2018) 'The mythical "Gerasimov Doctrine" and the language of threat', *Critical Studies on Security*, 7(2), pp. 1–5. doi:10.1080/21624887.2018.1441623.

GCHQ National Cyber Security Centre (2018) *Reckless Campaign of cyber attacks by Russian military intelligence service exposed*. https://www.ncsc.gov.uk/news/reckless-campaign-cyber-attacks-russian-military-intelligence-service-exposed, accessed May 6, 2020.

Gerasimov, V (2016) 'The value of science is in the foresight: new challenges demand rethinking the forms and methods of carrying out combat operations', *U.S. Army Military Review*. Translated by R Coalson, 96(1), p. 23.

Giles, K (2016) *Handbook of Russian information warfare*. Rome: NATO Defense College Research Division. http://www.ndc.nato.int/news/news.php?icode=995, accessed May 6, 2020.

Giles, K (2019) *Moscow rules: what drives russia to confront the west*. Washington, D.C.: Brookings Institution Press, pp. 35–58.

Giles, K and Hagestad, W (2013) 'Divided by a common language: cyber definitions in Chinese, Russian and English', *2013 5th international conference on cyber conflict*, Tallinn.

Greenberg, A (2017) 'The NSA confirms it: Russia hacked French election "Infrastructure"', *WIRED*. https://www.wired.com/2017/05/nsa-director-confirms-russia-hacked-french-election-infrastructure/, accessed May 6, 2020.

Haddick, R (2011) 'This week at war: Lessons from Cyberwar I', *Foreign Policy*. https://foreignpolicy.com/2011/01/28/this-week-at-war-lessons-from-cyberwar-i/, accessed May 6, 2020.

Hollis, D (2011) 'Cyberwar case study: Georgia 2008', *Small Wars Journal*. https://smallwarsjournal.com/jrnl/art/cyberwar-case-study-georgia-2008, accessed May 6, 2020.

Jonsson, O and Seely, R (2015) 'Russian full-spectrum conflict: an appraisal after Ukraine', *Journal of Slavic Military Studies*, 28(1), pp. 1–22. doi:10.1080/13518046.2015.998118.

Kishkovsky, S (2008) 'Georgia is warned by Russia against plans to join NATO', *The New York Times*. https://www.nytimes.com/2008/06/07/world/europe/07russia.html, accessed May 6, 2020.

Kramer, AE and Higgins, A (2017) 'In Ukraine, a malware expert who could blow the whistle on Russian hacking', *The New York Times*. https://www.nytimes.com/2017/08/16/world/europe/russia-ukraine-malware-hacking-witness.html, accessed May 6, 2020.

Lucas, E and Nimmo, B (2015) *Information warfare: what is it and how to win it?* Washington, DC: CEPA. http://infowar.cepa.org/Reports/How-has-Russia-Weaponized-Information, accessed May 6, 2020.

Ministry of Defense of the Russian Federation (2011) *Conceptual views regarding the activities of the armed forces of the Russian federation in the information space*. Moscow: Ministry of Defense. https://ccdcoe.org/uploads/2018/10/Russian_Federation_unofficial_translation.pdf, accessed May 6, 2020.

NATO Standardization Office (2019) 'Record 39182: cyberspace', *NATOTerm*. https://nso.nato.int/natoterm/Term.mvc/Display?termGroupId=39186, accessed May 6, 2020.

Office of the Director of National Intelligence (2017) *Assessing Russian activities and intentions in recent US elections, intelligence community assessment*. Washington, DC: Office of the Director of National Intelligence. https://www.dni.gov/files/documents/ICA_2017_01.pdf, accessed May 6, 2020.

Office of the President of the Russian Federation (2019) *National strategy for the development of artificial intelligence over the period extending up to the year 2030*. Edited by Konaev, M., Vreeman, A., and Murphy, B. Translated by Etcetera Language Group. Washington, DC: Center for Security and Emerging Technology, p. 490. https://cset.georgetown.edu/wp-content/uploads/Decree-of-the-President-of-the-Russian-Federation-on-the-Development-of-Artificial-Intelligence-in-the-Russian-Federation-.pdf, accessed May 6, 2020.

Oliker, O et al. (2015) *Russian foreign policy in historical and current context: a reassessment*. Santa Monica, CA: RAND Corporation. https://www.rand.org/pubs/perspectives/PE144.html, accessed May 6, 2020.

Pain, E (2000) 'The second Chechen War: the information component', *Military Review*. Fort Leavenworth: Department of the Army Headquarters, 80(4), pp. 59–69.

Paul, C and Matthews, M (2016) *The Russian "Firehose of Falsehood" propaganda model: why it might work and options to counter it*. Santa Monica, CA: RAND Corporation. https://www.rand.org/pubs/perspectives/PE198.html, accessed May 6, 2020.

RT (2017) '"Whoever Leads in AI Will Rule the World": Putin to Russian children on knowledge day', *RT*. https://www.rt.com/news/401731-ai-rule-world-putin/, accessed May 6, 2020.

Russian Federation (2000) *2000 information security doctrine of the Russian Federation*. Moscow: Russian Federation.

Russian Federation (2014) *The military doctrine of the Russian Federation*. Moscow. https://rusemb.org.uk/press/2029, accessed May 6, 2020.

Saeed, S (2017) 'US intelligence chief: Russia interfering in French, German elections', *Politico*. http://www.politico.eu/article/us-intelligence-chief-russia-interfering-in-french-german-elections/, accessed May 6, 2020.

Seddon, M (2014) 'Documents show how Russia's troll army hit America', *BuzzFeed News*. https://www.buzzfeed.com/maxseddon/documents-show-how-russias-troll-army-hit-america, accessed May 6, 2020.

Seely, R (2017) 'Defining contemporary Russian warfare', *The RUSI Journal*. Routledge, 162(1), pp. 50–59. doi:10.1080/03071847.2017.1301634.

Shea, TC (2002) 'Post-Soviet Maskirovka, cold war Nostalgia, and peacetime engagement', *Military Review*, 82(3), pp. 63–67.

Sieca-Kozlowski, E (2009) 'From controlling military information to controlling society: the political interests involved in the transformation of the military media under Putin', *Small Wars & Insurgencies*, 20(2), pp. 300–318. doi:10.1080/09592310902975430.

Sindelar, D (2014) 'The Kremlin's troll army', *Atlantic*. https://www.theatlantic.com/international/archive/2014/08/the-kremlins-troll-army/375932/, accessed May 6, 2020.

Socor, V (2008) 'The goals behind Moscow's proxy offensive in South Ossetia', *Eurasia Daily Monitor*. Jamestown Foundation, 5(152). https://jamestown.org/program/the-goals-behind-moscows-proxy-offensive-in-south-ossetia/, accessed May 6, 2020.

Stent, AE (2014) *The Limits of Partnership*. New Jersey: Princeton University Press, pp. 97–123. doi:10.2307/j.ctv7h0twn.11.

Thomas, T (2016) 'The evolution of Russian military thought: integrating hybrid, new-generation, and new-type thinking', *Journal of Slavic Military Studies*. Routledge, 29(4), pp. 554–575. doi:10.1080/13518046.2016.1232541.

Thomas, TL (2003) 'Information warfare in the second (1999-Present) Chechen War: motivator for military reform?', in Aldis, A and McDermott, RN (eds) *Russian Military Reform 1992–2002*. London: Frank Cass, pp. 209–233.

U.S. Department of Homeland Security (2016) *Joint statement from the department of homeland security and office of the director of national intelligence on election security*. Washington, DC: Department of Homeland Security. https://www.dhs.gov/news/2016/10/07/joint-statement-department-homeland-security-and-office-director-national, accessed May 6, 2020.

U.S. Joint Chiefs of Staff (2018) *JP 3–12 cyberspace operations*. Washington, DC: Department of Defense.

Whitmore, B (2008) 'Scene at Russia-Georgia border hinted at scripted affair', *Radio Free Europe/Radio Liberty*. https://www.rferl.org/a/Russia_Georgian_Scripted_Affair/1193319.html, accessed May 6, 2020.

9 US grand strategy and the use of unmanned aerial vehicles during the George W. Bush administration

Francis Okpaleke and Joe Burton

Introduction

Did the use of unmanned aerial vehicles (drones) during the Bush administration support or undermine US grand strategy? In answering this question, it is necessary to consider the foreign policy outlook of the Bush administration, the ways in which drones were used during this period, and the way they intersected with US grand strategy. The key argument in this chapter is that drone use during the Bush administration was closely related to a shift in US grand strategy from the defensive realist stance of successive US presidents in the post–Cold War period to an offensive liberalist posture in the aftermath of the 9/11 attacks. The chapter further argues that preventive and preemptive drone strikes served to undermine Bush's grand strategy, particularly in targeted states[1] such as Pakistan, Afghanistan, and Yemen by causing unintended civilian deaths, a broader escalation of anti-American sentiments, a rise in retaliatory strikes by radical insurgents, and through their propaganda value as a recruitment tool for extremists.

In building these arguments, the chapter is divided into 7 sections. The first section discusses the meaning of grand strategy. In the second section, the realist and liberalist interpretation of US grand strategy is examined. In the third section, the foreign policy of the Bush administration prior to and after the 9/11 terrorist attacks is analyzed. The fourth section examines the emergence of a post-9/11 grand strategy, including the role of neo-conservatism in that process. In the fifth section, the evolution of Bush drone warfare from 2000 to 2008 is assessed. The sixth section examines the role of and rationale for drone use during the Bush administration. In the last section, we examine how the use of drones undermined US grand strategy during this period.

Contextualizing grand strategy

The academic literature on grand strategy is extensive with a number of different definitions of the concept. For example Edward Luttwak (2016, p. 13) conceives it from a militaristic and expansive perspective, viewing grand strategy "as a confluence of hierarchical military interaction with varied external relations…", while other more narrow views consider grand strategy to be an instrument of statecraft.

For Paul Kennedy (1991, p. 22), grand strategy "represents an integrated scheme of interests, threats, resources, and policies". Likewise, for Peter Layton, (2012, p. 1) it is the "link between a country's ways, means, and ends". The essential purpose of grand strategy, as Inor Sloan (2012, p. 6) posits, "is to achieve a balance between means and ends: Combining the former to achieve the latter, but also adjusting the latter so as not to burden the former". In Christopher Layne's words, "it is the process by which a state matches ends and means in pursuit of security". (Layne, 2007, p. 5). This closely relates to Colin Gray's view that "grand strategy is purposeful and coherent set of ideas about what a nation seeks to accomplish in the world, and how it achieves it" (Gray, 2013, p. 4).

The definition put forward by Hal Brands (2014, p. 3) is perhaps the best applied in the context of this chapter. He defines grand strategy as, "the highest form of statecraft, which when reduced to its essence, forms the intellectual architecture that gives form and structure to foreign policy and the logic that helps states navigate a complex and dangerous world". That is, grand strategy is the compass that guides the articulation of a state's foreign policy. For Brands, an effective grand strategy: (i) Provides clarity on the nature of the international environment, (ii) articulates a country's highest goals and interests, (iii) identifies primary threats to these interests and, (iv) uses finite resources to deal with these threats. The overarching point here is that grand strategy should outline a country's interests, the threats to those interests, and the policies to defend them (Castillo, 2019).

US grand strategy: Realist and liberalist interpretations

This chapter adopts Bernard Miller's model for understanding US grand strategy in the post-9/11 security environment. The model analyzes contemporary versions of US grand strategy from realist and liberal perspectives and differentiates between offensive and defensive strategies. The model identifies 4 versions: Defensive realism, offensive realism, defensive liberalism, and offensive liberalism (see Table 9.1 below) (Miller, 2010).

Defensive realism presupposes that US power is in a decline and advocates a multilateral approach for dealing with international security threats (Schmidt and Williams, 2008). It asserts that the primary objective of the state is to ensure its own security (Miller, 2010) and views aggressive expansionism as an aberration in 2 ways – firstly, because it upsets a state's predisposition to conform to balance of power logic, and secondly, by undermining its capacity to challenge any hegemonic challenge to its power (Lobell, 2010). Intrinsic in defensive realism is the concept of deterrence, which should be achieved through the minimal use of force and multilateral efforts. Defensive liberalism is anchored by the multilateral use of hard and soft power as a compellence tactic to influence and change the objectives of rivals. It advocates the use of a multilateral approach for advancing peace (Miller, 2010).

Offensive liberalism is linked to the neo-conservative doctrine that emphasizes unilateralism, military assertiveness, primacy, and supporting the use of force by

Table 9.1 Realist and liberalist interpretations of US grand strategy

US Grand Strategy Interpretations	*Objectives*	
Versions	*Offensive*	*Defensive*
Realism	• Hard power • Military and economic strength • Survivalism • Great power competition	• US Declinism • Multilateralism • State security • Deterrence • Minimal use of force
Liberalism	• Unilateralism • Military assertiveness • Primacy • Hard power	• Intentional unilateralism • Compellence • Hard and soft power • Multilateralism

Source: Authors' adaptation.

the US to achieve its objectives, including to democratize non-democratic societies (Miller, 2010). The idea that offensive liberalism is 'liberal' lies in its goal of making the world safer for democracy. Offensive realism emphasizes the prevalent anarchy and 'relentless security competition' in the modern international system as the casus belli for conflict. For the US to survive in an anarchic environment, it must exercise its relative power advantages – military or economic – with the aim of becoming the global hegemon (Mearsheimer, 2001).

Bush foreign policy before and after 9/11

Since the US emerged as the sole superpower following the end of the Cold War, successive US presidents have sought to formulate a clear and consistent grand strategy in the absence of a peer competitor (Jervis, 2005). This applies to George H.W. Bush's "New World Order", Bill Clinton's "Engagement and Enlargement" strategy, George W. Bush's "Primacy" and, more recently, to Barack Obama's pursuit of "Restraint" and Donald Trump's "America First". Prior to 9/11, the foreign policy of the George W. Bush administration pursued a defensive realist strategy. As part of this, it showed little appetite for global democratic interventions.

Many analysts, such as Robert Art and Patrick Cronin (2003) and Michael Hirsh (2003), have argued that the foreign policy of Bush during this period was devoid of elements of offensive liberalism or, more precisely, of plans to forcefully project military into rogue states or those providing safe havens for terrorists. This was expressed by Bush's ambivalence before 9/11 to effect regime change or to order offensive military action against the Baathist government in Iraq. Likewise, his initial foreign policy orientation, or what has been described by Jervis (2003) as the "First Bush Doctrine" – was predicated on strengthening US military power and managing a great power competition with Russia and China. This stance was corroborated by Bush's objection to the deployment of US military power in open-ended democratic intervention missions, as stated during his 1999 presidential campaign (Bush, cited in Chollet et al, 2009, p. 62). In sum, prior to 9/11 the foreign policy of Bush can be summarized as one overly focused on changing the Clinton-era "enlargement and engagement" strategy to a more defensive realist approach predicated on multilateralism, limited interventions, and little inclination for offensive liberal proclivities (Miller, 2010).

The September 9/11 attack ushered in a new direction for US grand strategy by challenging the Bush administration's seemingly complacent view of threats in the international environment. The terrorist attack engendered new security concerns and a heightened fear that radical terrorist groups would acquire weapons of mass destruction (WMDs) to carry out coordinated attacks on US soil and interests (Jervis, 2003). This is mainly because the 9/11 attacks heightened the perception of a threat of a possible alliance between rogue states and radical militant groups in perpetuating further terror attacks against the US (Schmidt and Williams, 2008).

Consequently, the 9/11 attacks led to a shift in priorities and a new foreign policy direction for the Bush administration as encapsulated in the "Second Bush Doctrine" (Miller, 2010). The new doctrine was articulated in the 2002 *National Security Strategy* (NSS) and outlined 4 aspects of US strategy in the post-9/11 security environment. First, at the heart of the Bush doctrine was the idea of anticipatory self-defense – the requirement for the US to act unilaterally, with immediacy and offensively in dealing with global terrorist threats (Gaddis, 2005). This included preventive and preemptive wars against imminent threats from rogue states and terrorist groups seeking to develop WMDs (Jervis, 2003). Second, the strategy advocated US military primacy, arguing that a preponderance of US power could ensure peace and that the most serious threat to US primacy would be an "across-the-board" political, economic, and military challenger (Art, 2013, p. 4). The fundamental objective of the US was premised on a will to forcefully guard the international order against illiberal threats and the rise of a foreign power (Chollet et al, 2009). Fourth, it declared its goal for forcible democracy promotion, particularly in the Muslim world (Miller, 2010). The new approach encapsulated in the second Bush doctrine thus formed the main foreign policy focus of the administration and shifted the administration toward an offensive liberal approach to grand strategy.

Grand strategy after 9/11 and the role of neo-conservatism

Following 9/11, the Bush administration found that its existing foreign policy strategy was not suitable for addressing the threat of transnational terrorism and that a new direction was needed. More precisely, the 9/11 terrorist attacks demonstrated the flaws in defensive and offensive realist approaches that seemed ill-suited for dealing with non-state threats like suicide bombings, rogue states, or transnational terrorism (Leffler, 2011). In his assessment of the different versions of US grand strategies, Miller (2010, p. 33) argues that they were "incapable of offering any effective response for dealing with non-state threats". Implicit in defensive realism is the advocacy for the use of deterrence as a tactic to achieve its goals. While this may be useful against traditional state actors and against nuclear-armed states, deterrence was seen as an ineffective strategy to combat "irrational" terrorist or rogue states that were, or could be, willing to support transnational terrorist networks and use suicide bombs or WMD with little regard for their own lives (Kaufman, 2007).[2]

Furthermore, while preventive attacks could be allowed through an offensive realist strategy, and could potentially deal with the threats posed by rogue states and terrorist groups, it does not preclude the possibility of hostile regimes rebuilding infrastructure and conducting clandestine activities in hard-to-reach areas (Leffler, 2011). In the same vein, defensive liberalism, based on advocacy for multilateral diplomacy to advance non-proliferation, seemed to be a weak approach to adopt in dealing with regimes that were likely to hide their WMD programs or rogue states that could clandestinely support and finance transnational terrorism (Gurtov, 2006).

It was not possible to forge a grand strategy for dealing with the threat of Al Qaeda and rogue states via a rational, linear process. However, in the aftermath of 9/11, there appeared to be a symmetry between neo-conservative thinking and the immediate needs of the administration. Thus, the crucial issue for the Bush administration after 9/11 was formulating the most appropriate approach to deal with the rising threat of Islamic terrorism, and this served as the catalyst for the recrudescence of neo-conservative ideas for US foreign policy in the Bush administration.

Neo-conservatism had been a prominent part of the US domestic debate on foreign policy since the end of the Cold War and had a broader intellectual history stretching back at least to the 1960s. With the demise of the Soviet Union, neo-conservatives advocated that enforced democracy promotion based on an offensive liberal strategy should be the cornerstone of a new ideological American foreign policy (Halper and Clarke, 2004).

The neo-conservatives in the Bush's administration, such as Paul Wolfowitz, appeared to be the only ones ready with a detailed program that offered a credible response for dealing with the post-9/11 threats based on their long-standing offensive liberal agenda with regard to Iraq and the Middle East. Analyses by Robert Jervis (2003) and Robert Kaufman (2007) have shown the program included an explanation of the sources of terrorism (that states in the Middle East were aiding and abetting terrorism by inflaming American resentment and incentivizing the spread of radical Islamic ideologies), a program for forcible regime change in Iraq, and an action plan for dismantling terrorist operations and the rogue regimes that allegedly harbored them. Finally, Wolfowitz, an ardent neo-conservative under Bush, proposed replacing rogue regimes with liberal democracies, arguing that "the more democratic the world becomes, the more likely it is to be both peaceful and friendly to America" (Wolfowitz, cited in Kaufman, 2007, p. 26).

As a consequence, due to the availability and the seeming plausibility of the neo-conservative approach, together with the political and strategic shock of

9/11, key officials in the Bush administration, such as Dick Cheney and Donald Rumsfeld, who hitherto had not been seen as neo-conservatives, adopted this approach. Offensive liberalism thus became a key aspect of the Bush administration's post-9/11 foreign policy and informed the President's usage of the term, "Axis of Evil", during the January 2002 State of the Union address, and the articulation of the Second Bush Doctrine of September 2002. As Miller puts it,

> the offensive liberal strategy advocated by neo-conservatives, focused not only on changing the capabilities of rogue regimes, namely destroying their WMD and preventing them from acquiring more, but also on changing their ideological character by transforming the nature of their political system – namely, enforced democratization.
>
> (Miller, 2010, p. 53)

The offensive liberal strategy was perceived not only as necessary for US security but also as feasible, since it was expected that "the oppressed Middle Eastern peoples would welcome democracy if they could be helped to get rid of their oppressors" (Dorrien, 2004). The expectation of the neo-conservatives was that a successful regime change in Iraq that created a liberal democracy would have ripple effects across other states in the region, and would thus trigger democratization in the whole of the Middle East. Ultimately, while US power preponderance made this strategy possible, the high level of perceived threat post-9/11 provided the catalyst for its adoption, with neo-conservative ideology providing its intellectual underpinnings. The next section explores the phases of drone warfare under the Bush administration.

The evolution of drone warfare 2001–2008

Since the first lethal drone strike in Kandahar, Afghanistan in October 2001, the Bush administration's use of drone warfare consistently evolved in range and targets. This evolution is divided into 3 phases (see Table 9.2 below). The first phase (2002–2004) served as the testing period for the use of drones for the elimination of high-value targets (HVTs) in targeted states. The second phase (2005–2007) was characterized by slight increases in strikes on HVTs. The third phase (2007–2008) consisted of the escalation of drone strikes in targeted states.

The first phase of Bush drone warfare began mid-November 2001 and killed Mohammed Atef, an Al-Qaeda military commander. This was followed by

Table 9.2 Phases of Bush drone warfare (adapted from Hudson, Owens and Flames, 2011)

Drone Strikes by Phase	*Phase Strikes*	*HVTs Killed*	*Total Deaths*
(2002–2004)	2	2	11
(2005–2007)	6	2	53
(2007–2008)	48	5	333

another CIA drone strike in the Maarib province of northeastern Yemen. The strike killed six Al-Qaeda militants, including Salim Sanin al-Harethi, and it was the first drone strike to kill an American citizen, Kamal Derwish (Hudson, Owens, and Flannes, 2011). The next attack started in 2004, and killed Nek Mohammed, an influential member of the Taliban and a former Mujahedeen. The total number of HVTs killed in this period based on published data from the Bureau of Investigative Journalism is estimated to be five or six (BIJ, 2013).

The second phase occurred mainly in Pakistan. Drones strikes increased slightly during this period and targeted Al-Qaeda and Taliban HVTs in the remote Federally Administered Tribal Areas (FATA) region of Pakistan. Though there are conflicting reports that most of the targets killed during this phase were mainly low-value targets and not top members of the Al-Qaeda core (Boyle, 2013), the elimination of Hamza Rabia, the "number three" in the chain of command in Al-Qaeda, did occur during this phase. In the third phase, the use of drones for leadership decapitation became a manifest strategy, as more HVTs were eliminated following the intensification of lethal drone strikes in Pakistan, Afghanistan, Yemen, and Somalia (see Table 9.3 below).

The premise for the use of drones for leadership targeting is expressed in the 2003 *National Strategy for Combating Terrorism* (NSCT). The document highlighted the intersection between leadership decapitation and the collapse of terrorist organizations.

> The terrorist leadership provides the overall direction and strategy that links all these factors and thereby breathes life into a terror campaign. The leadership becomes the catalyst for terrorist action. The loss of leadership can cause many organizations to collapse. Some groups, however, are more resilient and can promote new leadership should the original fall or fail. Still others have adopted a more decentralized organization with largely autonomous cells, making our challenge even greater.
>
> (NSCT, 2003, p. 13)

Table 9.3 Drone strikes against high value targets (HVTs) in the third phase of the Bush administration's drone warfare. Author's adaptation (compiled from Bureau of Investigative Journalism and New America Foundation)

Drone HVT	*Terrorist Role/Affiliation*	*Date of Death*
Qaed Salim Sinan al-Harethi	Al-Qaeda leader in Yemen	November 3, 2002
Nek Mohammed Wazir	Senior Taliban Leader in Pakistan	circa 2004
Abu Laith al Libi	Senior militant/Al-Qaeda in Afghanistan	January 31, 2008
Abu Khabab al-Masri	Al-Qaeda Weapons Chief	July 28, 2008
Khalid Habib	Al-Qaeda Leader in Northwest Pakistan	October 16, 2008
Usama al Kini	Senior militant/Al-Qaeda in Pakistan	January 2, 2009

Thus, the main focus of drone strikes for the Bush administration was the elimination of leaders affiliated with global Al-Qaeda and its affiliates. As a counterterrorist tool, it demonstrated that the US could leverage the precision and lethal weapon capabilities of drone technology for leadership decapitation when targets were identified and intelligence matched the kill order (Boyle, 2013). The success of drones in eliminating HVTs like Salim Sinan al-Harethi – who was targeted for his involvement in the October 2000 USS Cole bombing and the October 2002 Limburg attack – and Nek Mohammed, a senior Taliban leader in Pakistan, gave the Bush administration the impression that if limited drone strikes were successful at leadership decapitation and terrorist disruption, more strikes would be even better. The logic behind this argument is that drone strikes create a climate of fear among terrorist targets by fracturing their organization and eventually leading to their collapse (Rae, 2014). In writings discovered after his death, Osama bin Laden lamented the impact of drone strikes and recommended that Al-Qaeda leaders flee Waziristan to safer terrain to avoid them (Minhas and Qadir, 2014).

The US process surrounding decapitation through targeted killings is worth noting here. According to Gallarotti, (2010) this involves (i) deciding if the target is a significant threat to US interests, (ii) recognizing state sovereignty issues, (iii) high confidence in the target's identity and (iv) ensuring that innocent civilians will not be harmed, and, finally, (v) engaging in an additional review process if the individual is a US citizen.

Explaining drone use during the Bush administration

The domestic rationale for lethal drones was predicated on the Authorization for the Use of Military Force (AUMF) joint resolution that was passed by Congress on September 18, 2001, which authorized the US president:

> To use all necessary and appropriate force against those nations, organizations or persons he determines planned, authorized, committed, or aided the terrorist attacks that occurred on September 11, 2001, or harbored such organizations or persons, in order to prevent any future acts of international terrorism against the United States by such nations, organizations or persons.
>
> (Grimmett, 2006, p. 3)

The AUMF not only expanded the executive powers of US presidents to use force for matters of national security but also served as the basis for the rationalization of lethal drone strikes for Bush's preemptive and preventive attacks in targeted states. The Bush administration further maintained that Article 51 of the UN Charter gave the US the authority to act in self-defense against high-level targets who were planning attacks, both in and out of declared theaters of war (Shaw & Ahkter, 2014).

Since their first lethal use, the precision, accuracy, and disruptive effects of drones on terrorist groups have helped to avoid the risks of conventional war – civilian and soldier deaths and collateral damage (Rae, 2014). This is premised on the idea that drones are more capable than human beings in gathering and processing information precisely, rapidly, and flexibly (Byman, 2013). Analogous arguments assert that drones provide the US with greater incentive and capacity to defend itself from external aggression through the projection and multiplication of the show of force (Hazelton, 2012).

At an operational level, one of the many US responses in the aftermath of the 9/11 attacks was the increase in the build-up of military and intelligence capabilities and the establishment of new military bases in Asia, including a new military command in Africa (Hudson, Owens, and Flames, 2011). This occasioned a rise in defense expenditures and emphasized the imperative of counterinsurgency initiatives for dealing with the new security threats in the aftermath of the 9/11 attacks. Under Bush, drones initially served intelligence gathering, surveillance, and reconnaissance (ISR) functions, with limited lethality in the battlefield, but by the end of the administration, they had become weaponized tools for counterinsurgency and counterterrorism operations that more directly and forcefully served the offensive aspects of the Bush doctrine.

Relatedly, as Shah (2016) recounts, America's experiment in Afghanistan and Iraq showed that a quick and decisive victory over the Taliban and Iraqi forces did not result in the complete destruction of the militant groups. In Iraq, armed groups mounted sustained attacks on US forces resulting in high military casualties.[3] In Afghanistan, US troops weakened but did not completely dismantle or eliminate the Taliban. The Afghan War caused Al-Qaeda's senior leadership to take safe haven in Pakistan's FATA following the fall of the Taliban (Shah, 2016). This made it difficult for the US to bring to bear the extraordinary advantage in conventional military power it had in Afghanistan and Iraq (Walsh, 2018) and compounded the pressure on the US to make quick and decisive progress in Afghanistan, which required dismantling Al-Qaeda's safe haven in Pakistan (Khan, 2014).

The expectation that American military involvement in Afghanistan and Iraq would be withdrawn as soon as stable and effective government institutions were created proved erroneous – meaning the US faced a choice between supporting a long-term presence in both countries or decreasing its military footprint and risking a protracted war and the overthrow of its local allies (Walsh, 2018). In addition, Al-Qaeda's clandestine transnational network became increasingly complex for the US to track, disrupt, and dismantle, especially as it established links with other armed groups and terrorist affiliations as far away as Southeast Asia, Northern and Eastern Africa, and the Middle East (Yousaf, 2017).

The onset of intense violence, the rising cost of war, mounting American soldier deaths, increasing uncertainty over the US democracy promotion mission in Iraq, and the perception that the administration lacked an exit strategy and a defined path to victory served to undermine President Bush's foreign policy and public support for it (Miller, 2006). Initially, the wars in Iraq and in Afghanistan received the support of a majority of Republicans and Democrats in Congress and from the majority of the American public. Pew Research research conducted when the Iraq War began in March 2003 showed more than seven-in-ten Americans (73%)

supported the use of force, including 93% of Republicans, 66% of Independents and 59% of Democrats (Pew Research, 2011). Toward the end of the George W. Bush administration, support for military action in Iraq had plummeted, with only 17% of Democrats and 73% of Republicans supporting the war (Pew Research, 2011). Thus, the use of drones under Bush, particularly in the latter stages of the administration, offered a way to continue the conflicts in the Middle East without sapping public support.

The Bush administration's experience in Iraq following the invasion of 2003 thus made the use of drone strikes a more attractive strategy (Gurtov, 2006). As its use in the latter part of the administration showed, drones became a manifestation of primacy and hard power empowerment for an offensive liberal grand strategy. Used in this way, drones fulfilled an important objective of aggressively countering terrorism as "an assertive and determined strategy to rollback threats that challenges US national security" (Nau, 2012, p. 59). In sum, the use of drones after 9/11 under Bush reflected not only America's technological and military power but also reiterated the commitment of the US to act alone in defeating Al-Qaeda and its affiliates. This pivotal role was expressed in its utility as a tool for the propagation of a form of "America First" multi-unilateralism – which advocates the use of brute force when US interest is threatened without compromising the benefit and necessity of multilateral cooperation (Boyle, 2013).

The dilemma of drone warfare for US grand strategy

Since the 9/11 attacks, drones have played an important yet controversial role in US counterterrorism strategy. Despite the acknowledged utility of drones for targeted killings and reducing troop deployment and casualty figures, the intensification of their use in targeted states had several negative effects. Ultimately, they worked at cross-purposes with the more liberal elements of the Bush administration's grand strategy and undermined efforts to promote democracy in Afghanistan, Pakistan, and Yemen. This was due to the countervailing democratic reactions engendered in the aftermath of drone strikes in these targeted states, such as local protests against their use, unintended civilian death (see Table 9.4),

Table 9.4 Drone strikes and casualty figures in the Bush era (Compiled from Bureau of Investigative Journalism and New America Foundation)

Bush Drone Years	
	Figures
Total drone strikes	51
Total reported killed	410–595
Civilians reported killed	167–332
Children reported killed	102–129
Total reported injured	175–277

anti-American sentiments and militant recruitment and violence. From this lens, rather than consolidating democratic promotion, drone use triggered the conditions that undermined it.

A number of scholars have taken up this line of reasoning. The use of drones against HVTs from 2002 to 2007 prompted what David Kilcullen and McDonald Exum (2009) termed "Death from above and outrage from below". They argue that drones engendered three effects that undermined their use. First, they created a siege mentality in targeted states, triggering a blowback effect for every unintended civilian death and collateral damage caused. Second, drone strikes triggered opposition from the population in target areas and increased anti-American sentiments due to the moral and territorial issues associated with its use. Third, drones externalize the burden of war. This approximates with Krieg and Rickli's (2018) explanation that drones are an instrument of "surrogate warfare" – which places the burden of war on human and technological surrogates, particularly in instances where the US judges no vital American interest to be at stake. This is analogous to what Shaw and Akhter (2014) call the "dronification of state violence", which highlights the shifting pattern of US state violence toward inclusion of weaponized drones.

More specifically, attempts to expand the offensive liberal strategy of the Bush doctrine beyond Iraq to Pakistan and Afghanistan appeared self-defeating and disempowering for the US. The reliance on drones as a tactic for targeted killing operations of terrorists increased the cause of anti-Western militancy and engendered a legitimacy crisis for governments in targeted states that supported drone operations (Gallarotti, 2010). Jervis (2005) contends that the way the US used it power in the war against terrorism increased American vulnerability by energizing terrorism and galvanizing support for anti-Western movements within the countries in which drones strikes occurred. Hazelton (2012, p. 31) adds, "drone strikes appeared to inflame existing enmities the US had prior to 9/11 by creating a web of fear and vulnerability in targeted states". The next paragraph supports the case that drone strikes inflames anti-American sentiments in targeted states. A recent study by David Jaegar and Zahra Siddique (2018) outlines the negative feedback in Muslim countries and the perception of the US as a military threat following drone strikes during the Bush era. The study reported the vengeance effects of drone strikes in Afghanistan and Pakistan based on an analysis of terrorist data attack on the Taliban from January 2005 to September 2011. The study showed that drone strikes induced further violence through vengeance by the Taliban (Jaegar and Siddique, 2018, p. 673).

A 2010 poll based on a sample of 1,000 residents in all seven FATA agencies showed that 76% of respondents opposed drone strikes; only 16% thought that such strikes accurately target insurgents; and 48% believed that the strikes largely kill civilians (Bergen and Tiedemann, 2011). Moreover, 60% believed that suicide attacks against the US were "sometimes" or "always" justified (ibid). Peter Bergen and Katherine Tiedemann who carried out the survey, have used these figures to claim that suicide attackers are widely popular across FATA, and that the main motivation for anti-American militancy "stems from anger at CIA-directed drone strikes at militants living in the area" (Bergen and Tiedemann, 2011, p. 15).

According to David Kilcullen, (2010, p. 18) "every one of these dead noncombatants represents an alienated family, a new desire for revenge, and more recruits for a militant movement that has grown exponentially even as drone strikes have increased". In short, Kilcullen implies that the aftermath of drones in targeted states drive anti-American sentiments, which is counterproductive for US objectives to win the hearts and minds of the population in targeted states (Kilcullen, 2010). As Ashan Iqbal of the Muslim League Party notes, the Islamist parties have used the pretext of the aftermath of drone strikes to mobilize thousands of followers throughout Pakistan in large protests in Punjab, North West Frontier Province, and Sindh (Iqbal cited in Williams, 2010). To this end, the excessive reliance on hard power (and drones) in dealing with terrorism under the Bush era rather than improve the US image as a promoter of democracy, actually undermined it.

Drone strikes have been argued to spur militant recruitment and retaliation in targeted states (Boyle, 2013). It is, however, hard to prove direct causation between drone strikes and militant retaliation – that the increase in suicide bombings was related to the emergence of insurgencies in targeted states that would probably have occurred even had drone strikes not been happening. However, drones clearly fuelled the conditions under which these insurgencies grew. Boyle, (2013) for example, argues that there is a substantive relationship between the increasing number of drone strikes and the increasing number of retaliation attacks. As data published by the Bureau of Investigative Journalism 2014 show, for every high-profile, purposeful attack by the US, many more low-profile attacks take place (BIJ, 2014). This position is supported by a report published by the CIA (2009) on HVTs:

> The potential negative effect of high level target operations include increasing the level of insurgent support, strengthening an armed group's bonds with the population, radicalizing an insurgent group's remaining leaders, creating a vacuum into which more radical groups can enter, and escalating or de-escalating a conflict in ways that favor the insurgents.
>
> (CIA, 2009, p. 3)

Drone attacks generate what David Kilcullen describes as the "accidental-guerrilla" phenomenon – which explains that drones incentivize the militarization of the locals and increase the propensity of reprisal attacks (Killcullen, 2009). Acknowledging the accidental guerrilla phenomenon, Aqil Shah writes:

> The new combatants unable to retaliate against the US within FATA, crossed over the border into Afghanistan, where US troops, NATO forces, and Afghan security forces are concentrated and present easily identifiable targets by joining the ranks of groups like the Pakistani Taliban, whose attacks within Pakistan destabilize the US-Pakistani alliance.
>
> (Shah, 2018, p. 57–58)

These "new combatants" contribute to the growth of terrorist cells which hinder US counterinsurgency operations. This implies that the use of drones as a decapitation strategy served as a propaganda tool for the creation of accidental guerrillas in targeted states, a consequence of which engendered a paradox of Bush counterinsurgency operations. Articulating this, Jessica Wolfendale stated, "counterterrorism policies that are intended to enhance security often have a counterproductive and paradoxical effect" (Wolfendale, 2016, p. 86). This corresponds to a modified version of security dilemma analysis which explains situations where the actions taken by a state to increase its own security cause reactions from other states, leading to a decrease rather than an increase in the state's security (Tang, 2009).

Data published by the Global Terrorism Database supports this (see Figure 9.1). As the trend in the graph below shows, retaliatory attacks in Pakistan increased with the intensification of US counterterrorism operations. The graph (2002–2008) corresponds with evidence from research on terrorist radicalization and recruitment between 2004 and 2008, which reported a steady rise in attacks by suicide bombers in Afghanistan and Pakistan – the two main centers for the Bush era drone strikes (Leffler, 2011).

Though a recent study by Shah (2018) based on interviews and surveys of "well-informed" respondents in FATA Pakistan discredits the credibility of

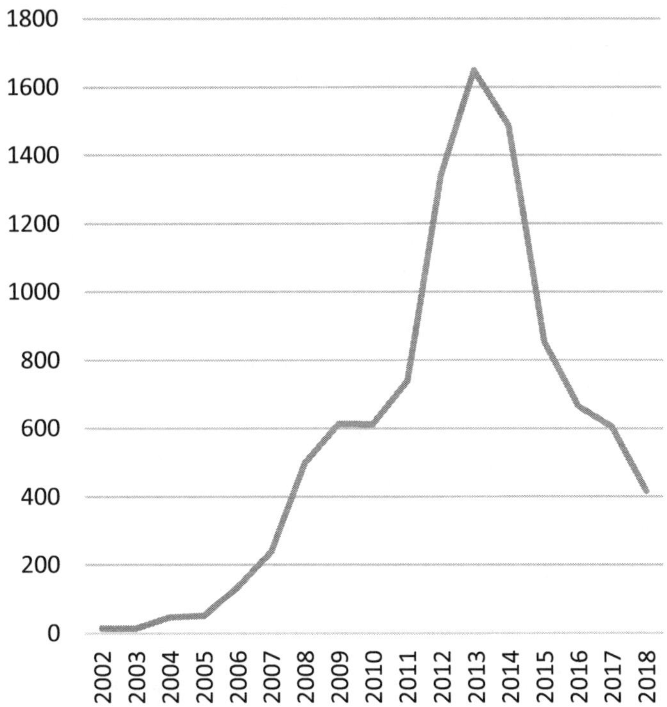

Figure 9.1 Trends in terrorist retaliatory attacks in Pakistan following US counterinsurgency operations. Note** 2002–2008. Source: Global Terrorism Database, University of Maryland

militant recruitment following drone strikes, his study still confirms drone strikes as contributing to the precarious security situation in Pakistan.

The use of drones as a counterterrorism instrument of statecraft under Bush came at a high price for US soft power proclivities. Drones emphasized US hard power, which complicated its strategic mission in Afghanistan, as well as affected its fragile relationship with Pakistan (Minhas and Qadir, 2014). The negative feedback of the new drone policy, particularly in targeted states, undermined the domestic and international soft power of the administration's foreign policy (Miller, 2006). According to Byman (2013, p. 35) drones caused "enormous pressure for governing structures in these countries while at the same time worsening social volatility in the target area with unpredictable outcome". This was the case in Pakistan, where the intensification of drone strikes culminated in a series of protests against the Pakistani government for aiding the US in killing Pakistanis (Boyle, 2013). Likewise, the US counterinsurgency mission in Afghanistan also became a victim of two forms of blowback. The first blowback arose from the feeling of asymmetric vulnerability from non-combatants on the ground in the target area. The feeling resulted in the desire to fight back and inflamed national sentiment against the use of drones. The second blowback is that drones potentially engender stiff resistance to local authorities in targeted states who are shown to be powerless (or even complicit) to stop drone strikes over their territories (Sadat, 2012). The point here is that the Bush administration's drone attacks served to further destabilize an already fragile nation by deepening divides between the citizenry that abhors the attacks and the government institutions that tolerate or facilitate them (Hudson, Owens & Flannes, 2011).

Conclusion

In this chapter, the role of drones in the grand strategy of the Bush administration has been critically assessed. One of the key arguments is that drones served as a tool for facilitating US offensive liberal strategies under Bush. The utility of drones for targeted killings, leadership decapitation, and as an instrument of statecraft in targeted states were also highlighted. The overarching argument in this chapter is that drone strikes in targeted states undermined rather than supported US grand strategy, especially toward the end of the Bush administration.

It should be noted that the use of drones under Bush created a precedent for successive US administrations. Under Obama, drones became a key instrument of his foreign policy as exemplified in the intensification of drone strikes in Afghanistan and Pakistan in 2009. The tactical reliance on drones under Obama also served as a double-edged strategy for facilitating the administration's defensive neo-realist hybrid grand strategy that sought to promote US exceptionalism, engagement, multilateralism, and restraint on the one hand, and its military assertiveness on the other. Furthermore, drones became the hallmark of Obama's restraint strategy in the Middle East, which favored a sharp rollback of the US military presence and a more austere foreign policy predicated on offshore balancing.

Under Donald Trump, drones have significantly expanded in use, target, and in lethality, especially in the Middle East, Africa, and South Asia (Rubin, 2020). Drones, as in previous US governments since 9/11, play an important role in the grand strategy of Donald Trump, which advocates a "nativist, protectionist, and nationalist-isolationist" vision for the US under the populist slogan "America First" and challenges post–Cold War US consensus on liberal hegemony (Brands, 2014).

In line with the themes of this book, the chapter suggests a number of other potential consequences for technology and international affairs. First, in devising strategies for the use of technology in war and conflict, due attention should be given to both the positive and negative effects and unintended consequences. This does not appear to have happened in this instance. Relatedly, grand strategy in this case was formulated according to a variety of factors, including external threats, the internal politics of the administration, and the weight of history (i.e. what had worked well in the past). Technology also arguably played a role in the formulation of strategy – during the early years of the Bush administration drones were a way to facilitate a light footprint and toward the end of the administration, they were a tool to deal with ebbing political support at home. In this respect, the strategic utility of the technology changed over the course of the administration. Furthermore, while drones seemed like an appropriate tactical and operational tool, they did not appear to serve US strategic goals particularly well. These could be important implications for other technologies, particular as the global security environment changes.

Finally, more broadly, the proliferation of drones may contribute to the emergence of new forms and types of warfare, such as the postmodern warfare identified by Burton in Chapter 1 of this book. There is the possibility that with current advancements in artificial intelligence that drone systems that are fully autonomous will be deployed in the future – that is, functioning entirely on programmed algorithms without human intervention. This also includes the development of lethal autonomous military robots that are capable of operating independently on land, air, and water – or underwater or in space – in attacking and protecting military targets. This would potentially engender further changes in the role of the state concerning technology due to the moral, ethical, and legal factors that may arise with these machines.

Notes

1 Targeted states here refer to states where the US is officially at war with and carries out drone strikes in. Non-targeted are states where the US is not officially at war with but continues to carry out drone strikes.

2 Alex Wilner argues that the Bush administration erroneously assumed that deterrence would be unsuccessful against terrorism when, in fact, various homeland security measures have proved effective at deterring subsequent attacks (Wilner, 2011).

3 According to the US Department of Defense, the US has lost 4,487 service personnel in Iraq since the start of Operation Iraqi Freedom on 19 March, 2003. By 31 August, 2010, when the last US combat troops left, 4421 had been killed.

References

Art, RJ (2013) *A grand strategy for America*. Cornell University Press.

Art, RJ and Cronin, PM (eds) (2003) *The United States and coercive diplomacy*. US Institute of Peace Press.

Bergen, P and Tiedemann, K (2011) 'Washington's phantom war: the effects of the US drone programs in Pakistan', *Foreign Affairs*, 90, p. 12.

Boyle, MJ (2013) 'The costs and consequences of drone warfare', *International Affairs*, 89(1), pp. 1–29.

Brands, H (2014) *What good is grand strategy? Power and purpose in American statecraft from Harry S. Truman to George W. Bush*. Cornell University Press.

Bureau of Investigative Journalism (2013) *Pakistan: reported US drone strikes 2013*. https://www.thebureauinvestigates.com/drone-war/data/obama-2013-pakistan-drone-strikes, accessed February 10, 2020.

Bureau of Investigative Journalism (2014) *US drone wars: 2014 in numbers*. https://www.thebureauinvestigates.com/stories/2015-01-07/us-drone-war-2014-in-numbers/, accessed February 11, 2020.

Byman, D (2013) 'Why drones work: the case for Washington's weapon of choice', *Foreign Affairs*, 92, p. 32.

Central Intelligence Agency (2009) *Making high-value targeting operations an effective counterinsurgency tool*. https://wikileaks.org/cia-hvt-counterinsurgency/WikiLeaks_Secret_CIA_review:of_HVT_Operations.pdf/, accessed 10 February, 2020)

Castillo JJ (2019) 'Passing the torch: criteria for implementing a grand strategy of offshore balancing', in *New voices in grand strategy*. Washington, DC: Center for New American Security.

Chollet, D, Chollet, DH and Goldgeier, J (2009) *America between the wars: from 11/9 to 9/11; the misunderstood years between the fall of the Berlin wall and the start of the war on terror*. Public Affairs.

Dorrien, G (2004) 'Benevolent global hegemony: William Kristol and the politics of American empire', *Logos*, 3, pp. 2–8.

Dueck, C (2004) 'New perspectives on American grand strategy: a review essay', *International Security*, 28(4), pp. 197–216.

Fukuyama, F (2004) 'The neoconservative moment', *The National Interest*, 76, pp. 57–68.

Gaddis, JL (2005) 'Grand strategy in the second term', *Foreign Affairs*, pp. 2–15.

Gallarotti, GM (2010) *The power curse: influence and illusion in world politics*. Lynne Rienner Publishers.

Gray, CS (2013) *War, peace and international relations: an introduction to strategic history*. Routledge.

Grimmett, RF (2006) *Authorization for use of military force in response to the 9/11 attacks (PL 107–40): legislative history*. Library of Congress Washington DC Congressional Research Service. January.

Gurtov, M (2006) *Superpower on crusade: the Bush Doctrine in US foreign policy* (vol. 35). Lynne Rienner Publishers.

Halper, S and Clarke, J (2004) *America alone: the neo-conservatives and the global order*. Cambridge University Press.

Hazelton, JL (2012) 'Drones: what are they good for?', *Parameters*, 42(4/1), p. 29.

Hirsh, M (2003) 'Bush and the world', *Foreign Affairs*, 81, p. 18.

Hudson, L, Owens, CS and Flannes, M (2011) 'Drone warfare: blowback from the new American way of war', *Middle East Policy*, 18(3), pp. 122–132.

Jaeger, DA and Siddique, Z (2018) 'Are drone strikes effective in Afghanistan and Pakistan? On the dynamics of violence between the United States and the Taliban', *CESifo Economic Studies*, 64(4), pp. 667–697.

Jervis, R (2003) 'Understanding the Bush Doctrine', *Political Science Quarterly*, 118(3), pp. 365–388.

Jervis, R (2005) 'Why the Bush Doctrine cannot be sustained', *Political Science Quarterly*, 120(3), pp. 351–377.

Kaufman, R (2007) *In defense of the Bush Doctrine*. University Press of Kentucky.

Kennedy, P (1991) 'Grand strategy in war and peace: toward a broader definition', *Grand Strategies in War and Peace*, 87(1), pp. 1–7.

Khan, WZ (2014) *US Drone Policy and Anti-American Sentiments in Pakistan (2001-2012)*. EduPedia Publications.

Kilcullen, D (2009) *The accidental Guerrilla: fighting small wars in the midst of a big one*. Oxford University Press.

Kilcullen, D (2010) *Counterinsurgency*. Oxford University Press.

Kilcullen, D and Exum, AM (2009) 'Death from above, outrage down below', *New York Times*, 16, pp. 529–35.

Krieg, A and Rickli, JM (2018) 'Surrogate warfare: the art of war in the 21st century?', *Defence Studies*, 18(2), pp. 113–130.

Layne, C (2007) *The peace of illusions: American grand strategy from 1940 to the present*. Cornell University Press.

Layton, P (2012) 'The idea of grand strategy', *The RUSI Journal*, 157(4), pp. 56–61.

Leffler, MP (2011) '9/11 in retrospect: George W. Bush's grand strategy, reconsidered', *Foreign Affairs*, 33(11), pp. 33–44.

Lobell, SE (2010) 'Structural realism/offensive and defensive realism', *Oxford Research Encyclopedia of International Studies*, 12, pp. 165–195.

Luttwak, E (2016) *The grand strategy of the Roman Empire: from the first century CE to the third*. JHU Press.

Mearsheimer, JJ (2001) *The tragedy of great power politics*. WW Norton & Company.

Miller, B (2010) 'Explaining changes in US grand strategy: 9/11, the rise of offensive liberalism, and the war in Iraq', *Security Studies*, 19(1), pp. 26–65.

Miller, SE (2006) 'The Iraq experiment and US national security', *Survival*, 48(4), pp. 17–50.

Minhas, Z and Qadir, A (2014) 'The US war on terror and the drone attacks in FATA, Pakistan', *Journal of the Research Society of Pakistan*, 50(4), pp. 17–20.

Nau, HR (2012) 'Realism', in *Routledge handbook of American foreign policy*. Routledge, pp. 77–90.

New America Foundation (NAF) (2013) *The drone war in Pakistan: America's counterterrorism wars*. https://www.newamerica.org/international-security/reports/americas-counterterrorism-wars/the-drone-war-in-pakistan/, accessed 8 February, 2020.

Nye Jr, JS (2003) *The paradox of American power: why the world's only superpower can't go it alone*. Oxford University Press.

Pew Research Center (2011) *Bi-partisan support for US drone strikes*. https://www.pewresearch.org/drone-strikes/, accessed 10 February, 2020.

Pew Research Center (2013) *A majority of Americans still support the use of drones despite questions*. https://www.pewresearch.org/fact-tank/2013/05/23/a-majority-of-americans-still-support-use-of-drones-despite-questions/, accessed 6 February, 2020.

Posen, BR (2014) *Restraint: a new foundation for US grand strategy*. Cornell University Press.

Rae, JD (2014) *Analyzing the drone debates: targeted killing, remote warfare, and military technology*. Springer.

Rice, SE (2003) *The new national security strategy: focus on failed states*. https://www.brookings.edu/research/the-new-national-security-strategy-focus-on-failed-states/, accessed 11 February, 2020.

Rubin, G (2020) 'Donald Trump, twitter, and Islamophobia: the end of dignity in presidential rhetoric about terrorism', in *Presidential rhetoric on terrorism under Bush, Obama and Trump*. Palgrave Pivot, pp. 105–128.

Sadat, LN (2012) 'America's drone wars', *Case Western Reserve Journal of International Law*, 45, p. 215.

Sauer, F and Schörnig, N (2012) 'Killer drones: the "Silver Bullet" of democratic warfare?', *Security Dialogue*, 43(4), pp. 363–380.

Schmidt, BC and Williams, MC (2008) 'The Bush Doctrine and the Iraq war: neoconservatives versus realists', *Security Studies*, 17(2), pp. 191–220.

Schulzke, M and Walsh, JI (2018) *Drones and support for the use of force*. University of Michigan Press.

Shah, A (2016) 'Drone blowback in Pakistan is a myth. here's why', *Washington Post*, 17.

Shah, A (2018) 'Do US drone strikes cause blowback? Evidence from Pakistan and beyond', *International Security*, 42(4), pp. 47–84.

Shaw, I and Akhter, M (2014) 'The dronification of state violence', *Critical Asian Studies*, 46(2), pp. 211–234.

Sloan, EC (2012) *Modern military strategy: an introduction*. Routledge.

Tang, S (2009) 'The security dilemma: a conceptual analysis', *Security Studies*, 18(3), pp. 587–623.

Walsh, JI, 2018. 'The rise of targeted killing', *Journal of Strategic Studies*, 41(1–2), pp. 143–159.

Williams, BG (2010) 'The CIA's covert predator drone war in Pakistan, 2004–2010: the history of an assassination campaign', *Studies in Conflict & Terrorism*, 33(10), pp. 871–892.

Wilner, AS, 2011. 'Deterring the undeterrable: coercion, denial, and delegitimization in counterterrorism', *The Journal of Strategic Studies*, 34(1), pp. 3–37.

Wolfendale, J (2016) 'The narrative of terrorism as an existential threat', in *Routledge handbook of critical terrorism studies*. Routledge, pp. 130–139.

Yousaf, F (2017) 'CIA drone strikes in Pakistan: history, perception and future', *CRSS*, https://ssrn.com/abstract=3160433 or http://dx.doi.org/10.2139/ssrn.3160433, accessed 28 August 2020.

Part III

The state, society, and non-state actors

10 Cyber autonomy

Automating the hacker – self-healing, self-adaptive, automatic cyber defense systems and their impact on industry, society, and national security

Ryan K.L. Ko

Introduction

In 2016, the Defense Advanced Research Projects Agency (DARPA) hosted the Cyber Grand Challenge (Song & Alves-Foss, 2015), a competition which invited participating finalist teams to develop automated cyber defense systems that can self-discover, prove, and correct software vulnerabilities at real time – without human intervention. For the first time, the world witnessed hackers being automated at scale (i.e. cyber autonomy) (Brumley, 2018). As the competition progressed, the systems were not only able to auto-detect and correct their software but also able to attack other systems (other participants' machines) in the network.

Even though the competition did not catch much mainstream media attention, the DARPA Cyber Grand Challenge proved the feasibility of cyber autonomy, stretched the imagination of the national and cyber security industries, and created a mix of perceptions ranging from hope to fear – the hope of increasingly secure computing systems at scale, and the fear of current jobs such as penetration testing being automated.

Since 2017, IT vendors started moving toward "security automation" (as evidenced by the recent rise in automation vendors at the annual RSA conference (RSA Conference, 2019) – the world's largest cyber security vendor trade show in San Francisco) – the first leap toward cyber autonomy. We have also witnessed the increased number of automated active cyber defense tools, including deception, penetration testing, and vulnerability assessment tools. Like other industries disrupted by automation, these trends have several implications for national security and private industry.

Many vendors are touting network automation programs with existing security information and event management (SIEM) tools as cyber autonomy. Others would label security "playbooks" – hardcoded heuristics that script responses according to triggers or a combination of triggers – as automation. An example of a "playbook" would be a pre-programmed workflow of actions responding to a variety of cyber-attacks (e.g. a response to a denial of service attack or a network port scan by an unknown source).

These examples are still a distance from the true potential and vision for cyber autonomy. The holy grail for cyber autonomy would be a capability that allows us to deter attacks and patch vulnerable computing systems at real time, at scale, *and* without disruption to normal operation. The crux of this is the assumption that a computing system handles abstract and virtual executions and, hence, has fewer physical limitations and boundaries for dynamic remediation.

However, from a practical implementation viewpoint, software systems, particularly those running critical infrastructure, emergency services, and 24/7 manufacturing, have very complex dependencies and do not have the luxury to be turned off and patched during downtime due to their operational demands. For example, the software running a nuclear power plant should not be shut down abruptly.

The dilemma between the need to patch system vulnerabilities and the need to maintain business or operational continuity also places pressure on software migration processes and time. Software migration (or modernization) is the current practice of modernizing software to a newer version. The interdependency of processes and software makes this a challenging change management process. Proponents of cyber autonomy would argue that with cyber autonomy, the need for systems (in particular, critical infrastructure systems) to be modernized would be reduced since the self-healing aspects of cyber autonomy will address vulnerabilities without disrupting business-as-usual.

In the author's opinion, it is important to note that cyber autonomy is differentiated against the concept of "hack-back" (Messerschmidt, 2013), which is linked to defense organizations proposing an autonomous system intentionally hacking back an attributed source of cyber-attack with the aim of decisively limiting or terminating future attacks. Cyber autonomy, with techniques such as dynamic binary instrumentation (Bernat & Miller, 2011), is more likened to the analogy of "fixing a car while driving it".

This chapter is organized into seven sections. The first covers the cyber autonomy challenge by discussing the scale and temporal challenges of cyber defense and the consequential need for cyber autonomy. Following this, the second proposes a novel framework of maturity levels required to achieve full cyber autonomy. The third part surveys current cyber defense automation approaches and the fourth section discusses the impact of automated cyber offense. The fifth segment covers the ethical and human rights challenges posed by full cyber autonomy. The sixth section discusses human resource trends and provides recommendations, before the chapter is concluded in the seventh part.

The crux of the problem

Why is cyber autonomy a turning point in the history of computing systems? One would be able to understand the implications better by first understanding the current "fire-fighting", unsustainable nature of cyber (and particularly, software) security.

Whenever a software is created and released, one can assume two factors occurring across most cases: (1) that the software will have some unknown bug or vulnerability despite best efforts in testing, due to the abstract nature of software development, and (2) that the software engineers working on the code have little knowledge or training in skills required to write code securely.

Until recent years, most programming language coding training at private training institutes or universities have seldom focused on writing code in a way which prevents weaknesses and logical errors which lead to cyber security vulnerabilities. In fact, several of the problems such as buffer overflow (Crispan et al, 1998) have been around for decades with no end in sight. Nevertheless, software is usually released before they are fully ready – in order to meet investment release deadlines or business targets.

These factors have contributed to software released to their users in a best-effort way. Even if the software has been formally verified for potential errors, up to the requirements of international standards such as the Common Criteria (Herrmann, 2002), the risks are only minimized at best. The recent Boeing 737 Max fatalities (Johnston & Harris, 2019) resulting from software bugs in the cockpit which automatically and repeatedly forced the aircraft to nosedive are testaments to that. In most cases, we are unable to be absolutely certain that there will not be new vulnerabilities discovered in the future. Contrast this to the building and construction industry standards and practices which assures reliability, safety, and accountability of the building industry. On this note, legal penalties for insecure software engineering practices have been disputed, but that is a separate discussion which will be discussed in the "Offensive cyber autonomy and its implications for national security and international regulations" section.

Reactive software vulnerability, remediation, and its challenges

Due to widespread insecure software engineering, the discovery of software vulnerabilities (commonly called "bugs") is a common occurrence and some bugs offer entry points for cyber criminals to enter software and access or breach sensitive data. Software released (ranging from operating systems (e.g. Windows, Linux, MacOS) to mobile phone apps) would have the following typical stages of vulnerability remediation (See Figure 10.1):

- X: Time from start of vulnerability discovery to the first public announcement/responsible disclosures
- Y: Time for vulnerability patch development
- Z: Time for patching the vulnerable software in deployed systems

Within the software industry, the time period X in Figure 10.1 would typically start from the discovery of vulnerabilities and conclude at the first public announcement by security professionals through disclosures or the appropriate software vendors. The vulnerabilities declared are usually included and categorized into MITRE's Common Vulnerability Enumeration (CVE) list (MITRE, 2005) and eventually into the National Institute of Standards and Technology

Current situation: X+Y+Z > A or B

Legend:

- X: Time from vulnerability discovery to public announcement/responsible disclosures
- Y: Time for patch development
- Z: Time for patching the vulnerable software
- A: Time for cyber criminals, based upon knowledge of publicly-disclosed vulnerabilities, to develop exploits to take over (i.e. pwn-ing) a system
- B: Time for cyber criminals, based upon knowledge of zero-day vulnerabilities to take over (pwn-ing) a system

Disaster: (X+Y+Z) > A

Figure 10.1 Current timeframes for software vulnerability, discovery, and remediation.

(NIST) National Vulnerability Database (NVD) to articulate the severity of the vulnerability (Booth et al, 2013).

The CVE list and the NVD serve as public service announcements, which enable security experts to understand the vulnerability and develop patches to remedy the problem or patch the vulnerability (i.e. time period Y in Figure 10.1). CVEs and the NVD are also resources for several cyber security tools and global cyber incident/emergency response teams.

Organizations and individuals would then update or patch their systems (i.e. Z in Figure 10.1). Unfortunately, sometimes Z can range from a few days to even months or years – depending on the complexity, processes, and dependencies of the systems. In 2018, statistics (Rapid7, 2019; Sheridan, 2018) showed that it takes an average of 39 days to patch a vulnerable system (i.e. time Z). It was reported that medium-severity vulnerabilities took an average of 39 days to patch while, interestingly, low-severity vulnerabilities took an average of 54 days. The oldest unpatched CVE (common vulnerability enumeration) took 340 days to remediate.

Even when the vulnerabilities are declared and X is short, like a double-edged sword, the CVE and NVD information may also serve as intelligence sources for cyber criminals to develop new ways and software scripts known as exploits to enter vulnerable systems which are not (yet) patched for the new known vulnerabilities. The time from knowing the new CVE entry to the development of an "exploit" (i.e. a way to automate the takedown of a machine via the known vulnerability) to the subsequent success cyber-attack is captured as "A" in Figure 10.1.

Compared to Z, the exact statistics for the length of X and Y are usually hard to measure. If a vulnerability was discovered by cyber criminals and not declared to the vendors, this would be known as a zero-day vulnerability. These cyber criminals may either choose to develop their own exploits or choose to sell their newly

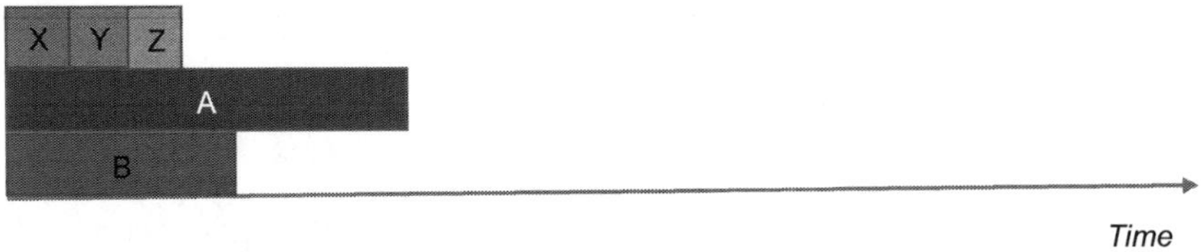

Ideal Situation: (X+Y+Z) < A or B

Figure 10.2 Ideal situation in software vulnerability, discovery, and remediation.

developed exploits (e.g. average price of bug bounties and exploits at around $3,400 (Lemos, 2019)) or vulnerability knowledge on anonymous platforms such as the Dark Web. It must be noted that even government state agencies purchase such knowledge. The time from the discovery of the vulnerability to the development of exploit would be captured in Figure 10.1 as "B". With reference to Figure 10.1, a zero-day vulnerability would also mean that the time period X is as long as it can be, until an individual or organization discloses the vulnerability publicly, or to the affected software vendor. Usually, B is shorter than X.

As shown in Figure 10.1, the current situation is that a typical organization would take (X+Y+Z) to patch their known vulnerabilities. The resiliency of an organization is increased when it is able to shorten Y and Z significantly. It would be disastrous for an organization if A<(X+Y+Z) or B<(X+Y+Z). In fact, these unfortunate scenarios are still happening today, as evidenced by the WannaCry ransomware attack (Mattei, 2017), where unpatched hospital equipment in NHS hospitals were unable to function after the attack.

The time taken to remediate the vulnerability also opens a window of opportunity for cyber criminals to take over systems. In some situations, Y and Z – both within the control of organizations – can take an extremely protracted time. An organization with a long Z for its production systems needs to work on the effectiveness, knowledge, and efficiency of its IT personnel and management. Therefore, our goal is to make X+Y+Z as short as possible (see Figure 10.2). This is why cyber autonomy – with its automated vulnerability detection and remediation (i.e. self-healing) – is such an attractive option.

Scale of the threat landscape

We have considered the situation from the temporal perspective. Let us now look at the situation from a scale perspective across a global context.

Rate of vulnerabilities discovered vs. rate of software produced – mobile, web applications

The statistics in Figure 10.3 show the average number of new Android app releases per day from the third quarter of 2016 to the first quarter of 2018. From

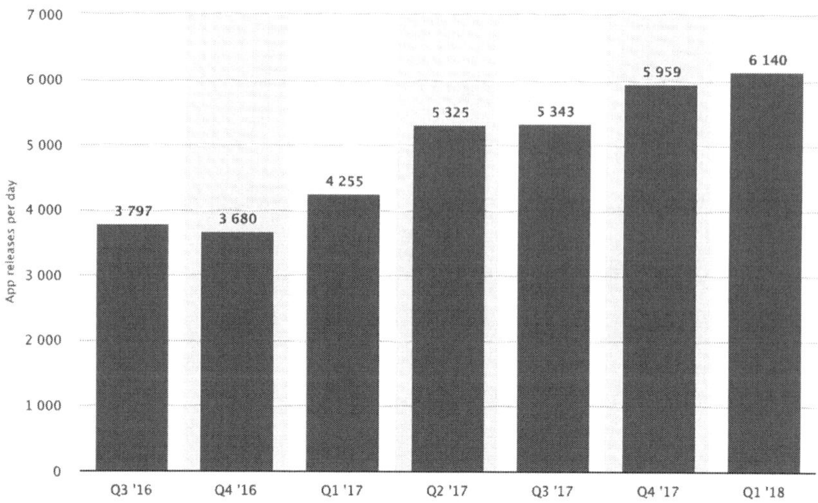

Figure 10.3 Average number of new Android app releases per day (Q3 2016–Q1 2018) (Clement, 2019).

these figures, it can be seen that during the last measured period, an average of 6,140 mobile apps were released through the Google Play Store each day. This is equal to about eight and one-half new apps released on Google's Android each minute. Such growth patterns are not expected to stop.

In contrast to the rate of new apps developed, the rate of discovery of vulnerabilities are lagging behind significantly. This builds a strong case for automation, especially because the software vulnerability identification and remediation processes can be done at rate faster than production. When the rate of such security automation increases at a higher rate than that of software produced, we will reach an equilibrium point, followed by the reduction of vulnerabilities exposed to potential criminals.

The best global indicators of vulnerabilities would be the number of vulnerabilities obtaining their relevant CVE numbers on the CVE list. As shown in Figure 10.4, since the advent of the CVE in 1999, the number of reported vulnerabilities to the CVE list has increased almost exponentially. It is important to point out that the increase could actually be even higher than this, as these records do not include zero-day vulnerabilities, which are used "in the wild", without the knowledge of manufacturers or users. Solutions for software security would need to be sustainable. Problems generated at a machine rate need to be matched with solutions formulated at a machine rate.

Rate of malware discovered

The AV-TEST Institute registers over 350,000 new malware and potentially unwanted applications (PUA) every day (AV-TEST – The Independent IT-Security

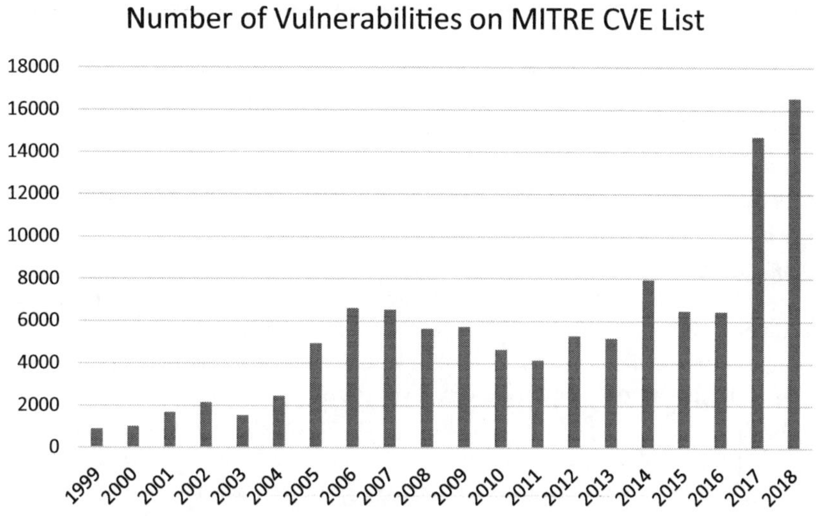

Figure 10.4 Number of vulnerabilities reported on MITRE CVE List. Src. (CVE Details, 2019).

Institute, 2019). These new malware or unwanted applications are then examined and classified according to their characteristics and saved. Visualization programs then transform the results into diagrams that can be updated and produce current malware statistics. This translates to about 4 new malware or PUA registered per second.

Comparing malware discovery rate to speed of skills training and education

When we look at the statistics of the threats generated, it is obvious that the best strategy to overcome the volume and veracity of the new vulnerabilities discovered should not be a linear one. With no intention of sounding cliché, there needs to be a strategy to "change the game". This could be achieved through new generations or types of weaponry to overcome adversaries or changing the battlefield into a "new chess board" to change the rules of engagement. History in the physical realms of security have shown us glimpses of how these strategies work. From the introduction of aircraft into wars, to the development and use of nuclear weapons in the Second World War, new responses seem to be the only viable solution to an unsustainable future.

Within the Web security space, we can see a simple way to "change the game" in responses to web spam visits by bots. CAPTCHA (Von Ahn et al, 2003), a randomly generated picture which can only be interpreted by human users, was introduced to web forms to cut down the number of automated spam bots which were crawling websites and sending spam information into organizations. The introduction of CAPTCHA, a simple additional feature, stopped web spamming. In a similar fashion, two-factor authentication techniques, which require users entering an additional authentication factor (e.g. a passcode sent to the user's mobile phone) on top of the usual password authentication, has also rooted out several phishing campaigns.

While there is a strong push for the cyber security skills gap to be addressed, and the author agrees with the need to educate more cyber security professionals across all skill levels, there is only so much training and education can do to fill the gap – which is growing at the rates shown above.

Logically and practically, it takes a much longer period of time for a person to be equipped with cyber security skills and experience relevant to manually finding vulnerabilities and remedying them. Typically, the time taken to approve the deployment of new patches into production systems, and the time taken to test new patches released, would not be able to catch up with the deluge of software and malware. The rate of threats and software released are at a much higher rate and pace, and there needs to be a shift from reactive to proactive defense. This is why active defense through full cyber autonomy is the only viable option we have so far.

The path to full cyber autonomy

So, what is *full* cyber autonomy? In its essence, this means a point or time when technical solutions match or overcome the scale and temporal challenges posed by malware and vulnerabilities described in the previous section. Full cyber autonomy is achieved when any computing user – regardless of his/her technical background – is able to help to protect himself/herself against cyber-attacks and attribute the attack sources. This is supported by automated tools allowing him/her to make judgement calls or plan cyber security strategies without the need to rely on technical experts or a need to actively and continuously observe visualizations for security events.

Achieving full cyber autonomy means the availability of tools which empower users to make strategies and implement them automatically. It is important to note that autonomy is not full automation, but more of a man–machine symbiosis (Licklider, 1960), which combines the strategic strengths of the human user with the scale and speed of software and automation solutions.

We propose four phases of maturity for full cyber autonomy (see Figure 10.5):

- **Phase 1: Data fusion** – At this maturity phase, tools are able to bring/fuse disparate data together to assist users with awareness of "what goes on behind the scenes" within and across machines – within, outside, and across the sovereign or organizational boundaries (e.g. cloud-computing services on Amazon Web Services or Microsoft Azure cloud). Most of the current solutions are based on historical data and not real-time data. Since the late nineties, the emergence of SIEM tools has assisted system administrators with abilities

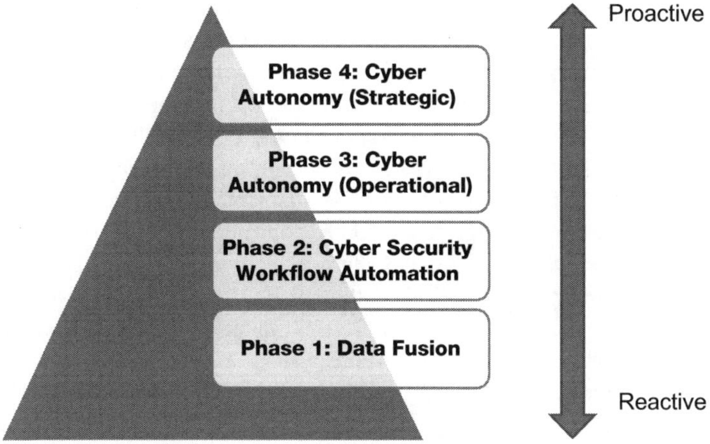

Figure 10.5 Cyber autonomy maturity phases.

to collate and correlate large volumes of log datasets from disparate devices, routers and computers across an organization's infrastructure landscape. The SIEM tools were able to present the risk levels (typically color-coded to traffic light colors, with red denoting "extremely serious" risk level and green denoting healthy systems) and automate remediation responses (e.g. blocking traffic from malicious IP addresses). Most current cyber security solutions are in Phase 1 as they automatically integrate disparate data sources but are unable to provide the capabilities of the following phase.

- **Phase 2: Cyber security workflow automation** – This phase involves cyber security tools integrating disparate data sources and automatically responding to simplistic scenarios, using workflows which can be pre-programmed as rules based on Boolean logic like *if this is true, then do x; otherwise, do y*. Since the early 2010s, the application of machine learning into (extremely) large datasets, epitomized by Google DeepMind's application of deep learning techniques using neural networks to perform so-called "predictions", to highly complex but bound problem sets, such as strategy games like Go (Gibney, 2016), has inspired a new generation of so-called "AI for Cyber Security" solutions. The abovementioned DARPA Cyber Grand Challenge in 2016 played a significant role in showing the feasibility of the then-mostly academic techniques (e.g. static and dynamic analysis of software, symbolic execution, dynamic instrumentation, fuzzing) for real-world applications. At the time of writing, the cyber security products and services industry is now moving toward Phase 2 at a steady pace, and other sectors, such as defense and technology, are at the cusp of moving into Phase 3 (at least on an aspirational level).
- **Phase 3: Operational cyber autonomy** – At this phase, tools will augment intelligence into mostly manual operational processes of identifying issues

in infrastructures at real time, thereby reducing the need to hire staff to audit software for vulnerabilities manually. The key difference with Phase 2 is that Phase 3 includes context awareness and the continuous awareness of organizational structure updates, processes, and policies. Phase 3 achieves the vision put forth in the eighties and nineties of "continuous auditing and monitoring" for accountability of actions (Kogan, et al, 1999), and users are empowered with true automation of lower-level tasks. Since there is no explicit way to model business processes from the lens of cyber security, the industry is still lacking description capabilities for automation and is naturally some distance away from Phase 3. Current techniques are mostly automating responses designed through the coding of heuristics rather than a truly intelligent plan automatically generated from artificial intelligence planning (a branch within artificial intelligence where systems take goals into consideration of constraints and generate new plans based on the reasoning algorithms they run on). When Phase 3 is achieved, systems will be resilient to adversarial machine learning attacks (i.e. garbage-in, garbage-out problems faced by machine learning algorithms) and will be able to handle both syntactic (i.e. raw string of characters) and semantic (i.e. meaning derived from the string of characters) levels of abstraction.

- **Phase 4: Strategic cyber autonomy** – At this phase, the user will be able to articulate *strategies* in plain language, supported by artificial *general* intelligence applied to cyber security scenarios. It is at this phase that we can *truly* claim that the user is able to help themselves protect against or respond to cyber incidents – achieving *full* cyber autonomy. At Phase 4, the systems will be able to handle pragmatics on top of semantics and syntactic levels of abstraction (Searle, et al, 1980). At Phase 4, the systems should also be able to handle ethical expectations and behavioral norms. Cyber norms (Finnemore, 2017) within and across jurisdictions would also be addressed.

As one looks at the state and future of cyber security tools through the lenses of the four maturity phases, it can be observed that the technologies display a shift from reactive security to proactive security as we progress from Phase 1 to Phase 4. Through this progression, we can also see an increasing degree of autonomy from the users' viewpoint. The main goal is to counter the current asymmetry where solutions are outweighed by the rate and variety of attacks with the aim of achieving parity between solutions and attack rates.

Automating cyber defense: Current approaches

From the above maturity phases, it is clear that the current solutions fall into Phase 2 (cyber security workflow automation) capabilities. Recently, some industry solutions labeled as "artificial intelligence for cyber security" products are trailblazing into Phase 3 (Operational cyber autonomy). To have an appreciation of the opportunities and challenges created from the road to full cyber autonomy, we will now look at the current approaches of Phase 2. While automatic discovery and patching of vulnerabilities (Bernat & Miller, 2012) take center stage in cyber autonomy, there are many more areas which complement the dynamic vulnerability discovery and patching process. These areas attempt to limit the capabilities of the hacker and outsmart the hackers through deception and a changing network topology.

Cyber deception

In several high-profile advanced persistent threats (APT) such as Stuxnet (Langner, 2011) or the Bangladesh bank heist (Kaspersky Labs, 2018), cyber criminals tend to covertly lurk within the network infrastructure for a long period of time – ranging from months to years – before they strike. The main goal of cyber deception technologies is to trap and defeat stealthy cyber attackers through luring or deceiving them into attacking seemingly vulnerable (but fake) infrastructure. This builds on earlier related but more simplistic network security technologies known as honeypots or honeynets (i.e. multiple honeypots) – which collect network traffic of attackers on machines intentionally exposed to potential network attacks. Cyber deception technologies aim to collect information about the movements and attack strategies of an attacker, rather than simply stopping access. The goal is to understand their modus operandi well enough to devise strategies and solutions so that the attackers cannot come back in again using the same techniques.

When deployed well, and supported by a strong team of experts, deception technologies can potentially reduce the asymmetry between solutions and attacks and turn the tables on attackers. That said, just like honeynets, attackers would soon understand that they are entering a deception technology environment after some time, rendering the deception technologies outdated and increasing the pressure on deception technology vendors to come up with more and more sophisticated cyber defense strategies.

Automated penetration testing/vulnerability assessment

As the number of vulnerabilities discovered increase exponentially over time, there is an increasing trend where companies use automated tools to conduct vulnerability scanning on their servers and applications. Like a building inspector or an auditor looking through a checklist of potential issues, the automated penetration testing and vulnerability assessment tools will systematically scan all programs on all relevant computers for known vulnerabilities informed by the NVD list. The current tools are good at flagging the known vulnerabilities and presenting the potential problems to the users but are clumsy when it comes to the next crucial step: Remediation of the vulnerabilities. The process is efficient but not very effective, since the results of such scans are often fraught with false positives. This means that there is still a necessity to rely on technical experts at some point in the process. As the tools get continuously refined, there are several further implications to the training and the skill levels of the workforce and these will be discussed in section seven.

Playbooks for security workflow orchestration

With solutions gaining momentum toward Phase 3 from 2017, solution providers and vendors are starting to offer add-on capabilities which allow security professionals to pre-program response workflows to initial attack symptoms, such as denial of service, port scanning, unauthorized or suspicious user privilege escalation, and so on. The steps for countering threats are programmed into workflows as so-called "playbooks". Playbooks are orchestrations of steps (think flowcharts) which will help to automate and save precious time, so that security professionals can focus on strategic tasks. Since mid-2019, the concept of playbooks has increasingly been adopted within incident response communities and organizations. In the author's opinion, playbooks are a necessary precondition for Phase 3 and beyond. For a playbook to work effectively, there is a need for human oversight to describe the workflows within the playbooks. This predefinition of a context is likened to the field of AI planning, where before an algorithm can come up with solutions in the form of a "plan", human operators need to first describe planning problem "domain" (e.g. a crane moving containers in a port), the domain's "atomic actions" (hold and release containers, up, down, left, right) and their "preconditions and effects" (Ghallab et al, 2004).

Moving target defense (MTD)

To level the playing field between attackers and defenders, moving target defense (MTD) techniques (Jajodia et al, 2011) will help cyber infrastructure landscape and settings (e.g. changing network topologies, evolving user account structures, etc.) to evolve. This will increase the level of effort needed by attackers to perform reconnaissance or lurk within the environments. The network topology learned by the criminal today will not be the same topology tomorrow, even though it is still the same target organization. In essence, MTD's goal is to constantly change the characteristics of the infrastructure so that attackers would become ineffective or discouraged from attacking their targets. Currently, the concept is appealing, but it is hard to implement at scale; it will be a few more years before this technology consolidates. That said, the concept of MTD promises to flip the asymmetric advantage toward the defenders.

Offensive cyber autonomy and its implications for national security and international regulations

As we progress toward full cyber autonomy, we begin to observe that the same capabilities used for autonomous *defensive* cyber security can also be used for *offensive* cyber security. The risk of weaponization of cyber automation also becomes more likely, and like other weapons, there needs to be clear legislation, regulations, and guidelines around their usage and disposal. Weaponization of cyber-related knowledge and software already exists. In fact, to a certain degree, weaponization is increasing through the thriving markets within the anonymous

Dark Web, including through the trading of zero-day vulnerabilities, "hacking as a service", and cyber-attack software – with buyers ranging from nation-states to mafia organizations (Radianti et al, 2009).

To the best of our knowledge, while efforts have been made to apply existing international law (e.g. Law of Armed Conflict) to cyber weapons (e.g. the Tallinn Manual 1.0 and 2.0), most current legal definitions and regulations of weapons do not cater to cyber weapons – a type of abstract weapon, which is agnostic to geographical boundaries and jurisdictions. The definitions and laws (both domestic and international) are also unable to handle complex cyber-attacks that blur the lines between crime and acts of war. In the words of Liivoja, Naagel, and Väljataga (Liivoja et al, 2019), "regulatory complexity increases as legal concerns become compounded". A good example is the February 2016 Bangladesh bank heist where hackers (attributed to the so-called Lazarus Group) attempted to illegally transfer close to US$ 1 billion out of the Bangladesh central bank's account with the Federal Reserve Bank of New York via the SWIFT network. While the actual cyber-attack was a transnational crime scenario, the crime had further geopolitical implications, since the "earnings" made by the Lazarus Group will eventually make it back to their paymaster, North Korea (Kong et al, 2019). Eventually, such funds and earnings will likely fund the North Korean leaders' lifestyles, nuclear enrichment programs, and ballistic missile tests.

The Bangladesh Bank heist example also brings out the critical need for government agencies to be able to coordinate together without being limited by their usual mandates or scope. In several countries, the role of protecting national critical infrastructure (e.g. banks) falls into a grey area between the usual responsibilities of defense forces, the police, and the signals or intelligence agencies. In the aftermath of a cyber-attack launched through autonomous technologies, who should lead the response and, if an interagency task force was already set up, how is it empowered and how effective can it truly be? It is critical that governments work proactively to answer these questions rather than be overtaken by events; being proactive will enhance their abilities to rapidly respond to cyber-attacks and reduce security and economic costs.

Another issue for national security and law enforcement is the (in)ability to prove or disprove intent. Cyber-attacks, especially those automated to the sophistication of Phase 2 maturity levels and above, will likely combine attacking machines owned by the attackers and machines owned by victims who are not even aware that their machines were controlled by attackers. In the 2016 Mirai botnet attack (Antonakis et al, 2017) which turned vulnerable unpatched devices into remotely controlled "bots" as part of a large-scale international botnet attack, several of the computers and devices launching the worldwide attacks were caused by ordinary unpatched devices ranging from medical devices to Internet-connected closed circuit television (CCTV) cameras. Several of the owners of these vulnerable devices were unaware that they contributed to the attacks and had no direct intention to launch or participate in the Mirai attack. Despite their ignorance or lack of intent, their devices nevertheless contributed to the inaccessibility of several prominent sites, such as Twitter and Netflix. Given this situation, would the owner of the vulnerable device(s) be complicit and accountable for the cybercrime? If so, how is intent proven?

Another challenge for regulation of cyber autonomy is the combination of the transnational nature of cyber-attacks and the existing difficulty in attributing the source of cybercrime. The global regulations currently do not facilitate or help address this challenge. The Budapest Convention on Cybercrime (Council of Europe, 2004) does exist, but non-member nations do not have the benefit of exchange of information derived by the 67 signatories. The effectiveness of the Budapest Convention on Cybercrime also has not been empirically measured or studied. At the time of writing, with 195 countries in the world, non-member nations make up the bulk of those that do not have access to the agreed cooperation and cybercrime information. They need to rely on other forms of communication for intelligence. While the usefulness of the Budapest Convention is recognized by most countries including non-member nations, many non-members view the requirement to change/align their respective national laws to align to Budapest Convention expectations as an obstacle for practical or timely implementation. For some countries, it may require much more effort to navigate through their political environments. As such, some alternative methods are the usage of their member nation status in the INTERPOL (190 member nations), or through one-to-one collaborative relationships with the countries they are seeking information from or with.

Cybercrime is increasingly automated and syndicated. The Blackhole Exploit Kit (Krebs, 2016) was a good example of criminals selling "hacking as a service" and providing fancy dashboards for their subscribers displaying the computers infected by their malware. As cybercrime grows in complexity and volume, the basic flow of information for investigation and crime-fighting purposes is also hampered by the difficulties of attributing the true source of cyber-attacks. For example, a distributed denial of service (DDoS) attack may be launched from computers in Country A, but these computers could be controlled by computers in Country B, or worse, controlled by Country C, and so on. When provided with the evidence of an attack through digital forensics and network logs, an investigator usually needs to find further clues that may debunk the assertion that attacks were from the said source. When trying to prove the source, some laws also contain technical requirements that mean that some of the critical digital evidence may not be usable in courts of law. For example, ordinary logs from cloud-computing servers have several duplicates across other cloud servers (a normal practice in large global clouds like Google or Amazon Web Service). These are transient in nature and do not lay a strong foundation for data integrity and tamper prevention, especially when compared to the traditional digital evidence of physical hard drives found in someone's computer. Unless an organization uses servers, which require high integrity or "chain of trust" from a cryptographic chip in the physical machines running the cloud instances, most logs collected do not satisfy strong evidentiary requirements.

Finally, because software is often released without being 100% assured, there is always a chance of the cyber automation tools (or weaponry) going out of control. Unintended consequences may occur and, even if the software is formally (i.e. mathematically) proven to a high level of assurance, there is still some chance of untested what-ifs. The recent example of the Boeing 737 Max cockpit software problems which caused several air crashes and hundreds of fatalities (Pasztor et al, 2019) was a good example of automation software going out of control in the form of unintended consequences. When we progress from Phase 1 to Phase 4 maturity levels, we need to develop policies and regulations that will address the liabilities and engineering standards imposed on software as an industry – and this goes beyond cyber automation software. In the case of a fully autonomous cyber software going rogue and taking down infrastructure because of a previously undiscovered bug, who or what can be held responsible and accountable? Making software engineers criminally responsible for badly engineered software has been a contentious and divisive topic, but there needs to be more analysis and discussion of this issue.

Implications for ethics and human rights

During the development of the technology and the progression from Phase 1 to Phase 4, adversarial nations may gain access to the automation through stealing intellectual property and knowledge. This could result in an increasingly fast-paced battle/war between the automation haves and have-nots. The chasm between the haves and have-nots will also widen, quite possibly forcing the situation where haves become the masters and the have-nots become subjects or slaves. For example, when every device you technically operate is under the control of a more powerful and Phase 4-ready organization or country, actors are likely to lose control over their activities, rights, and civil liberties. The recent shutdown of the Ukrainian power plants through cyber-attacks by an overseas state actor (Booz Allen Hamilton, 2019) demonstrated this possibility. The ethical boundaries and expectations around full cyber autonomy remains an open topic, but the prospects do not appear to be good. This conclusion does not emerge from a pessimistic viewpoint but, rather, comes from a consideration of the practical realities. In the face of adversaries with unequal terms of engagement or ethical expectations, the incentive to maintain the same ethical ground will naturally diminish over time – especially after a few high-profile incidents that will create pressure for decision-makers to act swiftly – often with compromises to privacy or ethics.

With more powerful and faster autonomous technologies for cyber defense and attack, there will naturally be more capabilities for individuals without technical expertise and organizations to scour the Web and other open source intelligence (OSINT) sources for reconnaissance purposes. In this context, ensuring personal privacy is likely to be an increasingly uphill battle. Much like the ethical and human rights expectations imposed on the usage of biomedical tools, such as gene editing, cloning, or stem cell research, there will need to be sustained advocacy around ethics, human rights, and norms of responsible behavior for the engineering and usage of these cyber autonomy technologies. The state-of-the-art in cyber security research suggests that ethical approval processes in the most developed nations consider mostly biological ethical risks. The same attention is not given to ethics for cyber security, data privacy, and data science technologies or experiments. A re-examination of ethics approval processes at the research and engineering stages could be a good baby step toward widely accepted global ethics and norms and the protection of civil liberties in autonomous systems for cyber defense.

Implications for human resource trends and recommendations

Full cyber autonomy will disrupt the global skills shortage problem and morph the skills gap problem from a focus on quantity to a focus on quality. With the current Phase 2 to Phase 3 transition, there is a stronger need for truly excellent penetration testers with actual bug bounty experience and tangible track records. The experience issue is a "Catch-22" situation, since it is impossible to train all fresh graduates to be the cream of the crop with a range of experience in bug- and threat-hunting. With the automation of vulnerability assessments and penetration testing tools, traditional penetration testing companies that depend on manual penetration testing face becoming a sunset industry.

The media plays an interesting and important role in shaping perceptions of job roles. When we observe the semiotics of current cyber security news articles, job brochures, or marketing material for tertiary qualifications, it is common to see images with locks and keys, handcuffs, a digital avatar wearing a hoodie jacket typing on a keyboard, dark images, and of course, backgrounds with 1s and 0s. The industry recently realized that this semiotics created a skewed demand for "hacking" jobs, instead of an actual cyber security profession. A recent design competition by the Hewlett Foundation (Sugarman & Wickline, 2019) to encourage new cyber security marketing material toward "non-hoodie" images was a sign that the industry has started to consider a more accurate representation of the diversity of the roles and jobs in cyber security. The move toward the diversity of roles and stakeholders in cyber security should also consider semiotics that encourage people to consider jobs as controllers of automation tools and a higher emphasis on strategic and analytical thinking. For all the advantages of full cyber autonomy, computers alone are insufficient to handle the increasingly complex cyber security attacks and challenges. We need skills and facilities, training, man–machine symbiosis, and to leverage the strategic skills of humans to monitor (and at times, prevent) mishaps from happening when automated tools perform unintended actions.

Generally, national skills policies aimed at increasing cyber hygiene and creating an awareness of career options are good, but they are not going to be sustainable for the future. There is currently too little emphasis on developing skills which allow professionals to constantly adapt to the new (and daily) challenges posed by cyber-attacks launched quickly by cyber autonomy tools. Skills such as research, data analysis, or the application of evidence-based techniques to solve complex issues are usually under-represented or under-championed. While industry stakeholders typically do not expect to hire cyber security job applicants with masters or PhD degrees, this could be a mistake from the macro viewpoint and in the longer term because they will miss out on the research thinking and mindsets of these cohorts. Cyber security skills and education policies must not only aim to develop a new generation of skills which are confined to penetration testing, they should encourage the training of researchers and toolmakers who can create automated cyber security tools.

Related to toolmaker training is the critical need for well-engineered cyber security software and tools. To put this into perspective, the buffer overflow problem commonly found in software is now decades old, but few schools actually teach coding with an emphasis to prevent such fundamental vulnerabilities. We need to advocate for a culture training new generations of prospective cyber security professionals with the ability to read and write secure code, backed by automation tools – supported by exposure to skillsets from other disciplines including but not limited to criminology, political science, and business management.

The road ahead

In this chapter, we illustrated the asymmetric scale and speed of attacks (i.e. four new malware per second) vs. the current mostly manual defensive techniques and the time it takes to conduct skills training. This sets the urgent context for cyber autonomy and highlights current gaps in the cyber security industry. To understand the current status in the context of *full* cyber autonomy, we proposed a novel framework with four phases of maturity for full cyber autonomy. It was observed that at the time of writing, the cyber security industry is currently at Phase 2 of our framework, while many organizations are moving toward Phase 3 (at least on an aspirational level). Phase 4 will be realized when technology achieves artificial general intelligence. We reviewed emerging cyber security automation techniques and tools, their impact on society, the perceived cyber security skills gap/shortage, and national security. We also highlighted the lack of global alignment using the Budapest Convention as an example and discussed several challenges, including the difficulty in proving criminal intent, since many owners of compromised devices do not even realize that their unpatched devices are contributing to global attacks.

Overall, the road toward full cyber autonomy has more positives than negatives but only if we can emphasize creating a mindset to engineer cyber defense tools rather than attack tools. Autonomous cyber defense tools should also focus on reversing the asymmetry between the rate of attacks and efficiencies of defense. The ethical principles around designing autonomous cyber defense tools should be well thought through and mirror industries such as the biomedical sector that regulate technologies, such as cloning and gene editing.

References

Antonakis, M et al. (2017) 'Understanding the Mirai Botnet', in *26th USENIX security symposium (USENIX security 17)*, pp. 1093–1110.

AV-TEST – The Independent IT-Security Institute (2019) *Malware statistics*. https://www.av-test.org/en/statistics/malware/, accessed December 24, 2019.

Bernat, AR and Miller, B (2011) 'Anywhere, any-time binary instrumentation', in *Proceedings of the 10th ACM SIGPLAN-SIGSOFT workshop on program analysis for software tools*. ACM, pp. 9–16.

Bernat, AR and Miller, BP (2012) 'Structured binary editing with a CFG transformation algebra', in *Proceedings of 2012 19th IEEE working conference on reverse engineering*, pp. 9–18.

Booth, H, Rike, D and Witte, G (2013) *The National Vulnerability Database (NVD): overview*. Gaithesburg: National Institute of Standards and Technology.

Booz Allen Hamilton (2019) *When the lights went out: Ukraine cybersecurity threat briefing*. http://www.boozallen.com/content/dam/boozallen/documents/2016/09/ukraine-report-when-the-lights-wentout.pdf, accessed December 24, 2019.

Brumley, D (2018) 'The cyber grand challenge and the future of cyber-autonomy', *USENIX Login*, 43(2), pp. 6–9.

Clement, J (2019) *Number of daily google play app releases worldwide 2016–2018*. https://www.statista.com/statistics/276703/android-app-releases-worldwide/, accessed December 24, 2019.

Council of Europe (2004) *Convention on cybercrime*. https://www.coe.int/en/web/conventions/full-list/-/conventions/treaty/185, accessed December 24, 2019.

Crispan, C et al. (1998) 'Stackguard: automatic adaptive detection and prevention of buffer-overflow attacks', *USENIX Security Symposium*, 98, pp. 63–78.

Finnemore, M (2017) *Cybersecurity and the concept of norms*. https://carnegieendowment.org/2017/11/30/cybersecurity-and-concept-of-norms-pub-74870, accessed December 24, 2019.

Ghallab, M, Nau, D and Traverso, P (2004) *Automated planning: theory and practice*. Elsevier.

Gibney, E (2016) 'Google AI algorithm masters ancient game of go', *Nature News*, 529(7587), pp. 445–556.

Herrmann, DS (2002) *Using the common criteria for IT security evaluation*. Boca Raton, FL: Auerbach Publications, CRC Press.

Jajodia, S et al. (2011) *Moving target defense: creating asymmetric uncertainty for cyber threats*. Berlin, Germany: Springer Science & Business Media.

Johnston, P and Harris, R (2019) 'The Boeing 737 MAX saga: lessons for software organizations', *Software Quality Professional*, 21(3), pp. 4–12.

Kaspersky Labs (2018) *Lazarus under the hood*. https://media.kasperskycontenthub.com/wp-content/uploads/sites/43/2018/03/07180244/Lazarus_Under_The_Hood_PDF_final.pdf, accessed December 24, 2019.

Kogan, A, Sudit, EF and Vasarhelyi, MA (1999) 'Continuous online auditing: a program of research', *Journal of information systems*, 13(2), pp. 87–103.

Kong, J-Y, Lim, JI and Kim, K-G (2019) 'The all-purpose sword: North Korea's cyber operations and strategies', in 2019 *11th international conference on cyber conflict (CyCon), IEEE*, 900, pp. 1–20.

Krebs, B (2016) *"Blackhole" exploit kit author gets 7 years*. https://krebsonsecurity.com/2016/04/blackhole-exploit-kit-author-gets-8-years/, accessed December 24, 2019.

Langner, R (2011) 'Stuxnet: dissecting a cyberwarfare weapon', *IEEE Security & Privacy*, 9(3), pp. 49–51.

Lemos, R (2019) *Bug bounties continue to rise, but market has its own 1% problem*. https://www.darkreading.com/vulnerabilities---threats/vulnerability-management/bug-bo

unties-continue-to-rise-but-market-has-its-own-1--problem/d/d-id/1335689, accessed December 24, 2019.

Licklider, JCR (1960) 'Man-computer symbiosis', *IRE Transactions on Human Factors in Electronics*, 1, pp. 4–11.

Liivoja, R, Naagel, M and Väljataga, A (2019) Autonomous *cyber capabilities under international law*. Talinn: NATO.

Mattei, TA (2017) 'Privacy, confidentiality, and security of health care information: lessons from the recent wannacry cyberattack', *World Neurosurgery*, 104, pp. 972–974.

Messerschmidt, JE (2013) 'Hackback: permitting retaliatory hacking by non-state actors as proportionate countermeasures to transboundary cyberharm', *Columbia Journal of Transnational Law*, 52, pp. 275–322.

MITRE (2005) *Common vulnerabilities and exposures*. https://cve.mitre.org, accessed December 24, 2019.

Pasztor, A, Tangel, A, Wall, R and Slider, A (2019) 'How Boeing's 737 MAX failed', *Wall Street Journal*. https://www.wsj.com/articles/how-boeings-737-max-failed-115536992 39, accessed December 24, 2019.

Radianti, J, Rich, E and Gonzalez, JJ (2009) 'Vulnerability black markets: empirical evidence and scenario simulation', in 2009 *42nd Hawaii international conference on system sciences, IEEE*, pp. 1–10.

Rapid7 (2019) *What's going on in production application security 2018*. https://blog .rapid7.com/2018/08/22/whats-going-on-in-production-application-security-2018/, accessed December 24, 2019.

RSA Conference (2019) https://www.rsaconference.com, accessed December 24, 2019.

Searle, JR, Kiefer, F and Bierwisch, M (1980) *Speech act theory and pragmatics* (vol. 10). Dordrecht: D. Reidel.

Sheridan, K (2018) *It takes an average 38 days to patch a vulnerability*. Dark Reading. https://www.darkreading.com/cloud/it-takes-an-average-38-days-to-patch-a-vulnerab ility/d/d-id/1332638, accessed December 24, 2019.

Song, J and Alves-Foss, J (2015) 'The DARPA cyber grand challenge: a competitor's perspective', *IEEE Security & Privacy*, 13(6), pp. 72–76.

Sugarman, E and Wickline, H (2019) *The sorry state of cybersecurity imagery*. https://he wlett.org/the-sorry-state-of-cybersecurity-imagery/, accessed December 24, 2019.

Von Ahn, L, Blum, M, Hopper, NJ and Langford, J (2003) 'Captcha: telling humans and computers apart automatically', *EUROCRYPT'03: Proceedings of the 22nd international conference on Theory and applications of cryptographic techniques*, pp. 294–311.

11 The international security implications of 3D printed firearms

Peter Cook[*]

Introduction

In May 2013, a US citizen by the name of Cody Wilson successfully manufactured and fired the first ever 3D printed gun (Walther, 2014). Named the "Liberator", this firearm was made from plastic and was developed so it could be "printed" on any entry-level home 3D printer. At the same time, computer files for the Liberator were made available online for anyone to download. Wilson has said his aim was to develop the Liberator, so it could be printed at home by anybody, out of readily available plastic (Wilson, 2016). A firearm made at home in this way, anywhere in the world, bypassing laws and regulations pertaining to licensing and distribution of firearms, has become a concern to law enforcement agencies around the globe.

The fact that the Liberator is made of plastic means it cannot be detected by metal detectors and, once printed, it could be smuggled onto aircraft, across borders and potentially into schools (Greig, 2013). Although it is crude and only allows for a single-shot to be fired (Walther, 2014), the technology around 3D printers is accelerating quickly, meaning more sophisticated weapons may be arriving sooner than anticipated, raising concern about its implications for national and international security. Despite this issue originating in the US, files for 3D printed firearms are available all over the world via the internet. To address this, in 2015 Australia's New South Wales parliament passed a law banning the possession of the 3D computer files for guns (Millsaps, 2015). So far it is the only country to introduce such legislation.

In recent years, 3D printing has become more mainstream with its rising use for benign purposes, such as industrial purposes like rapid prototyping of aircraft components (Wohlers Associates, 2017). Yet its use for manufacturing is still considered disruptive (Petrick and Simpson, 2013), and when 3D printers are used

* The ideas presented in this chapter are the conclusions and opinions of the author, not of the New Zealand Defence Force (NZDF) or the New Zealand government.

in the security context, it becomes very controversial and problematic. So, what is the significance of 3D printing and, in particular, 3D printed weapons? There are two viewpoints on this question: Pessimists believe it is a major concern right now, while optimists believe that although the technology does not yet pose a significant threat to public safety and international security, with technology advancing at a dramatic rate, it is likely to become a concern in the future and steps are required now to prepare for negative eventualities.

With the above in mind, this chapter contributes to the emerging literature that examines the threat level 3D printed firearms pose to international security. It considers whether or not the level of threat is legitimate, and which viewpoint is most valid. For this, data were obtained from a combination of government and technical literature, practical experiment, and interviews of experts in New Zealand working in relevant fields including: 3D printing, emerging disruptive technologies, firearms legislation, future force development, and border control. This chapter argues that since this technology is advancing so rapidly, it is important to be proactive across a range of issues, including the present threat assessment, what future development of the technology will offer to disruptive actors, and the consideration of regulatory change.

3D Printed firearm capabilities

According to Gerald Walther (2014), the Liberator is a one bullet handgun and consists predominantly of plastic components that can be printed by a 3D printer. The only non-plastic part is a hardware store nail, which acts as the firing pin. The US Bureau of Alcohol, Tobacco, Firearms and Explosives (ATF) conducted a test of the Liberator in 2013 to determine the capabilities of the firearm (Greig, 2013; Reilly, 2013; Walther, 2014). The conclusion was the Liberator posed a serious safety and security concern. Although not as powerful as most firearms in its class, the bullets fired from the firearm were able to sufficiently penetrate human skin to reach vital organs and puncture the skull, making it lethal.

Videos released by the ATF that show the Liberator being tested, determined the type of material used for the 3D print was critical to whether the weapon would function properly.[1] Several versions of the weapon were produced from either Visijet[2] or acrylonitrile butadiene styrene (ABS) plastics. Due to the explosive charge from the round, the Visijet plastic could not withstand it and exploded during the test. Therefore, the ATF warned, it would most likely kill the user (Greenberg, 2013b, November 14). The ABS version was more durable and was able to fire eight "single-shot"[3] rounds before the test was suspended. With an accurate range unlikely to extend beyond anything more than a few metres (News Inquiry 10, 2013), and being able to fire less than ten rounds before breaking, the Liberator is significantly less capable than virtually all firearms of the same class available either legitimately or on the black market. Add to that the fact it tends to explode if not manufactured properly, it places the user at as much risk of danger as the target, and thus the risk to public safety of this specific 3D printed firearm is considered low.

3D Printed firearm downloads

Although it is unknown how many times the schematics for the Liberator have been downloaded worldwide, approximately 100,000 instances of the Liberator pistol were downloaded from the Defense Distributed[4] website in the first two days after its release in 2013 before the website was shut down by US authorities (Greenberg, 2013a, May 8). The files were accessed mostly in the US, followed by Spain, Brazil, Germany, and the UK. Torrent Freak[5] has stated that once the Liberator files were available from Defense Distributed, the Pirate Bay[6] freely allowed distribution of the files for people to download (Ernesto, 2013). The files are now, and will remain forever, digitally available for anyone with the desire and means to find them.

Although Torrent Freak regularly collects data regarding the number of downloads of particular files for tracking purposes, they have not made the number of Liberator downloads available. At the height of the Liberator files availability in 2013, this information would have been widely accessible. However, Torrent Freak does not currently keep statistics for that period, as they periodically discard historical data. It was also confirmed that the locations the Liberator has been uploaded to cannot be physically determined, as that requires the collection of IP addresses.[7] Therefore, it cannot be determined how many have been accessed and where. So far, no Liberator is known to have been used in any murder since it was made available, suggesting very few (if any) downloads were for the sake of committing violence. The "novelty factor" of owning a 3D printed firearm, and the files to make one, may very well account for a good number of the recorded downloads.

3D Printed firearm manufacture

To determine how easy it would be to manufacture a 3D printed firearm at home, an experiment was conducted in 2018 at the New Zealand Defence Force (NZDF) to determine the projected cost and time required to "print" a 3D firearm. This experiment used downloaded Liberator files from the internet[8] and a mid-range 3D printer. The material used was ABS plastic, the only successful material tested by the ATF (Greig, 2013) and time to manufacture was approximately 12 hours; this, however, does not include download, machine setup, or assembly time. Therefore, depending on the availability of the person using the machine, manufacture should take less than one day. Weight of the material required to manufacture the Liberator was approximately 160 g and based on a cost of US$50/kg, estimated material cost to manufacture would be US$8. The machine cost around US$2,500; however, a simple internet search will show currently available printers capable of printing in ABS plastic in the US$650 range, making it more affordable. Due to legal restrictions in place by the NZDF[9] it should be noted no firearm was actually manufactured; the software running the 3D printer was used to determine the length of time and material required to successfully manufacture a gun, without the need to actually print it.

Many public libraries also have access to 3D printers that can be used by members of the public. This could potentially make things more affordable and easier to access without the need to spend on high machine costs to manufacture a firearm at home. Based on available documentation from the New Zealand National Library (n.d.), the approximate cost to use their 3D printer to print the Liberator would be US$15.[10] However, after consultation with the library's Team Leader of Operations, existing policy states 3D printers may only be used for lawful purposes. No one will be permitted to use the Library's 3D printers to create material presenting an immediate threat to the well-being of others, and, to the best of their knowledge, staff will refuse to print components that could be used to assemble a working weapon.[11] It should be noted the material used by the National Library of New Zealand is different to that specified for use by Cody Wilson (Wilson, 2016). Therefore, if used to successfully manufacture a Liberator, it will in all likelihood explode.

Global threat assessment

The threat level posed by 3D printed firearms can be considered as a national security, or criminal, issue. Based on the capabilities and access to 3D printed firearms discussed here, it is most likely the latter. However, transnational criminal and terrorist groups, or homegrown dissidents, could turn it into a national security issue by using 3D printers to smuggle firearms across borders or carry out lone-wolf attacks at locations where previously it would have been difficult, if not impossible, to access – for example, schools, government buildings, or airports (Johnston, Smith, and Irwin, 2018). That being said, transnational criminal groups would presumably have sufficient funds to access black market weapons without the need to 3D print firearms, and there have been no major global terrorist incidents whereby 3D printed firearms have been used that would give rise to this concern. From a societal perspective, criminality and public safety appear to be the main threats.

The threat of 3D printed firearms was considered even before the Liberator was ever developed, as Jensen-Haxel, who would fall into the pessimist camp, claimed in 2012, "3D printers promise a new industrial revolution" and "allow people with no technical expertise to produce firearms at home" (Jensen-Haxel, 2012, 447–448). The general consensus of virtually all literature relating to this subject is that due to the relatively low cost of home 3D printers and the speed at which technological developments are advancing, law enforcement agencies are concerned that 3D printed firearms pose a real threat to public safety and current firearms regulations (McCutcheon, 2014; Curtis, 2015). As plastic firearms can be undetectable in airports and other large public venues, such as stadiums or government buildings, the potential for smuggling and easy disposal is troubling.

In 2013, a reporter for the UK Mail on Sunday newspaper managed to "smuggle" a Liberator pistol on board a Eurostar train between London and Paris and then assemble it in the toilet (Murphy and Myers, 2013). As the reporter took neither the metal firing pin nor any bullets onto the train, the threat from the firearm was effectively non-existent. As this incident points out, limitations to the Liberator suggest concerns are not necessarily as high as the public would be led to believe (Walther, 2014). To this day 3D printed firearms still require a metal nail, to act as the firing pin, and ammunition to be fired (Daly and Mann, 2018), meaning they may not be as undetectable as some commentators suggest. The optimist view, espoused by Walther (2014) is that the sensationalism behind the Mail on Sunday story, in which the reporter claimed to have exposed a massive international security risk (Murphy and Myers, 2013) has been overhyped, as any real criminal would have also had to smuggle these two additional items on board – both of which would have been easily detectable. Despite advances in the technology, no plastic 3D printed firearm has yet been developed that does not require metal components to operate.

In November 2013, the ATF Public Affairs Division released a question and answer document on 3D printing of firearms (Bureau of Alcohol, Tobacco, Firearms, and Explosives, 2013). It states they are aware of, and concerned by, the phenomenon of 3D printing technology being used to produce firearms and make every effort to keep abreast of novel firearms technology and firearms trafficking schemes. When they receive credible information regarding the illegal possession of firearms, appropriate action will be taken. However, the ATF only enforces existing federal firearms laws and will only investigate cases in which technological advances allow individuals to avoid complying with existing statutes. It is not only US agencies concerned with the criminal use of 3D printers. New Zealand Police released a report on the emergence and expansion of 3D printing technology in March 2013 (National Intelligence Centre, 2013). The primary finding of this report was that, although there is no current evidence suggesting 3D printers have been used for criminal activity, decreasing price and, thereby, increasing potential for illegal purposes, it is likely that 3D printers would be put to criminal use at some point in the near future. Given the speed at which technological developments are occurring with 3D printers, it recommended police be aware of the criminal use of 3D printers. Additionally, defense forces are concerned this new technology could enable actors to bypass traditional import/export controls and increase firearm proliferation (Lott, 2015). Plastic firearms can also complicate attribution efforts as plastics can be melted down after use.

Despite law enforcement agencies' concerns about the threat 3D firearms pose, all personnel interviewed by this author on this topic agree the threat to public safety from a particular weapon system or technology is based on a combination of capability and intent. After the ATF conducted a test of the Liberator, the risk of criminals gaining access to the gun and using it was considered high. Yet because of the one-shot nature of the weapon (Walther, 2014), the risk to public safety was judged to be reasonably low. More advanced 3D printing techniques that can produce components out of metal (International Organization for Standardization, 2015) could be used for firearm production, increasing their capabilities. However, this would limit smuggling and disposal possibilities and, as Professor Johan Potgieter,[12] another optimist, claims, machines required for this type of manufacture are costly and usually designed for industrial purposes, putting them out of reach of the general public (Venuto, 2018). The cost of a 3D printer capable of creating a fully functional firearm remains too prohibitive to make a viable option for a person intent on killing others. Yet, he argues that given the technology is moving quickly and becoming more affordable, governments should remain vigilant. This is a view also held by the United Nations, who are concerned that while the reliability of 3D printed firearms is currently low, this may change in the future as technology progresses (Lupan, 2015).

Professor Potgieter believes only researchers, like himself, may have the capability and time to produce a 3D printed firearm, as it would require significant knowledge and design skills, as well as access to a 3D printer capable of printing large parts with suitable strength steel, aluminium, or titanium.[13] Such a machine would cost in excess of US$200,000 and is, therefore, out of reach of almost anyone attempting to print a firearm. Although there is a risk that criminals will try and make firearms on 3D printers, this is not the most pressing concern. As noted, CAD files can be downloaded by anyone trying to print a gun on a home machine, but this would be extremely unsafe and would most likely lead to injury. That being said, worrying signs are emerging. For example, since 2013, over ten incidents around the world have occurred in which police have seized 3D printed firearms and associated paraphernalia (Daly and Mann, 2018), including the first person in the UK to be convicted for using a 3D printer to manufacture a gun in 2019 (Noor, 2019).[14] Still, there is no evidence anyone has been killed or injured as a result of the production or use of a 3D printed firearm.

Future of 3D printing and 3D printed firearms

This chapter has shown the current risk from 3D-printed firearms is low. However, the speed of technology development demands an examination of the future risk. As the 3D Printing and Additive Manufacturing State of the Industry, Annual Worldwide Progress Report (Wohlers Associates, 2017) states, one day people may look back at 3D printing as historically significant and as important as the printing press, the automobile, or the computer. To date, 3D printing technology has been seen as a valuable tool for rapidly producing prototypes of digital parts and, as such, is relatively mature. For the past few years, however, the industry has been advancing to include many more production applications. This expansion includes the continued development of 3D printed firearm technology since 2013, although it still remains of limited quality. Although it remains far more difficult and complex to produce a high-quality 3D printed firearm than production of a basic prototype, there are opportunities for development in this area.

One area of concern is that the advance of 3D printing technology, plus associated lower costs, could significantly accelerate weapon proliferation and have drastic effects on all aspects of security – from international conflict and terrorism to everyday crime (Johnston, et al, 2018). Nations will face increasing threats to public safety, as everyone from protesters to members of criminal networks can potentially become capable of rapidly producing firearms in their own homes. Some futures expert interviewees[15] believe 3D printing technology is something that requires monitoring, as it provides the ability for civilian technology to be used for military ends. Historically, military technology has advanced faster than civilian technology. However, one of the interviewees explained that, in recent years, civilian technology has advanced faster than militaries have been unable to keep up. This does not solely include the development of 3D printed firearms but also includes the simultaneous advances in the development of Unmanned Aerial Vehicles (UAVs/drones) that are becoming cheaper, smaller, and more portable. As 3D printers get better and better, the manufacture of weapons will only become a greater issue in the future.

All interviewees agree that, with advances in process and component reliability and automation, future 3D printers will be able to produce increasingly sophisticated products without human supervision. With the collaboration of software and production, corporations have the potential to provide a greater disruption to the market. This makes the 3D printing industry susceptible to hacking by actors committing industrial sabotage (Kan, 2016; Wohlers Associates, 2017). In terms of firearms, one futures expert expressed their fear that this vulnerability will allow for the unmonitored manufacture of a vast number of quality weapons from the other side of the world – by state or non-state actors successfully committing industrial espionage – thus at low expense using the other state's own technology against them.

On the other hand, a firearms and 3D printing expert interviewee believes the media hyperbole surrounding the phenomenon of 3D printing firearms is only fuelling public fear. Despite the technology advancing at an unprecedented rate, the laws of physics will always prevail, and fully functioning and effective firearms cannot just come off a 3D printer. Currently, 3D printed firearms still require standard ammunition, and although projectiles can be 3D printed (Kleinman, 2013), the manufacture of entire bullets still requires components to be assembled with gunpowder, which cannot currently be 3D printed (Berkowitz, 2018). The interviewee cites an episode of the science fiction television series "Lost in Space" in which a character 3D prints a fully functioning firearm. Even at the point in time the series is set, this is impossible claims the interviewee, even though many people, including some journalists (Sloat, 2018) believe it is already possible. This discrepancy between those who claim certain technological developments are "impossible", and those declaring the same technology is already viable, suggests immense uncertainty over current 3D printing capabilities, and profound disagreement exists over the fundamental possibilities of 3D printing.

Another interviewee involved in future force development believes that in 20 years' time fully functioning 3D printed firearms will become easier and cheaper to manufacture and more accessible to the general public. However, the interviewee believes there will still be limits owing to electricity supply constraints and facilities requirements. More sophisticated manufacturing techniques require a significant amount of power and are bulky, and thus require sufficiently large buildings for storage. The requirement to purchase metal powder for the higher-end machines will also make it far more difficult to keep production secret and out of sight of law enforcement. Despite more sophisticated and exotic materials becoming increasingly commercialized and available, procurement of sufficient quantities would signal potential criminal activity to law enforcement agencies.

Proposals for legislation change

3D printing as a decentralized technology presents new challenges for development of legislation and enforcement of criminal law. There are already a number of approaches currently being implemented to respond to 3D printed firearms in various jurisdictions, including the UK, the US, and Australia (Daly and Mann, 2018). These include: Criminalizing manufacture of 3D printed firearms, registration of 3D printers, and new offenses for the possession of 3D printed firearm design files.[16] Although some commentators, such as Walther (2014), believe current international legislation, such as the Arms Trade Treaty (2013) and the EU Council Common Position (2008), is robust enough to cover the risks associated with 3D printed firearms, there are some gaps that require filling. Jensen-Haxel (2012, p. 448) claims, "3D printers will render current firearm regulations obsolete by allowing individuals to easily produce firearms", whilst Walther (2014, p. 1438) believes regulating 3D printers may not be possible or desirable, adding it "would hamper economic and technological progress". Katie Curtis (2015) claims new legislation should focus on what constitutes a 3D printed firearm and regulate the firearm itself, rather than 3D printing. Walther (2014) suggests regulation of 3D printed firearms could be accomplished by changing the coding of 3D printers to make them unable to print a functional firearm. However, he further posits an argument against regulation to be the difficulty of programming software to understand what a firearm is, and differentiate it from a toy gun, for example. Furthermore, 3D printers would need to be constantly updated to include new designs that are not allowed to be printed. This would though be an impossible task without participation of the 3D printing industry and would need significant collaboration between them and regulators.

Curtis (2015) has developed a legislative solution tailored to the American context, as it maintains the freedom to manufacture firearms for personal use by incorporating 3D printed weapons into existing legislation while still ensuring the safety of the population. She claims the US federal government should have the power to regulate 3D printed firearms meeting the definition of weapons regulated by the US National Firearms Act, and although 3D printed firearms have yet to match this description, "it is important for the legislative branch to proactively control such products before this occurs" (Curtis, 2015, p. 93). The Model State Wiki Weapon Act[17] would focus on safety, and would regulate 3D printed firearms through regulation of the firearms themselves and of explosive materials such as gunpowder.[18] Due to the fact that it is impossible to 3D print gunpowder for ammunition, Berkowitz (2018) recommends additional background checks for ammunition, which will always need to be procured for 3D printed firearms.

Due to the many legitimate and benign uses of 3D printers, most interviewees consider regulation of 3D printers would be impractical. For people or companies involved in manufacturing, art projects, or hobbies, the idea of regulating the use of a 3D printer would be overly prescriptive. Additionally, due to the number of 3D printers already in circulation that have not so far been registered, it would be virtually impossible to begin the registration process for 3D printers that would record where they are being used, or what for. Many machines and even subcomponents are available from online marketplaces allowing people to make their own machines at home (Wohlers Associates, 2017), making registration even more difficult. Some believe raising awareness of the problem is the most logical step in preventing misuse (Fey, 2017), and ensuring commercial sales of 3D printers to civilians and their use are monitored to identify unusual activities that could point to weapons manufacture. The expert on emerging disruptive technologies agrees and believes regulation of 3D printers to prevent the manufacture of 3D printed firearms is in fact possible, although as part of a broader context involving civil society including steps to counter criminal associations that could facilitate illicit and undesirable use of 3D printers. Regulation is not a matter of prevention but a matter of responsible use, making sure people act in a civil manner and do not endanger other people or wilfully cause damage. Legality of 3D printed firearms should be based on precursor activities regarding the person who downloaded them and whether or not they are of interest to law enforcement.

Since producing 3D printed firearms requires downloading files from the internet, one way to regulate their use would be to better control the transfer of files across the web. Professor Potgieter claims there is no way to prevent this, as the vast and growing amount of digital content makes regulation almost impossible.[19] On the other hand, these files originated in the US, so transfer of them across borders to other countries would come under the purview of the relevant Customs Service. Furthermore, domestic courts could also block files. For example, in the US, a Seattle judge in July 2018 issued a restraining order to prevent an activist's plans to release 3D-printed gun designs online. Specifically, the order prevents the international transfer of the files, not the domestic transmission of the files (Lopez, 2018). The New Zealand Customs Service interviewee suggested one way for this to occur would be for the government to classify these files as "objectionable material", similar to child pornography or extremist material. This way the transfer of the files can be considered illegal and be prevented from crossing borders. However, information regarding the files would have to be registered as objectionable and constantly monitored to ensure any changes or any newly developed files would come under the legislation. Additionally, this would have to be used in conjunction with precursor events and associations to specifically target the regulation against those with the intent to use them. For example, people with criminal backgrounds or a history of violence would be considered more of a public risk than researchers or gun enthusiasts. To monitor all transfers of files without applying some form of rigor would be overly cumbersome. Finally, export control authorities and customs officers could be trained and educated to recognize dangerous weapons or shipments of illicit 3D printed firearms (Fey, 2017).

While globalization and the expanse of the internet mean more and more countries are becoming connected and interdependent on each other, some assert distribution of material on the internet can still be regulated. The emerging disruptive technologies expert likens this issue to recent developments in the US regarding right-wing and conspiratorial radio host, Alex Jones. Jones's website and radio show, Infowars, was recently blocked by major social media companies[20] for promoting violence and hate speech (Hern, 2018). The interviewee believes if something available on the internet can be considered dangerous, there is a precedent for it to be censored. However, this is only related to the overt availability of files and material, and there would be ways of bypassing this form of censorship. As mentioned earlier, Liberator files have been made available through torrent website The Pirate Bay (Ernesto, 2013), making them downloadable even though Defense Distributed has been shut down.

Most interviewees believe the only way to legislate against the use of 3D printed firearms would be to introduce sanctions or punishments against those in possession of 3D printed firearms or the files required to manufacture them. Alexander Gillespie[21] believes looking overseas for best practice is the most logical perspective (Radio New Zealand, 2018). New South Wales in Australia already has legislation in place to punish those in possession of 3D printed firearms (Millsaps, 2015), and it is considered the best approach to take. Research carried out by David Wilson and others indicates policy decisions regarding events such as these are primarily made on information and arguments that can be quantified (Wilson, et al, 2010); without understanding the impact of 3D printed firearm use, preparedness for such impact is limited, and organizations and policy makers will continue to behave as they have always acted. Given this research, determination of what the legislation would need to entail would be difficult without an event having taken place.

Again, specific individuals' intent to use 3D printed firearms, plus their previous criminal and firearms history, will need to be considered when creating and applying new legislation; those with a legitimate firearm licence and with no criminal background or history of violence may well be exempt. Implementation would be up to legislators within government and would need to consider the threat assessment of 3D printed firearms and current regulations around firearm licencing. It must be recognized, however, that the ability to take steps to address the threat of 3D printed firearms is likely to be context-dependent; for example, there are debates in the US over gun control laws, and powerful lobby groups, like the National Rifle Association (NRA), endorse the use of 3D printed firearms and can be expected to fight any attempts to ban them (Macbride, 2018).

Conclusion

The aim of this chapter was to determine the threat and implications of 3D printed firearms for international security and public safety. Additionally, it was to determine how legislation could be updated to counter or mitigate any potential threat. Most of the current literature expresses concerns that 3D printed firearms will challenge law enforcement. However, as no events have occurred from 3D printed firearms (they have not been used for criminal activity or to harm others), early interest in the issue from government agencies has not been sustained; ascertaining the true level of threat 3D printed firearms pose is difficult to quantify. Based on initial media coverage and government agencies' reaction to this phenomenon, there is little doubt that, if a murder were to be committed with a 3D printed firearm in the future, there would be new efforts to try to assess the "new" level of threat they pose. Questions would also be asked about what could have been done to prevent the violent act, potentially spurring legislative reform and energizing agency efforts to devise ways to counter the spread and proliferation of 3D printed firearms and the files required to print them.

At present, given 3D printers are still an emerging technology and there is little evidence about the capacity in which they are being used for firearm manufacture and proliferation, this is a speculative space, and the level of threat 3D printed firearms pose can only be theorized. Nevertheless, a security event from a 3D printed firearm is likely to have an effect on the rest of the world. Despite concerns expressed by law enforcement agencies and militaries across the world, very little investigation, policy change, or legislation to address them have occurred or been implemented. This is unsurprising as, in light of the research on extreme and violent events, changes to legislation or regulations are unlikely to occur until after an event has occurred. Given the publicity from the time the Liberator was first manufactured, once an event does occur, calls from the public, media, and governments from around the world will be expected to ask why something like this was not prevented and to change legislation to prevent this from happening again in the future.

By examining the current capabilities of 3D printed firearms and challenges to using them and the related threat they pose, this chapter has argued the risk of an extreme event occurring in the near future as a result of a 3D printed firearm is low. There are, however, reasons to remain vigilant; technological advancements in the 3D printing industry and the capabilities of 3D printed firearms do mean the risk is likely to increase in the coming years and decades. Even so, there are limitations on what is possible, even in the future, and the risk is likely to remain low. As a result, and due to the difficulty and impracticality of regulating 3D printing and/or the transfer of 3D printable files, changes to regulation and/or legislation are neither practicable nor necessary at this time. This does not mean nothing can be done, however. Having a plan in place such that changes can be made in a relatively short timeframe in response to an extreme event could appease concerns of the public, media, and other interested parties, taking into account the context of rapidly taking steps to address the threat of 3D printed firearms. As regulation of 3D printing as a whole is not practical, any plan would most likely include legislating for sanctions against those in possession of 3D printed firearms, or the files to manufacture them, and to reduce proliferation. Additional policy changes in customs administrations, including customs agreements between countries, might also be required, including classifying the 3D files as objectionable material, in order to prevent the transfer of files across borders via the internet.

Notes

1 Available from ATF website (https://search.usa.gov/search/news?affiliate=atf&channel=4843&query=liberator).
2 A proprietary plastic made for 3D printing by 3D Systems, a leading manufacturer and distributor of 3D printers.
3 Single-shot refers to the fact that each round has to be loaded before firing; the chamber only holds one round at a time.
4 Cody Wilson's online company that manufactured and released the Liberator.
5 A website blog dedicated to reporting the latest news and trends on internet file sharing.
6 An online index for digital content and media.
7 E-mail: Torrent Freak, dated May 28, 2018.
8 It should be noted that, in accessing the files, no current laws were broken.
9 E-mail: NZDF Legal, dated July 23, 2018.
10 As the machines differ, there may be a slight difference in actual cost.
11 E-mail: National Library of New Zealand – Team Leader Operations, dated June 27, 2018.
12 An associate professor in mechatronics at Massey University.
13 E-mail: Johan Potgieter, dated June 29, 2018.
14 Tendai Muswere's conviction was as a result of UK law on firearm manufacture, rather than specifically for 3D printing.
15 Multiple futures experts were interviewed.
16 Many existing laws regarding unlicensed manufacture and possession of firearms may cover 3D printed firearms. However, these do not necessarily cover possession or distribution of 3D printed firearm design files.
17 A name given to the proposed act by Katie Curtis herself and not an existing act.
18 The single constant material required for a functioning firearm.
19 E-mail: Johan Potgieter, dated June 29, 2018.
20 Facebook, Apple, YouTube, and Spotify.
21 Law professor from Waikato University.

References

Berkowitz, J (2018) 'Computer-aided destruction: regulating 3D-printed firearms without infringing on individual liberties'. *Berkeley Technology Law Journal*, 33(1), pp. 51–84.

Bureau of Alcohol, Tobacco, Firearms and Explosives (2013) *Q&As: 3-D printing of firearms*. November 13. https://www.atf.gov/resource-center/docs/111313-hq-3-d-printing-technology-firearmspdf/download, accessed June 22, 2018.

Council Common Position on the Rules Governing Control of Military Technology and Equipment (2008) *Official Journal of the European Union*, 335, pp. 99–103. https://eur-lex.europa.eu/legal-content/EN/TXT/?uri=CELEX%3A32008E0944, accessed April 22, 2018.

Curtis, K (2015) 'A wiki weapon solution: firearm regulation for the management of 3D printing in the American household', *Rutgers Computer & Technology Law Journal*, 41, pp. 74–107.

Daly, A and Mann, M (2018) *3D printing, policing and crime*. Brisbane: Queensland University of Technology.

Ernesto (2013) *Pirate bay takes over distribution of censored 3D printable gun*. May 10. https://torrentfreak.com/pirate-bay-takes-over-distribution-of-censored-3d-printable-gun-130510/, accessed May 27, 2018.

Fey, M (2017) *3D printing and international security: risks and challenges of an emerging technology*. Frankfurt: Peace Research Institute Frankfurt (PRIF).

Greenberg, A (2013a) *3D-printed gun's blueprints downloaded 100,000 times in two days (with some help from Kim Dotcom)*. May 8. https://www.forbes.com/sites/andygreenberg/2013/05/08/3d-printed-guns-blueprints-downloaded-100000-times-in-two-days-with-some-help-from-kim-dotcom/#6355415110b8, accessed May 27, 2018.

Greenberg, A (2013b) *Forbes: 3D-printed gun stands up to federal agents' testfring – except when it explodes (video)*. November 14. http://www.forbes.com/sites/andygreenberg/2013/11/14/3d-printed-gun-stands-up-to-federal-agents-testfiring-except-when-it-explodes-video/, accessed October 8, 2018.

Greig, A (2013) *Daily mail: world's first fully 3-D printed gun is tested by ATF agents... and blows up in their faces*. November 15. http://www.dailymail.co.uk/news/article-2507654/The-worlds-fully-3-D-printed-gun-tested-Feds-blows-faces.html, accessed December 13, 2017.

Hern, A (2018) *The Guardian: facebook, apple, youtube and spotify ban infowars' Alex Jones*. August 6. https://www.theguardian.com/technology/2018/aug/06/apple-removes-podcasts-infowars-alex-jones, accessed September 30, 2018.

International Organization for Standardization (2015) *Additive manufacturing—general principles—terminology*. ISO/ASTM 52900:2015(E).

Jensen-Haxel, P (2012) '3D printers, obsolete firearm supply controls, and the right to build self-defense weapons under heller', *Golden Gate University Law Review*, 42(3), pp. 447–495.

Johnston, T, Smith, T and Irwin, J (2018) *Additive manufacturing in 2040: powerful enabler – disruptive threat*. Santa Monica: RAND.

Kan, M (2016) *PC world: hacked* 3D *printers could commit industrial sabotage*. July 12. https://www.pcworld.com/article/3094839/hacked-3d-printers-could-commit-industrial-sabotage.html, accessed July 29, 2018.

Kleinman, A (2013) *Huffington post: 3D-printed bullets exist, and they're terrifyingly easy to make*. May 23. http://www.huffingtonpost.ca/entry/3d-printed-bullets_n_3322370, accessed October 8, 2017.

Lopez, G (2018) *Vox: the battle to stop 3D-printed guns, explained*. August 29. https://www.vox.com/2018/7/31/17634558/3d-printed-guns-trump-cody-wilson-defcad, accessed October 2, 2019.

Lott, C (2015) *Additive manufacturing in the NZDF*. New Zealand Defence Force Joint Defence Services.

Lupan, V (2015) *Programme of action on small arms and light weapons second open-ended meeting of governmental experts – chair's summary*. New York: United Nations Office for Disarmament Affairs.

Macbride, E (2018) *Forbes: The NRA is tacitly endorsing 3D printed guns*. July 31. https://www.forbes.com/sites/elizabethmacbride/2018/07/31/when-it-comes-to-the-safety-of-3d-printed-guns-the-nra-is-conspicuously-silent/#4eee88ef622b, accessed October 2, 2019.

McCutcheon, C (2014) 'Deeper than a paper cut: is it possible to regulate three-dimensionally printed weapons or will federal gun laws be obsolete before the ink has dried?', *Journal of Law Technology & Policy*, 2014(1), pp. 219–250.

Millsaps, B (2015) *New South Wales, Australia: parliament passes law banning possession of 3D files for guns*. November 20. https://3dprint.com/106940/australia-ban-3d-files-guns/, accessed 8 October, 2017.

Murphy, S and Myers, R (2013) *Mail on sunday: how mail on sunday "printed" first plastic gun in UK using* a *3D printer – and then took it on board eurostar without being stopped in security scandal*. May 12. http://www.dailymail.co.uk/news/article-2323158/How-Mail-On-Sunday-printedplastic-gun-UK–took-board-Eurostar-stopped-security-scandal.html, accessed December 13, 2017.

National Intelligence Centre (2013) *The use of 3D printing technology in criminal activities*. March 15.

National Library of New Zealand (n.d.) *3D printing*. https://natlib.govt.nz/visiting/wellington/3d-printing, accessed June 26, 2018.

News Inquiry 10 (2013) *A gun fires a few meters from the Prime Minister*. July 7. https://www.10.tv/news/59590, accessed October 8, 2018.

Noor, P (2019) *The Guardian: London student convicted for making gun using 3D printer*. June 19. https://www.theguardian.com/world/2019/jun/19/london-student-convicted-for-making-gun-using-3d-printer-tendai-muswere, accessed September 1, 2019.

Petrick, I and Simpson, T (2013) '3D printing disrupts manufacturing', *Research-Technology Management*, November–December, pp. 1–6.

Radio New Zealand (2018) *NZ police monitoring developments around 3D firearms*. August 2. https://www.radionz.co.nz/national/programmes/checkpoint/audio/2018656403/nz-police-monitoring-developments-around-3d-firearms, accessed August 3, 2018.

Reilly, R (2013) *Huffington post: feds printed their own* 3D *gun and it literally blew up in their faces*. November 13. https://www.huffingtonpost.com/2013/11/13/3d-guns-atf_n_4269303.html, accessed October 7, 2018.

Sloat, S (2018) *Inverse:* 3D-*printed gun in "lost in space" isn't futuristic, it's already here*. April 19. https://www.inverse.com/article/43896-lost-in-space-3d-printed-gun, accessed July 27, 2018.

Venuto, D (2018) *New Zealand Herald: are 3D-printed guns a threat to NZ?* February 16. http://www.nzherald.co.nz/business/news/article.cfm?c_id=3&objectid=11995872, accessed March 4, 2018.

Walther, G (2014) 'Printing insecurity? The security implications of 3D-printing of weapons', *Science and Engineering Ethics*, 21, pp. 1435–1445.

Wilson, C (2016) *Come and take it: the gun printer's guide to thinking free*. New York: Gallery Books.

Wilson, D, Branicki, L, Sullivan-Taylor, B and Wilson, A (2010) 'Extreme events, organizations and the politics of strategic decision making', *Accounting, Auditing and Accountability Journal*, 23(5), pp. 699–721.

Wohlers Associates (2017) *Wohlers report: 3D printing and additive manufacturing state of the industry annual worldwide progress report*. Colorado: Wohlers Associates Inc.

12 Deepfakes and synthetic media

Curtis Barnes and Tom Barraclough

Introduction

This chapter introduces and defines synthetic media, a class that poses rapidly increasing risk to international security. It explains in general terms the connection of synthetic media to some existing technologies, particularly paradigms in artificial intelligence. It also provides emerged and emerging examples. The reader will then be introduced to an increasingly popular example of synthetic media, commonly called "deepfakes". With this in mind, the chapter then outlines the initial difficulties of regulating synthetic media technologies, while introducing the reader to the fundamental problems that synthetic media poses to international security. It discusses what work is being done to combat the threats of synthetic media, including legal and technical responses.

Defining synthetic media with examples

Increasingly realistic synthetic media artifacts can now be produced faster, cheaper, and with less technical skill (Barnes & Barraclough, 2019). This is in large part due to advances in the field of artificial intelligence (AI), as well as developers and programmers finding new applications for preexisting paradigms, like machine learning, neural networks, and facial recognition algorithms. Produced by these means, synthetic media are novel digital audiovisual artifacts synthesized from newly generated or preexisting digital information, such that the artifact which is produced is substantially different from the original data inputs, but often similar enough as to appear authentic to a naïve viewer (Maras & Alexandrou, 2019).

Furthermore, commercialization of these technologies is significantly reducing barriers to their use. Where formerly it would have taken significant technical skill and access to greater tools, resources, and time, everyday consumers can now generate realistic synthetic media artifacts with little effort using general consumer-level technologies, like smart phones and publicly available applications. The evolution and rapid speed of improvement in AI and consumer technologies will, likewise, lead to vastly improved digital audiovisual information over time.

We categorize synthetic media as a broad class of artifacts and technologies sharing common characteristics (Barnes & Barraclough, 2019). This class includes artifacts like deepfakes, which are one example of synthetic media that commentators already believe to be having a destabilizing effect on international security (Chesney & Citron, 2019). Synthetic media also includes a range of ordinary communications technologies that are designed and utilized for everyday civilian purposes, including augmented reality, virtual reality, computational photography, photo editing, and voice synthesis technologies.

The realistic quality of synthetic media artifacts will almost certainly continue to improve. So too will their accessibility to non-technical users. The main driver of this trend is seemingly limitless consumer and industrial demand. This demand is largely focused on benign and beneficial applications for synthetic media technologies, including entertainment, education, communication, customer service, and healthcare. Unavoidably, improvements in benign applications also lead to improvement in harmful uses too, because both tend to share the same architectural bases in computer science. Developing an effective and proportionate response to these technologies is challenging. The reasons for this are threefold.

1. It is debatable whether emerging synthetic media technologies are such a novel development from traditional methods of deception, misinformation, and disinformation that they truly justify the risks of intervention.
2. Identifying the correct regulatory targets for technical or legal intervention is difficult because the applications of the technologies are predominantly civilian and benign.
3. Disproportionate intervention in the use of synthetic media technologies risks undermining some of the pillars that are thought to support international security, such as effective democratic systems, free political engagement, and the free exchange of information. As a result, intervention may harm the very thing intended to be preserved.

Each of these points must be subject to ongoing debate in light of historical events, as well as foreseeable technological capabilities. To fully understand what emerging synthetic media technologies are capable of, those artifacts must be seen and heard. In lieu, we offer some simple descriptions of what emerged and emerging synthetic media technologies can do.

Voice synthesis

Digital data from voice recordings can be used to produce synthetic voices. Most importantly, these voices can be used to produce sounds, words, and phrases that were never actually captured on the original recordings. As such, synthetic voices are not the product of recordings being cut and stitched together. Rather, the synthetic voice is a digital asset that can be used and re-used to say new things. Because most modern audio information is recorded as digital data, these data can be analyzed and processed using AI techniques, like neural networks. Publicly available services are emerging which allow users to create their own synthetic voices by recording themselves saying random phrases (Lyrebird, 2019). Alternatively, synthetic voices can be made from recordings of other people, particularly where an abundance of data is publicly available. The quality of these voices is improving, while the amount of training data required is getting less and less. A market for synthetic voices has also resulted in commercial services offering pre-built voices in a variety of dialects and accents and licencing opportunities for well-known individuals (Replica Studios, 2019).

Facial synthesis

By very similar processes, human faces can also be synthesized. In other words, digital data representing one person's face can be manipulated by AI to look like another person's face. Facial synthesis is most often produced from preexisting visual information that is publicly available on the internet. However, increasingly, this process can also be performed in real time (i.e. data manipulation may take place contemporaneously with capture and display) (Thies et al, 2018). This means faces may be changed, altered, or swapped even in the course of a live video broadcast, within a person-to-person context or via mass media.

In terms of international security, there are already examples of this causing political confusion, including a case where members of the political opposition claimed that a video representing President Bongo of Gabon giving a public address was manufactured with facial synthesis. One week later, military figures attempted an unsuccessful coup, citing the video as justification (Harwell, 2019). Other examples show the potential for misuse, with developers giving real-time demonstrations that depict performers as figures like former US Presidents George W. Bush and Barrack Obama, President Vladimir Putin of the Russian Federation, Chancellor Angela Merkel of Germany, and others.

Moreover, several publicly accessible websites offer images of synthesized faces that present as though they were real people. For example, the website This Person Does Not Exist generates a novel and realistic image of a face each time the page is refreshed. (This Person Does Not Exist, 2018.) Some scholars believe state-level political disinformation, like that employed by Russia in and around the 2016 US Presidential election, strategically disseminated disinformation through agents presenting as ordinary individuals, rather than through the use of inauthentic news agencies as was previously typical (Saunders, 2018). This tactic acted upon Americans' historically low trust in mainstream media during the 2016 US presidential election (Swift, 2016). Increasingly, the synthetically generated face will be a tool that increases the persuasiveness of fake online profiles.

Unique synthesized faces that can be easily accessed and mass produced offer another layer of depth that may make an inauthentic social media profile appear authentic. Some scholars perceive such social media practices to constitute information warfare (Prier, 2017). Moreover, there are already cases where these same facial images have been used for political espionage (Satter, 2019).

Body animation

Technologies can make people appear to do things that they may never have actually done. By body animation we mean taking a static visual representation of a person and making it dynamic (i.e. making it move). In other words, a person who is standing still in one visual record can be used to create a record of that person moving and doing things that they never actually did. Demonstration videos show bad dancers represented performing the movements of skilled dancers, from ballerinas to Bruno Mars – what researchers call "do as I do" motion transfer (Chan et al, 2019). Movement may be transferred from a video of one person to the image of another, so that a performer can perform the action, while another person appears to be doing it.

Augmented reality and filters

Essentially, augmented reality is the addition of new digital assets to digital videos, often in real time. For example, consumer applications like Snapchat, Instagram, and Facebook Messenger allow a user to record themselves while adding objects to the recording that are not actually being captured by the camera on the user's recording device. Most often this is done for entertainment purposes, allowing people to represent themselves wearing things like glasses, beards, and different hair styles – all of which are designed to be deliberately unrealistic or cartoonish. However, assets added by augmented reality can be increasingly realistic, making a viewer believe that something present in the video must have been captured in reality by a recording device when, in fact, it was added to the visual record at a later stage. Augmented reality can also change the apparent conditions of a video record (e.g. make it appear like nighttime, though the video is being captured during the daytime).

The risk to international security may increase as augmented reality becomes more photorealistic and easier to use. The addition of certain objects in a video record may have far-reaching political ramifications, particularly where that record is conveyed and consumed rapidly through mass media and the internet. Of course, classical Photoshop examples of objects being added to images are by no means a novel problem in international security studies. However, augmented reality is an advancement on these by offering higher photorealistic quality as well as the capacity to add objects to videos instantly and in real time.

Rapid assisted image editing

We define rapid assisted image editing as the use of AI paradigms to alter visual records. For instance, a user can remove an object from a photograph, while machine learning algorithms then analyze the content photo and fill the emptied space with strikingly appropriate infill using freely available apps such as Adobe's Photoshop Fix for iOS.

Bearing in mind that the role of image editing in international security long predates even the Photoshop-era, the effect of advancements in the use of AI for digital editing cannot be overlooked. This sort of technology is now widely available for consumer download, often available as standard on new commercial smartphones, like the Apple iPhone 11. Historical examples of people being removed from photographs, as in the famously doctored Soviet images, like the removal of Nikolai Yezhov from F. Kislov's original photograph (Kislov, 1937), show the potential of these kinds of technology to be used for political purposes (King, 2014). They could also be used to manipulate audiovisual records as a means of insulating states and state actors from criticism for their actions captured on camera in the physical world, or for generating a basis for diplomatic confrontation between states, or removal of key personnel from sensitive inter-state relations.

Crowd animation

AI allows for realistic animation of large crowds of digital agents. In other words, programmers can realistically make it look like there was a crowd of people present somewhere, where there really was not. Rather than manually programming and animating each individual agent, as was previously the norm, the behaviors and actions of the agents can be guided by AI and algorithmic decision-making, so that each agent has its own relatively unique behaviors.

From the perspective of international security and geopolitics, we can imagine scenarios where this technique may be used to misinform or for counterintelligence purposes. For example, a modernization of ghost army tactical deception, like that which was implemented by the US during World War II (The Ghost Army, 2013). Alternatively, the tool might be used to enhance perception of the degree of public support for a political figure, by appearing to show much larger supportive audiences at events than what were present in reality.

Digital assets

Digital asset is a general term; however, in the context of synthetic media, we mean digital data files that can be used and reused to generate moldable audiovisual representations, not simply linear videos or audio files. For example, a recent high-profile film, *Gemini Man*, used a variety of different synthetic media technologies, traditional computer-generated effects, and AI techniques to produce a digital representation of the actor Will Smith at age twenty-three (Gemini Man, 2019). Preexisting footage of a younger Will Smith served as training data for training the AI algorithms, producing an asset which "doesn't take breaks or require the services of hair and makeup. And he doesn't need a trailer, since he lives on a hard drive" (King, 2019). This digital asset then played opposite the real Will Smith. The existence of this asset theoretically facilitates limitless future performances that appear as if done by Will Smith himself, with the asset able to be reproduced and animated in perpetuity.

From an international security perspective, highly realistic and persistent digital assets like this might have a key role in future disinformation campaigns. It is not uncommon for states to try and obfuscate the truth when a state official dies, falls ill, disappears, or is captured. By using digital assets, states could present a state agent as if he or she is perfectly well and performing normal activities. Another use for such assets is as part of state propaganda schemes. For example, Chinese news agency Xinhua has created AI-powered digital assets of two of its well-known anchors, providing it with persistent and tireless human-presenting assets as a mechanism to present state information (Xinhua News Agency, 2019).

Virtual reality

In essence, virtual reality is a technology for building digital spaces that present like physical environments. People use specialized goggles and controllers tracked by environmental sensors, so that their movements in the physical world are represented within a virtual world. This virtual environment can also contain other simulated agents that can be interacted with.

Virtual reality has enormous potential as a training simulation tool which allows the user to emulate environments, conditions, and situations that are otherwise difficult to recreate. It is being used to replace costly and spatially inefficient physical environments for military training (Saunders, 2018). Its effect on lowering cost barriers also increases the potential that it may be used as a training tool by insurgent groups. For example, a physical environment in the real world can be virtually simulated, like the inside of a building and its surrounding area. This would allow a person to train and become familiar with the virtual space before executing an action in the physical space, without ever actually going there – all at relatively little cost and from remote, mostly untraceable locations.

Natural language processing and text generation

Outside of audiovisual information, AI is also being applied to the synthesis of language. By using machine learning and natural language processing, computer systems can create convincing passages of text with little human guidance. This information is typically very well written, with few semantic or syntactic clues that it has been generated by a computer system. For example, Google's OpenAI project has "trained a large-scale unsupervised language model which generated coherent paragraphs of text ... and performs rudimentary reading comprehension, machine translation, question answering, and summarization" (OpenAI, 2019). Technology like this enhances the potential for large-scale misinformation or information saturation through written mediums and the expansion of "fake news" with its upwards of 32 attempted definitions (Jensen, 2018). It also improves the ability of computer systems to extract more meaningful information from data at scale by improving their comprehensive capacities and understanding of contextual information.

Understanding deepfakes

For the most part, "deepfake" is a term that refers to video records which incorporate facial synthesis technology. The etymology of the term dates to approximately

2016/2017, and relates to a user on the website Reddit operating under the pseudonym "Deepfake" while publishing pornographic videos produced using deep learning AI paradigms (Merrefield, 2019). These videos represented non-participant women as if they were participating in the original pornographic scene. "The typical cases involve porn movies in which the faces of famous actresses like Gal Gadot or Scarlett Johansson (this is regularly about women's faces) are used to replace the original faces" (Floridi, 2018). This remains their predominant use (Deeptrace Labs, 2019). In the most general sense, the word *deep* in deepfake means a deeper and more complex neural network. The word *fake* implies not only manipulation but also inauthenticity.

As such, deepfake is not a technical term. It is unstable and unreliable for regulatory purposes, in part because it infers that deepfakes are delineable and distinct from other synthetic media in the technologies and resources used to create them (Barnes & Barraclough, 2019). Rather, the technologies that underpin deepfake videos are often the same as those which underpin other artifacts, including facial recognition algorithms, deep learning, neural networks and computational photography present in ordinary smartphones. The shock value may be extraordinary, but the technologies are not.

Furthermore, whether an artifact is labeled a deepfake tends to depend not on the technical characteristics of the particular artifact in question but the content and the context in which it is presented. There may be little technical difference between a synthetic media artifact that is used to defraud somebody and one that is used to entertain a theater full of people. In spite of this, in reporting, the former is frequently referred to as a deepfake, while the latter is not. Another factor making the label "deepfake" unsuitable for regulatory use is its consistent misreporting and expansion. Put simply, things that are not deepfakes and which do not use emerging technologies at all, are increasingly being referred to as if they did. This includes audiovisual artifacts that are manipulated using conventional editing techniques, like the recent videos of Speaker Nancy Pelosi and journalist Jim Acosta – sometimes imprecisely referred to as "shallowfakes" by contrast (Johnson, 2019).

As such, a case-by-case basis analysis is necessary. It means focusing on the use of the synthetic media artifact, what it actually appears to depict, and what harm it causes. Necessarily, this analytical process must take place after the fact, at which point the synthetic media artifact may already have been disseminated and caused significant disruption to international security. Because such disruption is undesirable, many attempts to mitigate the harms of synthetic media focus on front-loaded detection or legal prohibition (i.e. stopping manipulated videos at the point of creation or distribution), rather than after they have been consumed. In the case of technical intervention, the question is whether harmful audiovisual content can be accurately sorted from benign audiovisual content, given both are digitally manipulated, potentially by the same techniques, thus minimizing the undesired detection and blocking of false positives. In the case of legal intervention, the question is whether legislation can be defined in such a way as to capture only harmful audiovisual content, without infringing unduly on civil rights and liberties.

High-level implications

Demand for detection and prevention services for synthetic media continues to grow as the public and policy makers become incrementally more aware of the technologies' deceptive capacities. As intimated above, this dialogue involves both legal regulators and forensic experts. At minimum, synthetic media technologies have four harmful implications:

1. That highly realistic audiovisual information can be used to seriously deceive at both the interpersonal and mass media scale
2. That the existence of these technologies undermines latent trust and reliability in digital audiovisual media
3. That degradation of trust in digital audiovisual media disproportionately harms authentic persons and benefits inauthentic persons, the so-called "liar's dividend" (Chesney & Citron, 2019)
4. That the human agents or agencies behind the deception may be difficult to identify and assign culpability

From the perspective of international security, these implications form the theoretical justification for any given method of intervention, whether legal or technical or both. We elaborate on several of these implications below.

Increasing capacity to deceive

By design, most synthetic media is deceptive. By deceptive, we mean that the artifact when viewed or heard makes it seem that something has happened when, in reality, it has not. This relatively simple concept is the common thread of all synthetic media but is complicated by the fact that *all* digital audiovisual information is inherently manipulated by digital processes, a fact that is often overlooked or misunderstood. For instance, simple photographic information produced by a digital camera is not merely captured; it is also constructed. The camera captures photon energy and converts this to digital data, often further altering this digital data to produce a better visual output for the consumer. The data is then interpreted by a computer back into an imperfect representation of what was originally captured, reconstructed, and output as photon energy by a display technology. Subsequent compressions and decompressions each time the digital file is accessed can cause further alterations to the data. Thus, data manipulation is inherent and present at every junction of this process, frequently deceiving the person that views the image as to the degree to which the visual record they are seeing represents an unaltered external reality.

However, user deception exists adjacent to data manipulation and often occurs to different degrees and in different contexts. Manipulation will be present even when deception is not. Frequently, the degree of deception is known, desired, and beneficial. Equally frequently, deception is undesired and harmful. The degree of manipulation may be unknown, but nevertheless beneficial, thus deception is acceptable. At the same time, the degree of manipulation may be known, but undesired, thus the deception is less acceptable. Context is immensely important in any given analysis.

Furthermore, the sheer volume of deceptive information is increasing, because the logical nexus of improved ease of use, quality, and availability is a significant increase in the amount of synthetic media that may be created, traded, and consumed. Indications exist to confirm this is happening for synthetic media. For example, researchers have found a greater than 84% increase in deepfake videos online within the last year (Deeptrace Labs, 2019). At the same time, quantitative analysis is hardly needed to see that the usage and quality of synthetic media is increasing: Augmented reality in social media applications, such as Apple's Animoji, is now ubiquitous; cinema blockbusters are using and improving the technologies that underpin deepfakes; one of the most recognisable people in the world has been reproduced as a persistent digital asset for repeated animation; and the latest smartphones are now shipped equipped with the sort of image and video editing capacities that would vastly outstrip what was previously possible unless the user was a well-resourced technical maven.

Concomitantly, the availability of data from which to produce synthetic media artifacts has drastically increased. Huge quantities of digital information are publicly available on internet platforms for video streaming and image caching. This means that even ordinary people are somewhat susceptible to being misrepresented. More importantly in the context of international security, this ample availability of data renders high-profile public officials and elected representatives extremely vulnerable because of the abundance of photos, videos, and audio depicting them that is available via the internet. These individuals regularly appear in public, on the radio, on television, and in podcasts. When recorded, the information from these appearances can be used to train synthetic media technologies that use deep learning and neural networks, and, as a general rule, the greater the training data in quantity and quality, the better the synthetic media artifact that can be built.

Undermining trust and reliability

The secondary effect of these technologies concerns the undermining of trust and reliability. By their mere existence they cast a shadow on the veracity of any given audiovisual record. This has at least two distinct effects: First, it generates plausible deniability for any person imputed in an activity ostensibly captured on an audiovisual device; and second, it damages the value of digital audio and visual mediums as a means of documenting fact (Gregory, 2019). By documenting fact, we mean that the audiovisual artifact evidences an event that actually took place somewhere in the physical world.

There are a number of possible impacts this may have on international security. On the one hand, there are numerous historical examples where audiovisual records have had an influence on international security, domestic politics, and global diplomacy. Consider, for instance, the purported photographic evidence of missile facilities in Cuba, Iraq, or North Korea and the relevance of these as catalysts for military intervention. Alternatively, consider the significance of audio records during the Watergate scandal or the recent telephone conversation between President Trump of the US and President Zelensky of Ukraine (Zurcher, 2019); the diplomatic fallout arising from visual evidence of atrocities by US soldiers at Abu Ghraib (Higham & Stephens, 2004); the murder of civilians by soldiers in Cameroon (Cameroon: Anatomy of a Killing, 2018); and the audio documentation of the killing of journalist Jamal Kashoggi inside the Saudi Consulate in Turkey (Corbin, 2019).

These situations might have played out very differently if those involved could credibly claim that the records were undetectably falsified – or even that such falsification was merely an omnipresent threat. Equally, states or rogue actors might utilize inauthentic synthetic media artifacts to proactively attack political adversaries, perhaps with no reliable technical method for proving the truth or falsity of the artifacts in question.

On the other hand, the true effects of a general loss of trust in the audiovisual medium is more difficult to encapsulate, given that audiovisual documentation is perceived to be one of the strongest tools available for revealing to the world the activities of states and state agents. These revelations may have untold effects on domestic politics and diplomatic relations, and widespread loss of faith in these revelations has the potential to undermine or enhance international security. With regards to the latter, it is necessary to consider how a lesser degree of faith in audiovisual records might actually enhance international security in certain scenarios. For instance, where the presence or absence of certain objects in photos, like missile sites and components, are posited as justification for a military intervention. A person abreast of developments in synthetic media technologies must be more sceptical as to the evidential value of such artifacts, and subsequently would require further information before assenting to any drastic action. This would prove prudent if the photos were, in fact, misleading.

Problems of agency and accountability

Several factors make it difficult to identify who is responsible for a synthetic media artifact that has caused a risk to international security. This could mean bad actors may be undeterred by the threat of being identified and held accountable. Moreover, it is not clear at this point what law would actually apply to the creation, dissemination, and consumption of synthetic media in any given case. For instance, consider the following scenario, which typically inspires significant concern: A deepfake-style video that is realistic and beyond forensic detection shows a state agent engaged in behavior that is highly damaging to the person's reputation and the reputation of the state they represent. There may be very few ways to definitively prove that the video is untrue, and the best anyone can say is that it plausibly may be false. In this situation, it is unclear exactly what laws may apply and how they will be enforced, particularly given the interjurisdictional nature of the internet (Barnes & Barraclough, 2019).

Ubiquitous access to the technologies that produce these kinds of synthetic media artifacts serves to further complicate the scenario. On the one hand, it increases the possibility that even a civilian or rogue agent could be the source. On the other hand, this fact makes it easier for states to conceal their activities and avoid accountability when they are complicit in the creation and strategic use of synthetic media technologies. It is already widely believed that states are using networks of small groups and individuals to conduct mass internet disinformation campaigns to strategically influence and destabilize the domestic politics of political adversaries (Jensen, 2018). This includes the use of falsified, photoshopped images. We believe the use of artifacts like deepfakes is simply a logical progression.

Another complicating factor is that the data from which synthetic media is trained may be drawn from many sources. For example, a dataset provided to a facial synthesis algorithm may contain frames from dozens of different videos and thousands of different photographs. These data are likely to have been created, been made public by, and belong to an array of different legal persons. The question then becomes: To what degree are each of these legal persons accountable for the use of their information to create a harmful synthetic media artifact? Many may be unaware of how their information has been used or that what it has been used for is feasible at all.

Forensic detection and political reactions

Policy concern around synthetic media has rapidly increased as the capabilities of technology have come into the public consciousness. Despite this, in our observation, many members of the public remain largely unaware of the capabilities of synthetic media technologies. They also tend to misunderstand the connection between harmful synthetic media and the ordinary technologies that they use every day.

As an indication of seriousness, it is worth noting that the US Defense Advanced Research Projects Agency (DARPA) has received approximately US$60 million to develop technological countermeasures to manipulated digital audiovisual information, like deepfakes. The Media Forensics (MediFor) project aims to "automatically detect manipulations, provide detailed information about how these manipulations were performed, and reason about the overall integrity of visual media to facilitate decisions regarding the use of any questionable image or video" (Turek, 2018). MediFor anticipates the risk that manipulated information can be used for "adversarial purposes, such as propaganda or misinformation campaigns" (ibid). Outside of DARPA, there are a number of private enterprises pursuing detection and forensic mechanisms for synthetic media. For example, some of the same researchers who are generating technologies of facial synthesis have also begun work on opposing detection technologies (Rossler et al, 2019). They state that:

> The rapid progress in synthetic image generation and manipulation has now come to a point where it raises significant concerns on the implication on the

> society. At best, this leads to a loss of trust in digital content, but it might even cause further harm by spreading false information and the creation of fake news.
>
> (Rossler, et al, 2019)

Synthetic media and, particularly, deepfakes have also inspired a number of political responses. In September, 2018, a bipartisan trio of Democratic and Republican Members of Congress wrote a letter to the Director of National Intelligence, asking for deepfake-style technologies to be investigated (Schiff et al, 2018). A month later, Senator Ben Sasse published an opinion editorial in *The Washington Post*:

> I spoke recently with one of the most senior US intelligence officials, who told me that many leaders in his community think we're on the verge of a deepfakes "perfect storm". First, this new technology is staggering in its disruptive potential yet relatively simple and cheap to produce. Second, our enemies are eager to undermine us. With the collapse of the Russian economy, Putin is trying to maintain unity at home by finding a common enemy abroad. He has little to lose and lots to gain – it's far easier to weaken US domestic support for NATO than to actually fight NATO head-on. Russia hasn't mastered these information operations yet, but China is running scout-team offense behind every play. China will eventually be incredibly good at this, and we are not ready.
>
> (Sasse, 2018)

Subsequently, US lawmakers across several domestic jurisdictions have proposed new laws to deal with the perceived threat of emerging synthetic media technologies. Some of this legislation has been directed toward the preservation of individual civil and political rights which are deemed to be under threat by the deceptive capacities of these new technologies, rather than the issue of domestic and international security. However, on June 13, 2019, the US House Intelligence Committee held a hearing on the National Security Challenges of Artificial Intelligence, Manipulated Media, and Deepfakes (US House of Representatives, 2019). Subsequently, the US House of Representatives introduced S.2065 – Deepfake Report Act of 2019, thus substantiating a growing sense of Congressional appreciation for the seriousness of deepfakes and their potential impacts on security. The purpose of the Act is "to require the Secretary of Homeland Security to publish an annual report on the use of deepfake technology, and for other purposes", including the underlying technologies, their evolution, how they can be used to create forgeries, the content of those forgeries, technical countermeasures, and so on. Most notably, the Act provided that each report should include:

(1) An assessment of how foreign governments, and the proxies and networks thereof, use, or could use, digital content forgeries to harm national security
(2) An assessment of how non-governmental entities in the US use, or could use, digital content forgeries (Deepfake Report Act, 2019)

Further considerations

The uses of audiovisual information for strategic misinformation, disinformation, obfuscation, and propaganda are nothing new to international security. While many commentators have become doomsayers for the end of truth, those working in international security require actionable understanding: What are the technologies? What can they do? Will they develop further? Can they be detected? Can harmful use be deterred?

The answers to several of these questions are all but known. The technologies can and will produce audiovisual artifacts that make it look like something happened, when it did not, and will be persuasive enough that anyone but the most sceptical individual will be unable to tell how far the artifact has been manipulated. Using existing technologies, the world's top digital forensic experts may be unable to tell if they are authentic or not. They will be disseminated and consumed at scale through the internet, and the means to create them will be widely accessible. If nation-states, along with all their resourcing, began producing and strategically utilizing the emerging technologies of synthetic media, they would be able to create misleading artifacts of a quality far beyond what has previously been possible.

With regard to forensic detection, matters are much less certain. Cumulatively, significant work is ongoing with the goal of rendering harmful synthetic media redundant as a strategic risk. Whether these programs will succeed is uncertain. It appears unlikely in the immediate future.

Beyond this, policy discussion must continue as to how to balance national and international security considerations against free speech protections and desirable market activity. This includes a continuation of the dialogue concerning the role of social media platforms as content hosts. Certainly, with a number of geopolitically significant elections looming, global preparedness for the growing prevalence of synthetic media is likely to be tested.

References

Barnes, C and Barraclough, T (2019) *Perception inception: preparing for deepfakes and the synthetic media of tomorrow*. Auckland: New Zealand Law Foundation.

Cameroon: Anatomy of a Killing (2018) [Film] Directed by Daniel Adamson, Aliaume Leroy. United Kingdom: BBC Africa Eye.

Chan, C, Ginosar, S, Zhou, T and Efros, AA (27 October 2019) 'Everybody dance now', in *International conference on computer vision*, Seoul, Korea (South).

Chesney, R and Citron, DK (2019) 'Deep fakes: a looming challenge for privacy, democracy, and national security', *California Law Review*, 107, pp. 1753–1819 at 1758.

Cole, S (2017) *AI-assisted fake porn is here and we're all fucked*. http://www.vice.com, accessed May 5, 2020.

Corbin, J (2019) *The secret tapes of Jamal Khashoggi's murder*. https://www.bbc.com, accessed May 5, 2020.

Deeptrace Labs (2019) *Mapping the deepfake landscape*. Amsterdam: Deeptrace Labs.

Floridi, L (2018) 'Artificial intelligence, deepfakes and a future of ectypes', *Philosophy & Technology*, 31(3), pp. 317–321.

Gemini Man (2019) Directed by Ang Lee. United States: Skydance Media.

The Ghost Army (2013) Directed by Rick Beyer. United States: Rick Beyer.

Gregory, S (2019) *Deepfakes and synthetic media: updated survey of solutions against malicious usages*. https://blog.witness.org/, accessed May 5, 2020.

Harwell, D (2019) *Top AI researchers on "deepfake videos": "We are outgunned"*. https://www.washingtonpost.com/, accessed May 5, 2020.

Higham, S and Stephens, J (2004) *New details of prison abuse emerge*. http://www.washingtonpost.com, accessed May 5, 2020.

Jensen, M (2018) 'Russian trolls and fake news: information or identity logics?' *Journal of International Affairs*, 71(1), pp. 115–124.

Johnson, B 2019) *Deepfakes are solvable—but don't forget that "Shallowfakes" are already pervasive*. https://www.technologyreview.com/, accessed May 5, 2020.

King, D (2014) The *commissar vanishes: the falsification of photographs and art in Stalin's Russia new edition*. London: Francis Boutle Publishers.

King, D (2019) *The game-changing tech behind gemini man's "young" will smith*. https://www.wired.com/, accessed May 5, 2020.

Kislov, F (1937) *Kliment Voroshilov, Vyacheslav Molotov, Stalin and Nikolai Yezhov walking along the banks of the Moscow-Volga Canal*. Tate.

Lyrebird AI. (2019) *Using artificial intelligence to enable creative expression*. https://www.descript.com/lyrebird-ai/, accessed May 5, 2020.

Maras, M-H and Alexandrou, A (2019) 'Determining authenticity of video evidence in the age of artificial intelligence and in the wake of deepfake videos', *International Journal of Evidence & Proof*, 23(3), pp. 255–262.

Merrefield, C (2019) *Deepfake technology is changing fast—use these 5 resources to keep Ip*. https://journalistsresource.org/, accessed May 5, 2020.

OpenAI. (2019) *Better language models*. https://openai.com/, accessed May 5, 2020.

Prier, J (2017. 'Commanding the trend: social media as information warfare', *Strategic Studies Quarterly*, 11(4), pp. 50–85.

Replica Studios (2019) *Create replica voices in seconds*. https://replicastudios.com//, accessed May 5, 2020.

Rossler, A et al. (2019) *FaceForensics++: learning to detect manipulated facial images*. [Preprint].

Sasse, B (2018) *This new technology could send American politics into a Tailspin*. https://www.washingtonpost.com/, accessed May 5, 2020.

Satter, R (2019) *Associated press news*. https://apnews.com.

Saunders, J. (2018) *Counter-terrorism police are now training with virtual terrorists*. http://theconversation.com/, accessed May 5, 2020.

Schiff, AB, Murphy, S and Curbelo, C (2018) *Letter to the honorable Daniel R. Coats*. https://schiff.house.gov/, accessed May 5, 2020.

Swift, A (2016) *Americans' trust in mass media sinks to new low Gallup*. September 14. https://news.gallup.com/, accessed May 5, 2020.

Thies, J et al. (2018) 'Headon: real-timereenactment of human portrait video', *ACM Transactions on Graphics*, 37, pp. 1–13.

This Person Does Not Exist (2018) *This person does not exist*. https://www.thispersondoesnotexist.com/, accessed May 5, 2020,

Turek, M (2018) *Media forensics (MediFor)*. https://www.darpa.mil/, accessed May 5, 2020.

U.S. House of Representatives (2019) *Hearing – national security challenges of artificial intelligence, manipulated media, and deepfakes*. Washington, DC.

Xinhua News Agency (2019) *Xinhua presents AI anchors at news agencies world congress*. http://www.xinhuanet.com/, accessed May 5, 2020.

Zurcher, A (2019) *Trump impeachment inquiry: dissecting the phone call behind it*. https://www.bbc.com/, accessed May 5, 2020.

13 Cyber threat attribution, trust and confidence, and the contestability of national security policy

William Hoverd

Introduction

Today, we regularly hear that Chinese company Huawei is a source of vulnerability for Western nations, and that North Korea and Russia are sources of cyber threats. Western populations are being asked to trust the word of intelligence agencies and world leaders that these unspecified emerging technological threats are real. The often-classified nature of a potential or emerging technological threat results in governments not being able to provide the public with an evidence base for threat attribution, and, even if such evidence were released to the public, how would they even begin to understand the origins of deleterious and malicious code? The challenges around cyber threat evidence create a situation for the public where government threat attribution discourse must be trusted (or not) by those receiving it.[1] Schluzke (2018, p. 954) recently argued that this form of cyber attribution creates an uncertainty in the mind of the public and that, given state knowledge is necessarily classified in this domain, care needs to be taken by cyber policy makers when crafting language around cyber-attacks. This chapter will argue that the threat potential posed by emerging technologies produces trust and confidence issues for Western populations, (Henschke & Brandt Ford, 2017, p. 82) as the public is asked to trust and have confidence that a government's particular technological threat attribution claim is accurate, and that robust classified evidence for this attribution exists and it is not being used selectively.

In recent years, narrative, counter narrative, spin, classified material, and "fake news" have politicized intelligence and cybersecurity discourse in such a way, that official security discourse is potentially contestable. Consequently, it is sensible for the public to ask: Whose security claim should be trusted and why? Likewise, given that there can be skepticism around this discourse, it seems a critical social responsibility for security policy makers and academia to first acknowledge this conundrum and then strive to develop frameworks to better understand the trust and confidence challenges around technological threat attribution.

The task of academia is to undertake constructive criticism by highlighting the nature of an issue of concern and then concluding by suggesting possible solutions. In a volume devoted to emerging technologies, a broader phenomenon has emerged where issues of trust and the potential for deception that results from new technology have significant ramifications for the social contract between the state and the population and how they are both secured. The age-old balance between human liberty and security is being continually addressed in response to the evolving technologies of attack and defense that are occurring globally across the cyber realm.

The chapter begins by outlining the recent Five Eyes (USA, UK, Canada, Australia, and New Zealand) condemnations of Russian and North Korean cyber policy as a sociological case study to illustrate where and if a technological threat attribution and trust and confidence challenge might be evident. Focus is given to the recent *WannaCry* (2017) and *NotPetya* (2017) cyber-attacks that have been attributed to these nations. This case study of the Five Eyes condemnation is used to sketch out how the contestability of national security strategy and government security discourse can present specific trust and confidence challenges around national security discourse for both the public and the government. To illustrate the challenges surrounding contemporary issues of technological threat attribution, it focuses on the smallest member of the Five Eyes alliance, the nation of New Zealand. The case study will compare and contrast the language of the government executive and that of the Government Communications Security Bureau (GCSB) when they attributed these two attacks to North Korea (*WannaCry*) and Russia (*NotPetya*).

This attribution discussion then returns to the broader context where, especially after the accusations of Edward Snowden, trust and confidence in the security actions and motives of the Five Eyes intelligence partners have been questioned. The chapter argues that trust and confidence challenges around the political nature of the Five Eyes cyber attribution is potentially exacerbated by two factors:

1) That the invisible nature of the cyber realm can only be communicated by metaphor. This naturally leads to conflict-laden hyperbole that creates notions of cyber war and cyber threat that may not be helpful for thinking about the intentionally deployed deleterious and malicious effects of code (Lawson, 2012)
2) That the development of the notion of fake news accusations (and other such accusations) brings into question the trustworthiness of the discourse of national security agencies

Lastly, the purpose of the chapter's conclusion is to move the content and goals of the chapter beyond simple problem definition to the initiation of a discussion of potential solutions to technological threat attribution, trust and confidence issues, and the contestability of national security policy when it comes to cyber attribution.

We now turn to review two recent global cyber-attacks, which the Five Eyes nations have publicly stated as deeply influencing their extant national security decision-making – especially around cybersecurity.

The *WannaCry* and *NotPetya* cyber-attacks and the Five Eyes attributions

The Five Eyes intelligence sharing network's condemnation of North Korea and Russian cyber policy concentrated on the threat and damage caused by *WannaCry*

(2017) and *NotPetya* (2017), two cyber-attacks attributed to these nations, respectively. Together these two attacks represent the two most costly cyber-attacks that the world has experienced to date (Snow, 2018).

WannaCry was a ransomware crypto-worm that, in May 2017, attacked computers using the Windows operating platform (Martin et al, 2018, p. 361). It utilized the Eternal Blue exploit developed by the NSA, which had been stolen by the organization or state that used it for this attack. Once loaded onto the platform, it would encrypt data and then demand payment in bitcoin to release the data. Europol (BBC News, 2017) claimed that it had infected some 75,000 computers across 99 countries within a day, including 48 UK National Health Service (NHS) trusts, leaving doctors unable to retrieve patient information (Martin et al, 2018, p. 361). Ultimately, Kaspersky Lab analysis noted that the attack affected some 200,000 computers across 150 countries (Snow, 2018) with much of this occurring within 24 hours (Jones, 2017). Microsoft was able to respond quickly to the attack by deploying emergency patches that prevented the crypto-worm's spread. Nevertheless, variants of *WannaCry* were still emerging in 2018, hitting Boeing (Mathews, 2018) in May and Taiwanese Apple chipmaker TSMC in August (Leswing, 2018).

What was unprecedented about the *WannaCry* attack was its indiscriminate global scale and its rapid spread, as well as its payload's ability to disrupt essential services and business. The cost of the disruption caused by the attack is significant, estimated at being between four and eight billion US dollars (Snow, 2018). Microsoft's president and chief legal officer, Brad Smith, stated that a global technological problem was emerging where governments were stockpiling cyber weapons that, if stolen, as in the case of Eternal Blue, create the potential for indiscriminate attacks like this to occur (Smith, 2017). He went on to argue that:

> The governments of the world should treat this attack as a wake-up call. They need to take a different approach and adhere in cyberspace to the same rules applied to weapons in the physical world. We need governments to consider the damage to civilians that comes from hoarding these vulnerabilities and the use of these exploits.
>
> (Smith, 2017)

The immense scale of the attack and the possibilities for further attack that *WannaCry* augured, indicated that globally a shift was required to address the perpetrators of such attacks.

Attribution for *WannaCry* was formally announced by the White House in December 2017 (Bossert, 2017). USA Homeland Security advisor, Tom Bossert, stated that:

> After careful investigation, the United States is publicly attributing the massive *WannaCry* cyberattack to North Korea. We do not make this allegation lightly. We do so with evidence, and we do so with partners. Other governments and private companies agree. The United Kingdom, Australia, Canada,

> New Zealand, and Japan have seen our analysis, and they join us in denouncing North Korea for *WannaCry*. Commercial partners have also acted. Microsoft traced the attack to cyber affiliates of the North Korean government, and others in the security community have contributed their analysis.
>
> (Bossert 2017, p. 1)

Usefully, here we see that commercial attributions for these attacks also exist to support government attributions. The global or, more specifically, US-led Five Eyes attribution for the *WannaCry* cyber-attack was made against North Korea, and here there was a stated (classified/private) evidence base for the attribution. During the question portion of the briefing, Bossert was asked about the evidence basis for the attribution and he stated:

> We took a lot of time to look through classified, sensitive information. What we did was rely on – and some of it I can't share, unfortunately – technical links to previously identified North Korean cyber tools, tradecraft, operational infrastructure. We had to examine a lot. And we had to put it together in a way that allowed us to make a confident attribution.
>
> (Bossert, 2017)

Above, we see Bossert assert that the US developed and shared its analysis with its partners, and on that evidence, made their attribution of the origin of *WannaCry* to North Korea. The often-classified nature of the potential or actual emerging technological threat results in governments not being able to provide the public with the evidence for cyber threat attribution.

NotPetya

Closely following the *WannaCry* cyber-attack, the *NotPetya* cyber-attack occurred on June 27, 2017. Kaspersky Lab reported infections in France, Germany, Italy, Poland, the United Kingdom, and the United States, but the majority of infections were occurring in Russia and the Ukraine where more than 80 companies were attacked (Snow, 2017). It mainly targeted businesses as one of the initial propagation vectors was through the financial software MeDoc (Snow, 2018). Kaspersky Lab argued *NotPetya* was the costliest cyber-attack the world has seen with its deleterious effects reaching some ten billion US dollars (Snow, 2018).

NotPetya was a variation of the 2016 *Petya* software, which was originally an e-mail-spread ransomware that encrypted hard drives and asked for payment (Snow, 2017, p. 1). However, Kaspersky Lab was quick to state that this attack was from a separate malware family, and utilized the Eternal Blue windows exploit (Robot, 2017, p. 1). Where *NotPetya* was radically different was that it masqueraded as ransomware but was, in fact, what is known as a "wiper" that permanently deleted material on a computer's hard drive (Robot, 2017). Moreover, it particularly affected businesses in Ukraine, with some 75% of incidents reported in that nation. It then spread globally to the business partners of those initially affected, with 9% of incidents occurring in Germany, and the Danish shipping company, Maersk, being particularly affected by the attack (Schwartz, 2018). Where *WannaCry* was global and relatively indiscriminate ransomware, *NotPetya* was a disguised, targeted attack on a particular nation-state's financial infrastructure.

In January 2018, the *Washington Post* stated that the CIA had attributed the *NotPetya* attack to Russian military hackers (Nakashima, 2018). Reportedly, its sources claimed "The GRU military spy agency created *NotPetya*, the CIA concluded with 'high confidence' in November, according to classified reports cited by USA intelligence officials" (Nakashima, 2018). On February 15, 2018, the White House released a brief press statement that attributed *NotPetya* to the Russian military:

> In June 2017, the Russian military launched the most destructive and costly cyber-attack in history… The attack, dubbed "*NotPetya*", quickly spread worldwide, causing billions of dollars in damage across Europe, Asia, and the Americas. It was part of the Kremlin's ongoing effort to destabilize Ukraine and demonstrates ever more clearly Russia's involvement in the ongoing conflict. This was also a reckless and indiscriminate cyber-attack that will be met with international consequences.
>
> (Statement from the Press Secretary, 2018)

By late February 2018, the *NotPetya* attack had been attributed by all Five Eyes nations as originating in Russia (Bussoletti, 2018). United Kingdom Foreign Office minister Lord Tariq Ahmad attributed the *NotPetya* cyber-attack to the Russian government, stating "The UK Government judges that the Russian Government, specifically the Russian military, was responsible for the destructive *NotPetya* cyber-attack of June 2017" (Foreign Office Minister condemns Russia, 2018). The rationale for this attribution was: "The decision to publicly attribute this incident underlines the fact that the UK and its allies will not tolerate malicious cyber activity" (Foreign Office Minister condemns Russia, 2018). Clear attribution for the attacks was leveled at the Russian military from across the Five Eyes nations, but, again, it is difficult to discern how and where evidence for this attribution arose and was then shared between these nations.

The April 2018 Five Eyes head of state cybersecurity meeting

The Five Eyes condemnation of the Russian cyber-attack occurred on April 18, 2018, when four heads of state (Australia, Canada, United Kingdom, and New Zealand) met at the United Kingdom's National Cyber Centre (Waldie, 2018). The meeting publicly attributed the *NotPetya* Ukraine cyber-attack to Russia. Canadian Prime Minister Justin Trudeau made the claim that Russia has been "using cyberwarfare as part of a wider effort to attack and undermine the international system" (ibid). In this meeting, British Prime Minister Theresa May described the Five Eyes alliance as "a unique security and intelligence-sharing partnership" that "has done much to protect our people from a range of threats" (ibid). In particular, with regard to Russian hacking, she added: "I have been clear to Russia that we know what it is doing. And we should be in no doubt that such cyberwarfare is one of the great challenges of our time" (ibid). Five Eyes's attribution discourse was universal in its condemnation. Another dimension to the trust challenges that the chapter has already outlined is that, in this example for both the Five Eyes governments and their populations, the attribution was denied by Russia (Marsh, 2018). Consequently, without presentation of the shared evidence, the cyber-attack attribution became contested, effectively creating a he-said/she-said situation between the various parties linked to ongoing great power competition.

The New Zealand attribution

The New Zealand government attributed both the *WannaCry* and *NotPetya* cyber-attacks to Russia and North Korea in line with their Five Eyes partners. However, I have not found a New Zealand government source stating that the nation has seen shared 'classified evidence' for these threat attributions. Official releases from the GSCB, in terms of both *WannaCry* (Government Communications Services Bureau, 2017b) and *NotPetya* (Government Communications Services Bureau, 2018), simply state the agency supports their international partners' condemnation. As we can see from Tom Bossert's statement above, he stated that attribution evidence for *WannaCry* was shared with its Five Eyes partners, specifically naming New Zealand. This means that the public must trust the government's discourse, and it is not clear whether its representatives have seen their partner's evidence of attribution or are simply repeating the attribution. We have to trust (or not) that they have seen evidence. To contextualize the New Zealand support, these attributions coincided with a significant 2018 government focus on cybersecurity as a key threat in the national security threat landscape (Little, 2018, p. 33).

The 2018 Department of Prime Minister and Cabinet (DPMC) *Cyber Security Strategy Refresh* stated that "New Zealand recognizes that the relevance of cybersecurity concerns now extends across multiple branches of government" (Department of Prime Minister and Cabinet, 2018a, p. 6). The primary responsibility for cybersecurity is given to the Government Communications Security Bureau, which delivers signals intelligence. A close look at the government's cyber discourse in 2018 shows a discourse had emerged across the national security sector where a cyber technological threat was being attributed to Russia and North Korea. This threat attribution discourse is usually accompanied by an argument that New Zealand is not immune from this threat, and that additional cyber infrastructure development is required. For example, in February 2018, Andrew Hampton, the director general of the GCSB (Government Communications Security Bureau, 2018), stated, "In terms of cyber threats, the GCSB noted a 15% increase in serious incidents affecting New Zealand in the year to June 2017", and that "The GCSB's international partners have today attributed the *NotPetya* cyber-attack to the Russian Government". Here we see that the attribution for *NotPetya* is sourced from Five Eyes partners, and it is not made clear whether the statement was made simply in support of, or after reviewing classified evidence supplied by their partners.

Similarly, in the 2018 winter issue of *Line of Defence* the Right Honorable Andrew Little, Minister Responsible for the GCSB and NZSIS stated:

> In February, the Government added New Zealand's voice to the international condemnation of the *NotPetya* cyber-attack which international partners have now attributed to the Russian Government. It targeted Ukraine but had a global impact – including affecting supply chains in New Zealand. In December, New Zealand also expressed concern about international reports which link North Korea to the major *WannaCry* ransomware campaign.
>
> (Little, 2018, p. 33)

A similar type of attribution claim can also be found in the July 2018 Ministry of Defense Strategic Defence Policy Statement which states that:

> Physical distance is no protection in cyberspace, and New Zealand is subject to a growing cyber threat from state sponsored and other malicious actors… Cyber blurs boundaries between conflict and peace, and public and private. (...)
>
> Russian 'active measures' in the 2016 United States Presidential election brought to light 'cyber enabled information warfare' as a disrupter in liberal democracies. (...)
>
> North Korea has a substantial store of chemical and biological weapons, a significant cyber capability (which it has shown a willingness to use).
>
> (Ministry of Defence, 2018, pp. 18–21)

Across these three examples, and more recently in the 2019 New Zealand cybersecurity strategy (Department of Prime Minister and Cabinet, 2019), we clearly see attributions of cyber threat to Russia and North Korea. We also see in all these examples that the threat attribution for cyber-attacks occurring across nation-states is linked explicitly to a governmental claim that there is a growing cyber threat environment impacting adversely on New Zealand's national security and, by implication, that there is an increased need for investment in the nation's cyber infrastructure. The implication is that cyber-attacks represent a form of virtual domain of threat that circumvents the nation's geographic isolation and that while it may not be a direct target of any such future attack, both *WannaCry* and *NotPetya* demonstrated the ability these attacks have to deliver collateral damage on a global scale. Nevertheless, we still need to better understand what might motivate such a small nation to make explicit attribution claims.

Cyber threat attribution

Cyber attribution is generally understood to be a difficult technical process (Kello, 2013, p. 7; Lin, 2016, p. 75). Commonly, the challenge of attributing a cyber-threat has created problems for nations trying to establish deterrence against foreign actors (Lupovici, 2016, p. 322; Lindsay, 2015, p. 53). However, it is the political construction of cyber threat attribution in national security that is useful here. Myriam Dunn Calvety (2013, p. 105) argued that "The link between cyberspace and national security is often presented as an unquestionable and uncontested 'truth'. However, there is nothing natural or given about this link: It had to be forged, argued, and accepted in the (security) political process". She argues that cybersecurity discourse is, in fact, constituted by a variety of authority figures in governments. For Dunn Calvety, the political nature of cyber discourse is further complicated by the fact that the very building blocks of cybersecurity language employ analogies or metaphors to describe and explain the effect of unsolicited changes in code across networks. Similar to a conceptualization of disease as an attack (Sontag, 1978), this metaphorical language of cyber "attack" and "weapons" is potentially distorting and results in a tendency to describe the cyber landscape as unruly, dangerous, or threatening, when, in fact, much of the cyber environment is benign and intended to be enabling.

One consequence of the political construction of cyber discourse is that there are real risks when it comes to generating accurate attribution claims. Thomas Rid and Ben Buchanan (2015, p. 4) argue, "Doing attribution well is at the core of virtually all forms of coercion and deterrence, international and domestic. Doing it poorly undermines a state's credibility, its effectiveness, and ultimately it's liberty and security". When it comes to cybersecurity, the attribution debate is evolving surprisingly slowly, and this slowness remains the case today (Eichensehr, 2020, p. 19). Rid and Buchanan (2015, p. 4) provide a detailed technical account of the attribution process, where they conclude that developing the evidence for cyberattack attribution is a difficult and costly process which inevitably only indicates the likely source of an attack. Moreover, this evidence cannot easily determine motive or the political gain that may have initially motivated an attack. They conclude with the finding that *attribution is what states make of it*, and public attribution is a political act used by states to legitimate security positioning. If we are to follow Rid and Buchanan's logic, we could understand the Five Eyes attribution of cyber threat to Russia and North Korea as a broader function of what is at stake politically in terms of the broader relations between these nations.

In the New Zealand case study, the combination of technical evidence challenges and political nature of attribution claims is made explicit when we look closely at the GCSB's National Cyber Security Centre Annual report, where it states:

> Publicly reporting attribution is a significant [political] decision and is not made by the NCSC alone. Public attribution is one way to reduce the efficacy of malicious cyber actors by revealing their tools or increasing the reputational costs of illegitimate activity. However, it also carries risk for New Zealand and is considered alongside our other national objectives, including the need to maintain our ability to protect the networks that are of importance…
>
> (Government Communications Service Bureau, 2017a, p. 12)

Given that there is academic and local technical agreement that cyber threat attribution is ultimately a political act, why might this government be making these threat attribution claims – why now and what might be the risks inherent in this attribution?

There are three possible explanations for this attribution. Firstly, there is a clear discussion in other sections of these documents that New Zealand is part of the Five Eyes network and is speaking in solidarity with its partners to increase the reputational costs of these activities. The challenge here is that when there is no unclassified evidence to support these attribution claims (there was commercial evidence for *WannaCry* but none for *NotPetya*), the public must trust the veracity and motives of these partners, and they must trust that their politicians are maintaining an independent policy stance. For example, Reuben Steff's chapter in this volume (Chapter 3) explains that small states are traditionally considered to be "consumers" of security technology supplied by larger states, given they often cannot develop it on their own. This means that it is necessary for New Zealand to align itself with the US on key technological security issues, and the case suggested here is that this could extend to technological threat attribution too, in order to secure its relations to the US and maintain access to critical technology advances and supply. If this is the case, it fuels critical questions about the independence of the nation's foreign affairs policy (Small, 2018, p. 1), which, if undermined, would be a potential blow to public trust.

A second explanation is sourced from DPMC's National Cyber Policy Office release of April 20, 2018, which takes the form of a letter written by the Right Honorable Claire Curran, then-minister of Broadcasting, Communication, and Digital Media. The topic of the letter is the *Refresh of New Zealand's Cyber Security Strategy and Action Plan* which resulted in 2019 Cybersecurity Strategy.

Claire Curran stated:

> The clear trend is an upward trajectory of cyber security threats… New Zealand's geographical location does not exempt us from this threat... We will need to consider the mechanisms available to us to dissuade or deter malicious cyber activities, particularly where it is state-sponsored, or state condoned. This includes the option of publicly attributing malicious cyber activity as a way of holding states to account.
>
> (Department of Prime Minister and Cabinet, 2018b, p. 4, p. 9)

Clearly, in the above document, the government has made an active political decision to publicly attribute what it describes as malicious cyber activity to particular nation-states as part of "expanding its international cyber efforts" (ibid). The challenge is that the source and reliability of classified evidence for the threat and the rationale for public attribution are unable to be made transparent, and the public is being represented by this claim. Here, the public is asked to trust that the correct decisions and attributions are being made as part of being a good international citizen and that the government is not being selective in who it calls to account.

A third explanation comes from the Copenhagen school, which developed the securitization theory that focuses on "speech acts" and the significance these acts can have upon political agenda settings and political relations (McDonald, 2008, p. 568). Securitization theory suggests that when a threat is identified and a "speech act" identifying it is utilized, this discourse prioritizes the threat on the political agenda in such a way that it necessitates the development of urgent mitigation measures that could potentially extend even to the encroachment of privacy, the need for secrecy, and the utilization of force (Kassab, 2014, p. 65). For the Copenhagen school, securitization speech acts emphasize the dangers of the cyber threat environment and legitimate additional government funding for the construction of various infrastructures designed to protect the public from the threat. The challenge here, again, is that the speech acts have the power for change, not any classified evidential basis that may sit behind the act. When it comes to cyber threats, the public has no choice but to trust that those who are making the speech act are fully informed by subject matter experts and are making the best national security decisions possible.

Taken together, these explanations suggest that the various 2018–2019 New Zealand Cybersecurity attribution speech acts appear designed to:

1. Align New Zealand closer to its Five Eyes partners
2. Call out certain acts of cyber aggression where New Zealand can then position itself as a good international cyber citizen
3. Be used as a mechanism to dissuade or deter malicious cyber activity by publicly attributing the activity to hold states to account
4. Create, justify, and legitimate the development of additional domestic cybersecurity infrastructure

Together, these four purposes suggest that the attributions are being utilized for both international and domestic positioning in terms of alignment with our partners, being a good international citizen, and creating global and domestic cyber infrastructures to further New Zealand's interests and safety.

Public trust and national security discourse

In a civil society environment where the evidence for attribution is unavailable and the integrity of state institutions such as the GSCB (Rogers 2015, 46) or the Minister for Broadcasting's judgement has recently been called into doubt (Edwards, 2018), it is possible that future trust and confidence issues may arise around the efficacy of this language and the effectiveness of current government cyber initiatives. In 2015 and 2016, similar trust and confidence issues occurred in the justification of counter terrorism funding (Key, 2014), where evidence of a domestic Muslim radicalization threat has never eventuated and the claim of the "Jihadi bride" risk was found by the media to have little evidentiary basis and to have been exaggerated (Vance, 2015) and shifted attention away from the right wing. It is clear that, at least within the New Zealand Fourth Estate (News

Media), there is a tendency to critically examine government security discourse rather than trust it implicitly, and it is this view that underpins their reporting to the population (Jancic, 2019). Serious issues around trust in government security discourse and action resulted in recent wholesale reform of the New Zealand intelligence agency legislation with the creation of the *Intelligence and Security Act 2017* governing the NZSIS and GSCB (Rogers, 2018, p. 657). Further destabilization of public trust in security agencies is also evident with ongoing "inquiries" into the New Zealand Defence Force's actions in Afghanistan (*Operation Burnham Inquiry* 2020) and the Christchurch Terror attacks (*Christchurch Attack Royal Commission* 2020), where the integrity of national security apparatus and discourse has been brought into contention by investigative journalists (Hagar & Stephenson, 2018), expert commentators, or community groups (Ainge Roy, 2019), creating a situation where it is impossible to know whose claim is trustworthy (Hoverd, 2019, p. 17). All the national security organizations (NZDF, GCSB, DPMC, and NZSIS) under investigation in these two inquiries are the same agencies producing the cybersecurity attributions. While clear distinctions need to be made between these individual one-off events and the general effectiveness of these agencies' everyday business and their successful operations, these examples show that they are regularly and recently seen by the public to have their integrity brought into doubt.

These types of one-off trust challenges to national security agency integrity are also seen at the international level. In 2013, *WikiLeaks* released an electronic dispersal of NSA classified material of some 1.7 million documents that had been leaked by Edward Snowden. These documents suggested strongly that the NSA, and perhaps also its Five Eyes partners, were engaged in regular global surveillance of their domestic populations (Verbal, 2014, p. 14). Again, it was impossible to know the truth of this claim based on open-source information, but part of the multifaceted collateral damage caused by this leak is a tarnishing of the reputation of the Five Eyes's agencies within their own populations where trust and confidence in those intelligence agencies is diminished in the eyes of the public.

In 2016, an additional social media technological challenge emerged for government discourse by the name of *fake news* (Allcott & Gentzkow, 2017, p. 211). Social media allows content to be passed between providers with no verification of its authenticity (ibid). Morgan (2018, p. 39) asserts that this verification problem allows disinformation and what is described as fake news to manipulate the public sphere and has the effect of "declining levels of trust in institutions and experts". There are significant implications for democracy and free speech emerging from this phenomenon. When it comes to intelligence agency claims, the discussion of fake news is particularly pertinent when it comes to the relationship between US President Donald Trump and his intelligence agencies (McCarthy, 2018). Over a number of highly publicized national security issues related to Iran, ISIS, North Korea, and Russian interference in the 2016 Presidential election, the president has called the discourse of his intelligence agencies into question. Donald Trump effectively stated that the intelligence agency claims are fake news and that the integrity of their discourse needs to be stringently questioned.

Conversely, one must ask why should the public trust Donald Trump, who has a conflict of interest because of investigations into his own actions, over politically neutral actors? Consequently, it is sensible for the public to ask whose security claim should be trusted and why? When it comes to Russian interference in the US election, *Thirdway.org* data show that the Donald Trump/James Comey counterclaims have created significant confusion about whose security discourse to trust (Eoyang et al, 2018). The broader ramification of such a claim is that all intelligence agency discourse, including that of cyber threat attribution, could be questioned and further magnifies the question over how Five Eyes partners, and their respective public, know who they should trust.

An essential assumption of my argument is that a problem of trust exists today when it comes to the public receiving national security agency discourse. How does one define trust? In political science, it has been traditionally assumed that "Trust is relational; it involves an individual making herself vulnerable to another individual, group, or institution that has the capacity to do her harm or betray her" (Levi & Stoker, 2000, p. 476). Trust can be understood as dichotomous in that one trusts or one does not, but trust relations can also be a sliding scale of degrees of trust. For example, Western populations give their trust to their national security agencies to be protected from cyber threats (Shulzke, 2018, p. 954). That trust is essential for cyberspace to function and for governance and security of that critical infrastructure (Azmi et al, 2018, p. 258).

Henschke and Ford (2017, p. 82) have argued that since the Snowden releases there has been a breakdown in the public's trust of national security agencies.

> The cost here is in the public's willingness and capacity to trust in state mechanisms of national and international security, especially those intelligence programmes involving cyber-surveillance. The loss of trust in this area is particularly detrimental to the effective functioning of national security.
>
> (Henschke & Ford, 2017, p. 82)

For Henschke and Ford (2017, p. 82), there is a twofold element of trust in the relationship between national security agencies and the public: 1) That surveillance is necessary and 2) that intelligence activities must be kept secret from the public. The challenge of Snowden (and likewise the GCSB's illegal spying on Kim Dotcom) (Rogers, 2015, p. 46) is that the Five Eyes nations were "caught out abusing this trust risk losing the goodwill of the public. If the public's trust in this area is lost and the government is judged as untrustworthy, then effective governing can become more difficult" (Henschke & Ford, 2017, p. 84). Pew data since the Snowdon releases reveal that the American public's approval for surveillance of citizens has declined as has their confidence that their government can protect their private data (Geiger, 2018). *Thirdway.org* suggests that in relation to contentious national security issues, most Americans did not know whose claim to trust (Eoyang et al, 2018). The poll data are not conclusive in showing a loss of trust in national security discourse, but it certainly demonstrates a public confusion and a lack of implicit unquestioned trust in national security discourse.[2]

Given that recent Pew data also suggest the public believe there will be increasing cyber threats to elections and national security infrastructure (Poushter & Fetterolf, 2019), it is essential that trust is maintained. Ultimately, the loss of public trust is counterproductive for effective national security (Poushter & Fetterolf, 2019) because it leads to increasing media scrutiny and hyperbole, which then adversely impacts security agency transparency in terms of comfort levels with communicating to the public and not over classifying internal documentation. Thus, when it comes to making cyber-security policy, Ball et al (2019, p. 103) have argued that "Cybersecurity policymaking at the national level should be attentive to the central role of trust for maintaining resilient cyber systems – and it should have plans in place to repair trust". Trust therefore is an essential element of cyber-governance. To create trust with the public, governments must be seen to be active, careful, and deliberate in the policy and infrastructure they create and sustain to keep cyberspace safe, both domestically and globally. All these activities (including attribution) will be more effective if trust has not been reduced by earlier state failings and when/if a government can be transparent.

Conclusion: Addressing the contestability of attribution discourse

In this chapter, my intention has been to argue that, when it comes to cyber threat attribution claims made by governments, it is not always clear whose claim to trust or to what extent. There are inherent challenges in cyber threat attribution discourse: a) The evidence for attribution is not publicly verifiable, b) attribution is inherently political, c) the public must choose to trust attribution discourse or not, and d) attribution discourse often comes from institutions whose integrity has been brought into question elsewhere (this decreases trust). Given these concerns, we can say that there is a trust and confidence challenge that must be addressed by governments when it comes to the production of such discourse in the future.

As a social scientist, I traditionally look for empirical evidence to back a claim; without it, we are, at best, left with circumstantial evidence or, at worst, simply a truth claim. Nevertheless, we must acknowledge that the public do not (and perhaps should not) have access to classified empirical data to make assessments about the efficacy of cybersecurity decisions and discourse (Kello, 2013, p. 7). This creates a grey zone where cyber-attacks and their attributions live in a contested space of discourse between evidence and trust, where maintaining trust in government action is essential. I have demonstrated that attribution claims are inherently political and are contestable by both the accuser and the accused. While the effects of cyber-attacks are very real, communication about their effects and the intentions and motives behind an attack exist in a land of metaphor and discourse. This makes national security discourse around cyber-attacks contestable; a fact that is exacerbated by an inclination across social media, traditional media, and even the president of the US (with his conflict of interest) to question the integrity of these attribution claims. Therefore, a social scientific crisis emerges where, because of the classified nature of the evidence, the public is asked to trust and have confidence in a technological threat attribution claim without any further assurance. Consequently, it is then sensible for the public to ask whose security claim should be believed and why.

In modern democracies, trust in government is a key indicator of political legitimacy and stability (Marcinkowski & Starke, 2018, p. 87). Key questions for governments and populations are *how* and *when do* we trust the national security discourse of our government. A lack of trust inherently undermines and frustrates the positive intent of governments to keep their citizenry safe. In a contemporary environment where the national security discourse of a nation such as New Zealand (not to mention many other states) can be contested, it seems a critical social responsibility for security policy makers and academia to first acknowledge this trust and confidence conundrum and then strive to develop frameworks to better understand the trust and confidence challenges around threat attribution and apply it to this emerging cyber discourse. How to ensure that there is trust and confidence in the government's ability to identify, communicate, and respond to technological national security threats will be essential 21st-century governance questions for democracies.

Unfortunately, civil society cannot be relied upon to meet such a challenge. A civic challenge exists for both government and academia to consider how to best address the inherent weakness (that they are political and contestable) that exists in national security threat attributions. Ideally, probably the most effective manner in which cyber attribution could be separated from the individual interests of nation-states or commercial entities is if some form of international legal attribution agency was formed.[3] However, realistic solutions to the attribution problem must come from individual governments, who should be sufficiently motivated to increase the efficacy of its security apparatus by decreasing potential criticism of its discourse.

The producers of national security discourse must acknowledge that these weaknesses exist in their discourse and that it is their responsibility to find ways to mitigate these challenges. This is critical because *WannaCry* and *NotPetya* have shown us that cyber issues are changing the national security landscape indelibly. While care is required around cyber hyperbole and metaphor, the potential scale, spread, and damage of future attacks could be catastrophic. Governments need to find better ways to effectively communicate with each other, their opponents, and their own populations when it comes to mitigating these challenges.

It is important to conclude by asking what might be possible solutions to respond to the technological cyber-attack attribution challenge? First, cyber discourse makers must accept that national security discourse justifications that arise from classified sources cannot always provide a persuasive public evidence basis for their assertions. Second, cyber discourse makers should acknowledge that their claims are inherently political and therefore potentially contestable. Such acknowledgment would allow the discourse to be more reflexive and to self-regulate its claims in a more nuanced manner. Nuanced cyber threat discourse lessens the possibility that a government institution or representative will be subsequently called to account for making an incorrect or exaggerated claim.[4] Governments should look to employ these observations in the communication tools that they already possess. Cybersecurity strategy refreshes offer a real opportunity to engage with the public. They can focus both on private enterprise and the public. They can offer an education capacity and more transparency around the nature of various cyber threats. One place where a degree of transparency and limited evidence can exist for the actual nature and prevalence of cyber threats is in the annual reports from their National Cyber Security Centers (see Government Communications Security Bureau, 2017a). These carefully constructed and communicated reports and data could and should be more widely distributed, publicized, and utilized as an evidence basis for future claims. And lastly, one additional way to address the confidence challenge and limited trust around secrecy in national security discourse is to regularly craft language that clearly notes that a fine balance exists between secrecy and transparency. This balance relies heavily on building a trusting and transparent relationship between government officials and the public while working to repair any past breaches of confidence.

Notes

1 The chapter does not claim that cyber-attack attribution evidence is lacking; simply, it argues that government claims about possessing robust attribution evidence must be trusted by the public.
2 Trustworthiness is based on the ability to understand the process and the evidence presented to you. If one lacks the technical knowledge and/or when there are competing claims around national security, mediated by the Fourth Estate, a great deal of confusion will be at play in the mind of the public. This may mean that cybersecurity discourse, by being difficult to understand and hyberbolic in nature, makes establishing trust in that discourse problematic from the outset.
3 The Christchurch Call is one example of State/Private Sector policing of social media threat. Bodies such as the United Nations and Interpol, in collaboration with states and private sector, could potentially develop such a complex mechanism.
4 One place this could occur is in a careful differentiation between the prevalence of cybercrime events and state sponsored cyber-attacks to ensure that the actual prevalence and nature of cybersecurity threats are properly represented.

References

Ainge Roy, E (2019) *Christchurch shootings real issues not being heard*. https://www.theguardian.com/world/2019/aug/08/christchurch-shootings-real-issues-not-being-heard-by-inquiry-critics-say, accessed April 13, 2020.

Allcott, H and Gentzkow, M (2017) 'Social media and fake news in the 2016 election', *Journal of Economic Perspectives*, 31(2), pp. 211–236.

Azmi, RA, Tibben, W and Than Win, K (2018) 'Review of cybersecurity frameworks: context and shared concepts', *Journal of Cyber Policy*, 3(2), pp. 258–283.

Ball, K, Degli Esposti, S, Dibb, S, Pavone, V and Santiago-Gomez, E (2019) 'Institutional trustworthiness and national security governance: evidence from six European countries', *Governance*, 32(1), pp. 103–121.

BBC News (2017) *Cyber-attack: Europol says it was unprecedented in scale*. https://www.bbc.com/news/world-europe-39907965, accessed April 13, 2020.

Bossert, T (2017) *Press briefing on the attribution of the WannaCry Malware attack to North Korea*. https://kr.usembassy.gov/121917-press-briefing-attribution-wannacry-malware-attack-north-korea/, accessed April 13, 2020.

Bussoletti, F (2018) *All five eyes countries have blamed Russia for the NotPetya cyber attack*. https://www.difesaesicurezza.com/en/cyber-en/all-five-eyes-countries-have-blamed-russia-for-the-notpetya-cyber-attack/, accessed April 13, 2020.

Department of the Prime Minister and Cabinet (2018a) *Refresh of New Zealand's cyber security strategy and action plan*. Wellington: New Zealand Government.

Department of the Prime Minister and Cabinet (2018b) *Refresh of New Zealand's cyber security strategy and action plan*. https://www.dpmc.govt.nz/sites/default/files/2018-04/ers-18-paper-refresh-of-new-zealands-cyber-security-strategy-and-action-plan_1.pdf, accessed April 13, 2020.

Department of the Prime Minister and Cabinet (2019) *New Zealand's cybersecurity strategy 2019*. Wellington: New Zealand Government.

Dunn Cavelty, M (2013) 'From cyber-bombs to political fallout: threat representations with an impact in the cyber-security discourse', *International Studies Review*, 15(1), pp. 105–122.

Edwards, B (2018) *Political roundup: Clare Curran's agonising downfall*. https://www.nzherald.co.nz/nz/news/article.cfm?c_id=1&objectid=12121270, accessed April 13, 2020.

Eichensehr, K (2020) 'The law & politics of cyberattack attribution', *UCLA Law Review*, 67, 19–36.

Eoyang, M, Freeman, BG, and Wittes, B (2018) *Confidence in government on national security matters*. April 2018. https://www.thirdway.org/polling/confidence-in-government-on-national-security-matters-april-2018, accessed April 13, 2020.

Geiger, A (2018) *How Americans have viewed government surveillance and privacy since snowden leaks*. https://www.pewresearch.org/fact-tank/2018/06/04/how-americans-have-viewed-government-surveillance-and-privacy-since-snowden-leaks/, accessed April 13, 2020.

Government Communications Security Bureau (2017a) *National cyber security centre: unclassified cyber threat report*. Wellington: New Zealand Government.

Government Communications Security Bureau (2017b) *New Zealand concerned by North Korean cyber activity*. https://www.gcsb.govt.nz/news/media-release-new-zealand-concerned-at-north-korean-cyber-activity/, accessed May 5, 2020.

Government Communications Security Bureau (2018) *New Zealand joins international condemnation of NotPetya cyber-attack*. https://www.gcsb.govt.nz/news/new-zealand-joins-international-condemnation-of-notpetya-cyber-attack/, accessed April 13, 2020.

Hagar, N and Stephenson, J (2018) *Hit & run: the New Zealand SAS in Afghanistan and the meaning of honour*. Nelson: Craig Potton Press.

Henschke, A and Brandt Ford, S (2017) 'Cybersecurity, trustworthiness and resilient systems: guiding values for policy', *Journal of Cyber Policy*, 2(1), pp. 82–95.

Hoverd, W (2019) 'The changing New Zealand national security environment: new threats, new structures, and new research', *National Security Journal*, 1, pp. 17–34.

Jancic, B (2019) *SIS apologises to hager over unlawful spying*. https://www.nzherald.co.nz/nz/news/article.cfm?c_id=1&objectid=12275352, accessed April 13, 2020.

Jones, S (2017) *Timeline: how the WannaCry cyber-attack spread*. https://www.ft.com/content/82b01aca-38b7-11e7-821a-6027b8a20f23, accessed April 13, 2020.

Kassab, H (2014) 'In search of cyber stability: international relations, mutually assured destruction and the age of cyber warfare', in Frederik Kremer, J and Müller, B (eds)

Cyberspace and international relations: theory, prospects and challenges. Berlin: Springer.

Kello, L (2013) 'The meaning of the cyber revolution: perils to theory and statecraft', *International Security*, 38(2), pp. 7–40.

Key, J (2014) *Speech to NZ institute of international affairs*. https://www.beehive.govt.nz/speech/speech-nz-institute-international-affairs-0, accessed April 13, 2020.

Lawson, S (2012) *Putting the "war" in cyberwar: metaphor, analogy, and cybersecurity discourse in the United States*. https://firstmonday.org/ojs/index.php/fm/article/view/3848/3270, accessed April 13, 2020.

Leswing, K (2018) *One of apple's key chip manufacturers was hit with a virus that targets windows computers—and it took nearly 3 days to recover*. https://www.businessinsider.com.au/apple-tsmc-wannacry-virus-2018-8?r=US&IR=T, accessed April 13, 2020.

Levi, M and Stoker, L (2000) 'Political trust and trustworthiness', *Annual Review of Political Science*, 3, pp. 475–507.

Lin, H (2016) 'Attribution of malicious cyber incidents: from soup to nuts', *Journal of International Affairs*, 70(1), pp. 75–137.

Lindsay, J (2015) 'Tipping the scales: the attribution problem and the feasibility of deterrence against cyberattack', *Journal of Cybersecurity*, 1(1), pp. 53–67.

Little, A (2018) 'Andrew little addresses the national security conference', *Line of Defence Magazine*, 1(8), pp. 33–35.

Lupovici, A (2016) 'The "attribution problem" and the social construction of "violence": taking cyber deterrence literature a step forward', *International Studies Perspectives*, 17(3), pp. 322–342.

Marcinkowski, F and Starke, C (2018) 'Trust in government: what's news media got to do with it?', *Studies in Communication Sciences*, 18(1), pp. 87–102.

Marsh, S 2018. *US joins UK in blaming Russia for NotPetya cyber-attack.* https://www.theguardian.com/technology/2018/feb/15/uk-blames-russia-notpetya-cyber-attack-ukraine, accessed April 13, 2020.

Martin, G, Ghafur, S, Kinross, J and Hankin, C (2018) 'WannaCry—a year on investment is important, but a culture change is crucial', *British Medical Journal*, 361.

Mathews, L (2018) *Boeing is the latest WannaCry ransomware victim.* https://www.forbes.com/sites/leemathews/2018/03/30/boeing-is-the-latest-wannacry-ransomware-victim/#771f28026634, accessed April 13, 2020.

McCarthy, T (2018) *Why is Donald Trump attacking the US intelligence community?* https://www.theguardian.com/us-news/2018/aug/18/why-is-donald-trump-attacking-the-us-intelligence-community, accessed April 13, 2020.

McDonald, M (2008) 'Securitization and the construction of security', *European Journal of International Relations*, 14(4), pp. 563–587.

Ministry of Defence (2018) *Strategic defence policy statement.* Wellington: New Zealand Government.

Morgan, S (2018) 'Fake news, disinformation, manipulation and online tactics to undermine democracy', *Journal of Cyber Policy*, 3(1), pp. 39–43.

Nakashima, E (2018) *Russian military was behind "NotPetya" cyberattack in Ukraine, CIA concludes.* https://www.washingtonpost.com/world/national-security/russian-military-was-behind-notpetya-cyberattack-in-ukraine-cia-concludes/2018/01/12/048d8506-f7ca-11e7-b34a-b85626af34ef_story.html?utm_term=.26d65cb883f8, accessed April 13, 2020.

Nakashima, E and Timberg, C (2017) *NSA officials worried about the day its potent hacking tool would get loose. Then it did.* https://www.washingtonpost.com/business/

technology/nsa-officials-worried-about-the-day-its-potent-hacking-tool-would-get-loose-then-it-did/2017/05/16/50670b16-3978-11e7-a058-ddbb23c75d82_story.html?noredirect=on&utm_term=.7f27f2e1b9b4, accessed April 13, 2020.

Poushter, J and Fetterolf, J (2019) *International publics brace for cyberattacks on elections, infrastructure, national security*. https://www.pewresearch.org/global/2019/01/09/international-publics-brace-for-cyberattacks-on-elections-infrastructure-national-security/, accessed April 13, 2020.

Rid, T and Buchanan, B (2014) 'Attributing cyber attacks', *Journal of Strategic Studies*, 38(1), pp. 4–37.

Rogers, D (2015) 'Extraditing Kim Dotcom: a case for reforming New Zealand's intelligence community?', *Kotuitui: New Zealand Journal of Social Sciences Online*, 10(1), pp. 46–57.

Rogers, D (2018) 'Intelligence and security act 2017: a preliminary critique', *New Zealand Law Review*, 4, pp. 657–692.

Schulzke, M (2018) 'The politics of attributing blame for cyberattacks and the costs of uncertainty', *Perspectives on Politics*, 16(4), pp. 954–968.

Schwartz, M (2018) *NotPetya: from Russian intelligence, with love*. https://www.bankinfosecurity.com/notpetya-from-russian-intelligence-love-a-10589, accessed April 13, 2020.

Small, Z (2018) *Jacinda Ardern doubles down on New Zealand's independent foreign policy*. https://www.newshub.co.nz/home/politics/2018/10/jacinda-ardern-doubles-down-on-nz-s-independent-foreign-policy.html, accessed April 13, 2020.

Smith, B (2017) *The need for urgent collective action to keep people safe online: lessons from last week's cyberattack*. https://blogs.microsoft.com/on-the-issues/2017/05/14/need-urgent-collective-action-keep-people-safe-online-lessons-last-weeks-cyberattack/#sm.00000xl4qcz818edarjiw7w28w6qj, accessed April 13, 2020.

Snow, J (2017) *Petya ransomware eats your hard drives*. https://www.kaspersky.com/blog/petya-ransomware/11715/, accessed April 13, 2020.

Snow, J (2018) *The five most notorious cyber attacks*. https://www.kaspersky.com/blog/five-most-notorious-cyberattacks/24506/, accessed April 13, 2020.

Sontag, S (1978) *Illness as metaphor: AIDs and its metaphors*. New York: St Martins Press.

Robot, M (2017) *New Petya / NotPetya / ExPetr ransomware outbreak*. https://www.kaspersky.com/blog/new-ransomware-epidemics/17314/, accessed April 13, 2020.

Vance, A (2015) *John Key and SIS boss made an ambulance-chasing pitch about Jihadi brides*. https://www.tvnz.co.nz/one-news/new-zealand/andrea-vance-john-key-and-sis-boss-made-an-ambulance-chasing-pitch-about-jihadi-brides, accessed April 13, 2020.

Verbel, J (2014) 'The NSA and Edward Snowden', *Computers and Society*, 44(3), pp. 14–20.

Waldie, P (2018) *Trudeau and key allies condemn Russia's role in cyberattacks, poisoning*. https://www.theglobeandmail.com/world/article-trudeau-and-key-allies-condemn-russias-role-in-cyberattacks/, accessed April 13, 2020.

Unauthored websites

Foreign office minister condemns Russia for NotPetya attacks. https://www.gov.uk/government/news/foreign-office-minister-condemns-russia-for-notpetya-attacks, accessed April 13, 2020.

Operation burnham inquiry. https://operationburnham.inquiry.govt.nz/, accessed April 13, 2020.

Statement from the press secretary. https://www.whitehouse.gov/briefings-statements/statement-press-secretary-25/, accessed April 13 2020.

The royal commission of inquiry into the attack on Christchurch Mosques. https://christchurchattack.royalcommission.nz/, accessed April 13, 2020.

14 Disrupting paradigms through new technologies

Assessing the potential of smart water points to improve water security for marginalized communities

Nathan John Cooper

Introduction

Today a billion people lack access to safe drinking water and 3.6 billion live in areas with potential water scarcity (WWAP, 2018). Growing human population and dwindling groundwater reserves are conspiring to make water security among the most urgent and profound of global challenges. For decades, conflicts over water resources have been predicted across the world and, in particular, around transboundary lakes and river basins in the Middle East, South America, Central Asia, and Africa. Violent altercations over water access between tribes on the Kenya-Ethiopia border are commonplace, while the significant and sustained conflicts in Darfur, western Sudan, which raged for over a decade, are now accepted as having been largely driven by water scarcity (Polgreen, 2007). This combination of climate change and human pressure on water resources is straining the global hydrological cycle, leading to warnings that by 2050 an additional half a billion people will be subject to water stress, leading to further damaging interventions in water systems (including water table depletion through shallow boreholes and ad hoc river diversions) (Stockholm Resilience Centre, undated).

Advancements in water technology offer the possibility of achieving reliable, sustainable, and equitable water services for users in marginalized communities, but, at the same time, they represent a disruption to established relationships vis-à-vis water management. This chapter reflects on three geographical areas in order to consider: First, the paradigm-shifting legacy of the "water war" in South America; second, the consequences of pursuing cost-recovery through prepaid water meters in South Africa; and third, the potential effects on water security of smart pumps in The Gambia. The chapter considers the effects of technological interventions to help achieve local water security, and theorizes on the complex inter-relational and institutional dynamics that are necessarily affected by its arrival.

Water scarcity as a threat to security – disrupting water governance paradigms

The connection between secure access to sufficient, safe water and healthy and productive life is obvious, as is the inverse connection between insecure and insufficient water access, ill health, and poverty. What may be a less obvious, but no less present, connection is that between water security and social and political stability. This connection is illustrated well by events in South America, which culminated in the so-called "Cochabamba Water War", where water services were controversially privatized, then renationalized, in the Cochabamba region of Bolivia.

A case from Argentina serves as a useful point of departure for understanding events in Cochabamba. The case of *Compania de Aguas de Aconquija (AdA) v Argentine Republic*, (ICSID Case No. ARB/97/3) involved a 30-year water service contract signed in 1993 between AdA (Vivendi Universal), a French multinational water services company, and Tucuman province, Argentina (Assies, 2003). The water service contract contained no provisions to ensure stability of water prices and water quality. Prices subsequently rose, and there were two specific incidents that gave rise to allegations of poor water quality. As a result, in 1996 the provincial government attempted to renegotiate the contract with the intention of securing reduced water prices and more acceptable water quality for its residents. The renegotiation failed, resulting in repudiation of the contract by AdA along with a claim for US$300 million in damages.

The contract stipulated that litigation was subject to the jurisdiction of the local Tucuman Court. However, the relevant French-Argentinian Bilateral Investment Treaty directed that jurisdiction lay with the International Centre for the Settlement of Investment Disputes (ICSID). This was confirmed by ICSID arbitrators, and the case was duly heard at the ICSID. AdA's claim for damages was awarded, which was interpreted as a victory for corporate capital and a reassurance to private water service providers that their interests were well-protected, despite considerable public opposition and civil unrest.

In 1999, in the Bolivian city of Cochabamba, a similar story began to play out. The previously state-owned water supply company, Semapa, was privatized. The privatization contract went to Aguas del Tunari, a joint venture with the US multinational construction company, Bechtel. Immediately water rates were raised by 35%. Consequently, many residents could not afford to pay for their water, becoming disenfranchised from what they considered their inherent right (affordable access to sufficient water) (Assies, 2003, p. 15). This led to large-scale, sustained public protest.

As in the previous case, the water service provider, Aguas del Tunari, also sought to enforce the contract through the ICSID, claiming between US$25 and US$100 million in damages. But unlike in the previous case, the Bolivian government withdrew from the ICSID in response to overwhelming domestic pressure, including civil unrest which claimed two lives. As a result of withdrawal from the ICSID, local jurisdiction over the dispute was reasserted. Eventually, Aguas del Tunari was forced to agree to an out-of-court settlement for a symbolic payment of two bolivianos (Bolivia's national currency) (ibid, p. 30). On the April 10, 2000, less than five months after the privatization contract had been signed, the Bolivian government reversed the privatization and returned Cochabamba water to state ownership (ibid).

One important legacy of the Cochabamba Water War is that it was here that issues of water rights (articulated since Cochabamba as "water justice") and community-based resistance to the commodification and privatization of water were first located within broader struggles for global justice, including global environmental justice (Mehta et al, 2014, p. 161). It is also important to note that the Cochabamba Water War was influenced by non-Western, indigenous worldviews and communal ways of life that emphasized the importance of keeping access to water available for all, through an understanding of water as part of the "commons". The concept of commons encompasses notions of communality, manifested through structures of community deliberation and decision-making regarding resources conceived of as collective, corporately held, or common, and including water (Assies, 2003, pp. 16–17). As such, Cochabamba represents a bold challenge to the dominant paradigm of water-as-commodity. By resisting water commodification that results in disenfranchisement, and by restating water as part of the (global) commons, a paradigm of "water as commons" can be discerned, the disruptive effects of which continue to be seen in contemporary water governance. Indeed, extending beyond the context of water services, commons logic represents a challenge to atomized individualism – embodied in the dominant neo-liberal narrative of citizens as consumers – and a counterpoint to Westphalian state sovereignty, independence, and isolation.

Another important legacy of Cochabamba is its influence on the 2010 United Nations General Assembly resolution on the right to water and sanitation, which is the most recent international affirmation of water as a universal human right. A brief summary of this human right now follows.

A human right to water in the *Anthropocene* – national obligations and international cooperation within a context of social-ecological security

Access to sufficient water is a basic requirement for life. Although it is not mentioned explicitly as a human right in the Universal Declaration of Human Rights (UDHR, 1948) nor the International Covenant on Economic Social and Cultural Rights (ICESCR, 1966), access to sufficient water has progressively been recognized internationally as a human right since the 1977 UN Water Conference in Mar del Plata. In 2003, the General Comment No. 15, issued by the Committee on Economic, Social, and Cultural Rights (CESCR) reemphasized water as a prerequisite for the realization of other human rights and restated that access to water was itself a human right. In July 2010, the UN General Assembly adopted a resolution recognizing access to clean water and sanitation as a human right (UNGA Resolution 64/292, 2010). This further entrenched access to sufficient water as an internationally accepted human right to which the obligations of states party to the ICESCR apply. Indeed, the resolution was introduced by the Bolivian Ambassador to the UN, Pablo Solon, acknowledging the importance of events in Cochabamba in securing this international commitment.

Part II of General Comment No.15 begins as follows:

> The right to water contains both freedoms and entitlements. The freedoms include the right to maintain access to existing water supplies necessary for the right to water, and the right to be free from interference, such as the right to be free from arbitrary disconnections or the contamination of water supplies. By contrast, the entitlements include the right to a system of water supply and management that provides equality of opportunity for people to enjoy the right to water.
>
> (CESCR, 2003, para 10)

Such freedoms and entitlements reflect the negative *and* positive obligations on states: To respect, to protect, and to fulfil.

Part IV of the General Comment sets out actions and omissions which amount to a violation of the right to water. Violations can occur through acts of omission, corresponding to a failure to realize "positive" obligations to protect and/or fulfil. Violations can also occur through acts of commission, which are contrary to the "negative" obligation to respect. For instance, adoption of retrogressive measures incompatible with the right to water, arbitrary or unjustified disconnection, pollution or diminution of water resources may also count as violations by commission (CESCR, 2003, paras 42, p. 44). The degree to which introducing new water technology in the form of prepaid water meters can be characterized as one such retrogressive measure is discussed in the following section.

Declared as a universal human right, access to water, and the challenges that it poses, must be applied to the particular context(s) in which it operates. While geographical, climatic, and socioeconomic conditions conspire to create distinctive challenges, the emerging concept of social-ecological security (SES) offers an important general and contemporary context within which to situate the human right to water, which is itself contextualized by the exigencies of the Anthropocene.

Social-ecological security (SES) attempts to better articulate the multifarious challenges to the security of the human (and nonhuman) environment. In particular, it emphasizes that social and human security cannot be separated from ecological security. It thus reflects a changing notion of security toward a broader, more interconnected, and contingent understanding, better placed to problematize and address the changing nature of ecological and developmental threats faced. Such threats have been defined as:

> an action or sequence of events that (1) threatens drastically and over a relatively brief span of time to degrade the quality of life for the inhabitants of the Earth, or (2) threatens significantly to narrow the range of policy choices available to the international community, governments of states or to private, nongovernmental entities. (persons, groups, corporations).
>
> (Ebbesson, 2014, p. 77)

Applied to water, this means that not only must water services be "socio-sustainable" (sustainable from the perspective of human development) but they must also be "eco-socio-sustainable" if such services are to withstand the manifold crises of planetary degradation (Crutzen & Stoermer, 2000, pp. 17–18). While traditionally threats to water, food, and energy security have been addressed in isolation, SES offers not only a more holistic framework for understanding and responding to these challenges, but also gains relevance when set against the backdrop of the Anthropocene. The Anthropocene has been unofficially proposed as a new geological epoch (ibid). Formally still, the *Holocene* remains the present epoch, as it has for 10,000–12,000 years. It has been characterized in the main by stable environmental conditions, which have supported the enormous growth of the human population, and the development of modern societies. But the global human imprint on the biosphere has become so significant that the Earth is moving into a critically unstable and inharmonious state (Steffen et al, 2007, p. 615). It is asserted that humankind has played a central role in moving the planet toward this critically unstable state, characterized by less predictable and less harmonious Earth systems (ibid). So central is this role, that humanity can be considered a discrete geological force capable of moving Earth systems outside their natural range of variability and into a new and unstable epoch.

The effects of the Anthropocene on water are not yet fully understood. But salination, drought, and heavy rainfall are all consequences of the less predictable weather patterns experienced globally. Furthermore, so crucial is fresh water for life, that pressure on water supplies adversely affects aspects of human life ranging from food security, sanitation, health, and economic development (Meisch, 2014, p. 427).

Unsurprisingly, SES and the Anthropocene are becoming part of the discourse around environmental law and governance,[1] giving new impetus and urgency to principles including sustainable development, while offering new perspectives with which to interrogate the juristic interventions that must ultimately be better able to respond to the exigencies of the Anthropocene – now and well into the future.

The negative consequences of new water technologies

In light of the importance of achieving and maintaining universal access to clean water, for social, political, economic and ecological security, and aware of states' obligations to do the same, the chapter now turns to examine some of the consequences of imposing one instance of new water technology onto existing water supply structures. In South Africa, beginning in 2007, the City of Johannesburg and its water company, Johannesburg Water, began to install prepayment meters onto standpipes in the greater Soweto area. This came as a response to acute water losses in Soweto as a result of corroded pipes, an inaccurate tariff system (that meant more water was used than was predicted to be necessary), and a "culture of non-payment" for water services that had "arisen originally as part of the resistance to apartheid local government" (*Mazibuko and others v City of Johannesburg and others*, 2009, para 166, hereafter *Mazibuko* (CC)). In short, prepayment meters were heralded as necessary technological interventions to improve stewardship of water and to ensure a greater level of participation (through payment) in this improved system.

It is important to note the particular context within which this disruption to the status quo took place and the implications of this for the success or failure of the technological intervention. Such context is aptly provided in the opening paragraphs of the case of *Mazibuko*, which is discussed:

> Although rain falls everywhere, access to water has long been grossly unequal. This inequality is evident in South Africa. While piped water is plentifully available to mines, industries, some large farms and wealthy families, millions of people, especially women, spend hours laboriously collecting their daily supply of water from streams, pools and distant taps ... despite the significant improvement in the first fifteen years of democratic government, deep inequality remains and for many the task of obtaining sufficient water for their families remains a tiring daily burden. The achievement of equality, one of the founding values of our Constitution, will not be accomplished while water is abundantly available to the wealthy, but not to the poor.
>
> (*Mazibuko* (CC), para 2)

Shortly after such meters were installed, a group of residents from the Phiri area of Soweto began court proceedings to challenge the legality of these meters in light of the country's constitutional right of access to sufficient water (Constitution of the Republic of South Africa 1996, s. 27(1)(b)). In the 2008 High Court case of *Mazibuko and others v City of Johannesburg and others* (*Centre on Housing Rights & Evictions as amicus curiae*), hereafter *Mazibuko* (W), it was contended that since prepayment water meters, by design, require users to pay for water in advance, access to sufficient water is curtailed if users cannot afford to prepay. Such a situation was commonplace for Phiri residents and was raised as incompatible with the constitutional right of access to sufficient water. Quantification of sufficient water, in the Water Services Act (WSA) as a minimum standard of 25 liters per day (lpd) was directly challenged in this case on the basis that what is a sufficient quantity of water depends on the requirements of users in particular social circumstances. For instance, people using waterborne sanitation require a greater volume of water to support life and personal hygiene than those using pit latrines.[2] The decision of the High Court put great emphasis on the need to redress past injustices (as a result of apartheid policies) and the dire social and material state of many Phiri residents, described as "poor, uneducated, unemployed and ravaged by HIV/AIDS" (*Mazibuko* (W), para 5).

In determining the applicants' grounds, the High Court looked to the CESCR General Comment No. 15. Applying the General Comment, the court's view was that "[T]he State is under an obligation to provide the poor with the necessary water and water facilities on a non-discriminatory basis" (*Mazibuko* (W), para 36).

Moreover, the progressive realization of the constitutional right of access to sufficient water meant that:

> Retrogressive measures taken by the state are prohibited. If such retrogressive measures are taken, the onus is on the state to prove that such retrogressive measures are justified with reference to the totality of the rights provided for in the Covenant (ICESCR, 1966). The state is obliged to respect, protect and fulfil the right to water.
>
> (*Mazibuko* (W), para 37)

The installation of prepayment meters was held to be just such a retrogressive step, preventing residents from access to sufficient water that they had previously enjoyed (before the prepayment meters, Phiri residents had access to a constant supply of water – despite many accruing arrears as a result).[3] The retrogressive step was taken without adequate justification.

It was held that, given the particular needs of the Phiri community (including the need to use waterborne sewerage) a volume of 50 lpd would be a more appropriate quantification of sufficient water than the statutory 25 lpd limit. Satisfied that the respondent could provide this increased amount "without restraining its capacity on water and its financial resources" (*Mazibuko* (W), para 181), the High Court decided wholly in the applicants' favor.

The City of Johannesburg and Johannesburg Water appealed to the South African Supreme Court of Appeal in February 2009 (*City of Johannesburg & others v Mazibuko & others* (*Centre on Housing Rights & Evictions as amicus curiae*), hereafter *Mazibuko* (SCA)). Here the quantity amounting to sufficient water for Phiri residents was reduced on appeal to 42 lpd. But the High Court's approach was otherwise upheld, and the City of Johannesburg and Johannesburg Water were directed to formulate a revised water policy accordingly (*Mazibuko* (SCA)).[4]

Mazibuko in the High Court and Supreme Court of Appeal was heralded as an important milestone in socioeconomic jurisprudence in South Africa (Jansen van Rensberg, 2008, p. 434). It showed the courts' willingness to push the legislature toward concrete manifestations of constitutional rights and not to allow the "progressive realization" of these rights to result in unconstitutional policies. The impetus to promote and fulfil the right of access to sufficient water was clearly discernible (particularly in Tsoka J's High Court judgment in *Mazibuko* (W)) in the acceptance of the need for sufficient water to be a quantity that promotes dignity and goes beyond the minimum of Free Basic Water already set (*Mazibuko* (W), para 1). The potential implications of *Mazibuko* for people living in similar situations to the Phiri residents were significant. Both decisions demonstrated the courts' engagement with polycentric matters in order to help realize socioeconomic constitutional rights more quickly and more explicitly than would otherwise be the case.

However, in September 2009 the Phiri residents appealed to the Constitutional Court (unhappy with the Supreme Court of Appeal's order to reduce the amount of water deemed to be sufficient from 50 to 42 lpd). This was the first time the Constitutional Court had considered the proper interpretation of the right of access to sufficient water. The orders made by the High Court and Supreme Court of Appeal, respectively, were set aside.

The Constitutional Court held that the City of Johannesburg's Free Basic Water policy was not in conflict with Section 27 of the Constitution or Section 11 of the Water Services Act[5] and the installation of prepaid water meters was lawful. The court was satisfied that, while the Free Basic Water Policy was flawed, it was consistent with the constitutional right of access to sufficient water (*Mazibuko* (CC), para 163). Consequently, the applicants' appeal was dismissed, and the installation of prepaid water meters in Phiri was affirmed as compatible with Section 27 of the Constitution (*Mazibuko* (CC), para 169). Indeed, installation of prepayment meters has continued.

The final *Mazibuko* decision has spawned much criticism from activists and academics alike. Although it might be defended as a constitutionally deferent and pragmatic judgment, it must be acknowledged as profoundly disappointing for the original claimants and the millions of other South Africans in similar situations.

To those Phiri residents now denied a quantum of water commensurate with their needs, and necessary for their dignity, their right to water rings hollow. Their experience of new water technology has been one of disenfranchisement and dislocation from the most basic of necessities. While the aims of water suppliers in the country to improve sustainable water supply and optimize its use are laudable, it would appear that the policy of imposing prepayment meters lacks the sensitivity necessary to simultaneously promote water justice and protect people's dignity. For these reasons, prepayment meters in South Africa represent a negative experience of new water technology on people's access to water. The next section contrasts this with something rather more positive, although no less complex in its layering of socio-cultural-political-technological-legal considerations. Here the focus turns to The Gambia and to the potential of "smart pumps" to improve access to water for people living in rural areas.

The positive potential of new water technologies to improve access to water

The Gambia is a small country in West Africa with a population of 2 million. It is a popular tourist destination for Europeans in search of winter sun. Other main industries include fishing and peanuts. Average annual national income per capita is just over US$1,500, and life expectancy is 61 for men and 63 for women. In keeping with much of the region, The Gambia suffers from many symptoms of poverty, including high infant mortality rates and a significant portion of the population using unimproved drinking water sources and sanitation facilities (UNICEF, 2013).

The author had the privilege of visiting The Gambia in 2017, as part of an experimental water technology team. We were invited by the Department of Water Resources, within the newly created Ministry of Water and Fisheries, in partnership with a UK-based NGO, the GLOVE Project. Our task was to install prototype monitoring devices to existing hand-operated water pumps in villages on the edge of Serekunda. These devices were designed to monitor the flow (performance) of pumps and to communicate this information remotely by means of mobile phone short message service (SMS) technology.

In the Serekunda area, most hand pumps are between 25 and 30 years old. Many hand pumps have become ineffective (requiring multiple depressions of the pump handle to draw any water), while others have ceased working altogether. This pattern of disrepair and obsolescence is familiar and well-documented across the continent. Studies report that between 20% and 65% of hand pumps installed in various African countries are broken or out of use (RWSN, 2010). It is estimated that approximately 61.8 million people across the continent are served by broken water pumps. All types of water pumps will deteriorate and exhibit worsening performance with age (De Palencia and Pérez-Foguet, 2011). But when such infrastructure malfunctions, local communities will often resort to using less-protected water sources, increasing their exposure to a wide range of water-related diseases.

If broken pumps remain in a state of disrepair, and as established pumps get older, this could represent a regression in people's access to water, contrary to the states' obligations toward *progressive* realization of the human right to water (as discussed earlier). Furthermore, broken pumps constitute a capital loss in terms of the investment that is represented by this infrastructure. It is reported that over the last 20 years, broken hand pumps in Africa have represented between \$1.2 and \$1.5 billion of lost investment, with 30%–40% of rural water systems failing prematurely (USaidwaterckm, 2016).

The operational and reliability problems of such pumps have been attributed to a wide range of factors including insufficient local financial resources to fund necessary repairs, limited access to spare parts, limited technical capacity within the user community, inappropriate project implementation and/or technology choice, and limited post-construction monitoring and support from external agencies. It is reported that less than 5% of WASH (water, sanitation, and hygiene) projects are visited after installation, and, as such, broken infrastructure frequently goes undetected or is not addressed by relevant stakeholders (USaidwaterckm, 2016).

Clearly the problems of deficient and broken water pumps in The Gambia are broadly in keeping with continent-wide challenges. Specifically, in Serekunda villages, one hand pump serves around 80 compounds (extended households). Generally, these pumps were found to be poorly maintained, despite being crucial to the water needs of so many people. While problems with bearings and other mechanical aspects of the pumps resulted in reduced flow (and therefore the need for more pumping to draw the required amount of water), a more serious symptom of disrepair was the contamination of the water source (usually a fairly shallow borehole) during the rainy season, once the concrete foundation and casing for the pumps deteriorated and allowed seepage.

Improved post-construction monitoring of remote water projects by so-called "smart pumps", which can remotely monitor operational performance in place of physical site visits, could potentially solve some of these challenges and reduce some of the heavy time and resource demands on stakeholders that are characteristic of traditional monitoring strategies. It was this potential that our team tried to test and which was met with mixed success: Installation of the smart devices into some hand pumps proved impossible, because the pump aperture was too narrow. Some other devices failed to send SMSs once installed. Data *were* generated by some correctly-fitted and operational devices, until their battery packs ran down – providing a useful picture of some pump performance for a limited time period, all of which has been fed back into the research and development process. But any effective addition of new technology to established modes of supply must be attempted with an appreciation of the social, cultural, legal, and other contexts to which it connects and inevitably disrupts.

Responsibility for, and governance of, water sources in the Serekunda villages visited, is complex. A village development committee, comprised of a district chief, village chiefs, and elected representatives (usually covering more than one village) takes overall responsibility for the water pumps, along with other infrastructural issues. Within this is a smaller water committee, responsible for collecting residents' water payment subscriptions and for organizing and paying for repairs. At both levels, there was an explicit commitment made to equal representation of both genders in written documentation. But in reality, membership of each of the committees we interviewed was overwhelmingly male. This contrasts with the fact that collection of water from the hand pumps is accepted as being the exclusive preserve of women and young children. In discussions with residents, this imbalance was the source of considerable tension, since it is often women who have to work extra hard to draw water when the (male-dominated) committees fail to maintain the pumps. There were also occasional problems with some residents who failed to pay their subscription. But a combination of charity and collective responsibility within the villages appeared to be sufficient to overcome such problems. While financial resources earmarked for repair were never large, there was some money available. Indeed, a commons type approach to water resources (mentioned earlier in this chapter) was clearly discernible in the communal approach to questions of payment and repair, which stands in contrast to the dominant paradigm of water-as-commodity, so prevalent in some other cultural contexts.

The primary problem facing communities appears to be that committee members and residents often struggled to identify which part(s) of the pumps needed repair or replacement. Because of this, they are reliant on itinerant mechanics and inspectors from the relevant government department, whose occasional visits may not correspond with times when repairs are needed and who, on arrival, may not have the necessary parts or experience to resolve the matter without further delays.

Such delays undoubtedly have the potential to force residents into using water sources that are less safe (including shallow, self-dug wells and rivers), and also to raise tensions within and between communities. Into this context the potential of smart hand pumps, that can self-identify faults and then send this notification immediately and directly to those able to make repairs, is obvious. While it would not solve all issues, there is scope to significantly increase the speed with which faults are identified, responded to, and repaired. This could translate into material improvement in people's experience of water security. Questions remain, not least around who the appropriate recipients of such notifications should be. Representatives of the Ministry of Water were quick to emphasize their own limited resources and the challenges of a dispersed, rural population, in a poor country.

One possibility is that smart pump data form part of monitoring initiatives by stakeholder NGOs (like the GLOVE Project) in order to inform and direct their development priorities. Additionally, this could enhance public engagement and awareness efforts by feeding directly into infographic resources which show real-time information on the status of hand pumps in an area and highlighting hot spots with a high density of problems. This could also complement efforts toward realization of Sustainable Development Goal 6 on water and sanitation and its associated monitoring requirements (UN Water).

Certainly, any permanent (or long term) application of such smart pumps would also require an ongoing dialogue with community stakeholders and decision-makers while showing cultural sensitivity to these governance structures. But as an early experiment in helping to improve secure access to water through new technology, it appears to be broadly positive. To-date, fieldwork continues.

Conclusions

Life and clean water are inseparable. Yet water is becoming increasingly scarce, and ever more people are becoming subject to water stress. Such an existential threat must surely lead us to pursue all appropriate avenues in order to reach the one billion people who currently live in water poverty, and to secure safe, sufficient, and sustainable water access for present and future generations. This goal becomes even more imperative in light of the growing number of violent confrontations over water and the potential for water scarcity to become a common flashpoint for conflict and political unrest, as evinced in Bolivia.

Advancements in new water technology which seek to improve people's access to water offer one such avenue toward water security. As such, they must be thoroughly researched and understood within the social, cultural, political, and legal context, in which they may be deployed, if they are to realize their potential to optimize existing water resources without contributing to regressive dynamics of disenfranchisement and disconnection, as seen in South Africa.

Such technology offers the possibility of achieving reliable, sustainable, and equitable water services for users in marginalized communities, if applied sensitively, and in partnership with existing community governance structures (however imperfect they may be), as illustrated by recent trials in The Gambia. The intervention of technology does represent a disruption to the established social, cultural, political, and legal relationships governing water, but the effects of this disruption need not be negative. Rather, the positive improvements in people's access to water are there to be realized.

Notes

1 For a list of the more recent publications, see among others: Rakhyun, K. and Bosselmann K. (2013) International Environmental Law in the Anthropocene: Towards a Purposive System of Multilateral Environmental Agreements. *Transnational Environmental Law,* 2, 285–309; Scott, K. (2013) International Law in the Anthropocene: Responding to the Geoengineering Challenge. *Michigan Journal of International Law*, 34(2) 309–358; Robinson, N. (2012) Beyond Sustainability: Environmental Management for the Anthropocene Epoch. *Journal of Public Affairs*, 12(3) 181–194; Kotzé, L. J. (2014) Human Rights and the Environment in the Anthropocene. *The Anthropocene Review,* 1(3) 252–275; Kotzé. L. J. (2014) Transboundary Environmental Governance of Biodiversity in the Anthropocene. In Kotzé L. J. and Marauhn, T (eds), *Transboundary Governance of Biodiversity*. Leiden: Brill Nijhoff, 12–33; Kotzé, L. J. (2014) Rethinking Global Environmental Law and Governance in the Anthropocene. *Journal of Energy and Natural Resources Law,* 33(2) 121–156; Ebbesson, J. (2014); Philippopoulos-Mihalopoulos, A. (2014) *Spatial Justice: Body, Lawscape, Atmosphere*. London: Routledge.

2 This is particularly pertinent to the interpretation of sufficient water in the Constitution since section 27 links food and water: "Everyone has the right to have access to … sufficient food and water". Also, since sanitation is not listed in section 27 of the Constitution, but is recognized as a right in the Water Services Act (section 3(1)), the volume of water that is sufficient must depend on the type of sanitation system being used.

3 *Mazibuko,* (W) Note: Prior to installation of prepayment meters and the associated improvements made to water pipes as part of the City's water services improvement project in Soweto, "Operation Gcin'amanzi", water services were poor, but the volume of water available was unlimited (except when affected by intermittent technical problems).

4 *Mazibuko*, (SCA) Note, because the SCA found that 42 lpd was the quantity of sufficient water, not 50 lpd as decided by the High Court, the appeal was upheld.

5 Water Services Act 1997 (108 of 1997), the duty on the part of the Water Services Authorities to provide access to water services is clearly spelled out in section 11(1): "Every water service authority has a duty to all consumers or potential consumers in its area of jurisdiction to progressively ensure efficient, affordable, economical and sustainable access to water services."

Note: While this duty is subject to a number of conditions including inter alia the availability of resources and the duty of consumers to pay reasonable charges (11(2)), the Water Services Act entrenched this duty by stating in section 11(4) that a water services authority may not unreasonably refuse to give access to water services to a consumer or potential consumer in its area of jurisdiction. Further in section 11(5), the act states that in emergency situations, a water service authority must take reasonable steps to provide basic water supply and basic sanitation services to any person within its jurisdiction and may do so at the cost of that authority.

References

Assies, W (2003) 'David versus Goliath in Cochabamba: water rights, neoliberalism, and the revival of social protest in Bolivia', *Latin American Perspectives*, 30(3), pp. 14–36.

City of Johannesburg & others v Mazibuko & others (Centre on Housing Rights & Evictions as amicus curiae) [2009] JOL 23337 (SCA).

Compania de Aguas de Aconquija (AdA) v Argentine Republic ICSID Case No. ARB/97/3.

Constitution of the Republic of South Africa, 1996.

Crutzen, PJ and Stoermer, EF (2000) 'The anthropocene, global change newsletter', *International Geosphere–Biosphere Programme (IGBP)*, 41, pp. 17–18.

De Palencia, AJF and Pérez-Foguet, A (2011) 'Implementing pro-poor policies in a decentralized context: the case of the rural water supply and sanitation program in Tanzania', *Sustainability Science*, 6(1), pp. 37–49.

Ebbesson, J (2014) 'Social-ecological security and international law in the anthropocene', in Ebbesson, J, Jacobsson, M, Klamberg, MA, Langlet, D and Wrange, P (eds.) *International law and changing perceptions of security: liber amicorum Said Mahmoudi*. Leiden: Brill Nijhoff, pp. 71–92.

GLOVE Project. https://www.gloveproject.life/water/hand-pump, accessed November 2019.

Jansen van Rensburg, L (2008) 'The right of access to adequate water [discussion of Mazibuko v the city of Johannesburg case no 13865/06]', *Stellenbosch Law Review Stellenbosch Regstydskrif*, 19(3), pp. 415–435.

Kotzé, LJ (2014a) 'Human rights and the environment in the anthropocene'. *The Anthropocene Review,* 1(3), 252–275.

Kotzé, LJ (2014b) 'Rethinking global environmental law and governance in the anthropocene'. *Journal of Energy and Natural Resources Law,* 33(2), 121–156.

Kotzé, LJ (2014c) Transboundary environmental governance of biodiversity in the anthropocene. In Kotzé, LJ and Marauhn, T (eds), *Transboundary Governance of Biodiversity*. Leiden: Brill Nijhoff, 12–33.

Mazibuko and others v City of Johannesburg and others [2009] JOL 24351 (CC).

Mazibuko and others v City of Johannesburg and others (Centre on Housing Rights & Evictions as amicus curiae) [2008] JOL 21829 (W).

Mehta, L, Allouche, J, Nicol, A and Walnycki, A, (2014) 'Global environmental justice and the right to water: the case of peri-urban Cochabamba and Delhi', *Geoforum*, 54, pp. 158–166.

Meisch, S (2014) 'The need for a value-reflexive governance of water in the anthropocene', in Bhaduri, A, Bogardi, J, Leentvaar, J and Marx, S (eds) *The global water system in the anthropocene*. Switzerland: Springer International Publishing, pp. 427–437.

Philippopoulos-Mihalopoulos, A (2014) *Spatial justice: body, lawscape, atmosphere*. London: Routledge.

Polgreen, L (2007) 'A godsend for darfur, or a curse?', *The New York Times*, July 22. https://www.nytimes.com/2007/07/22/weekinreview/22polgreen.html, accessed November 2019.

Rakhyun, K and Bosselmann, K (2013) 'International environmental law in the anthropocene: Towards a purposive system of multilateral environmental agreements'. *Transnational Environmental Law*, 2, 285–309.

Robinson, N (2012) 'Beyond sustainability: environmental management for the anthropocene epoch'. *Journal of Public Affairs*, 12(3), 181–194.

RWSN (Rural Water Supply Network) (2010) *Myths of the rural water supply sector, supply network perspectives no. 4*. St. Gallen: RWSN.

Scott, K (2013) 'International law in the anthropocene: responding to the geoengineering challenge'. *Michigan Journal of International Law*, 34(2), 309–358.

Steffen, W, Crutzen, PJ and McNeill, JR (2007) 'The anthropocene: are humans now overwhelming the great forces of nature?', *AMBIO: A Journal of the Human Environment*, 36(8), pp. 614–622.

Stockholm Resilience Centre (undated) *The nine planetary boundaries*. Stockholm: Stockholm University. https://www.stockholmresilience.org/research/planetary-bound

aries/planetary-boundaries/about-the-research/the-nine-planetary-boundaries.html, accessed November 2019.

United Nations Committee on Economic, Social and Cultural Rights (CESCR) (2003) *General comment no. 15: the right to water (Arts. 11 and 12 of the Covenant)*, 20 January 2003, E/C.12/2002/11. https://www2.ohchr.org/english/issues/water/docs/CESCR_GC_15.pdf, accessed November 2019.

UNICEF (2013) *At a glance: Gambia*. UNICEF. https://www.unicef.org/infobycountry/gambia_statistics.html, accessed November 2019.

United Nations General Assembly (1948) *Universal declaration on human rights*. (217 A (III)) 10 December 1948.

United Nations General Assembly (1966) *International covenant on economic, social and cultural rights*. 16 December 1966. (2200A (XXI)) Entered into force 23 March 1976.

United Nations General Assembly (2010) *Resolution 67/97, The human right to water and sanitation A/RES/64/292* (28 July 2010). https://undocs.org/en/A/RES/64/292, accessed November 2019.

UN Water. https://www.unwater.org/launch-of-sustainable-development-goal-sdg-6-data-portal/, accessed November 2019.

USaidwaterckm (2016) *USAID solicitation – sustainable WASH systems* [blog]. January 29. https://sanitationupdates.blog/2016/01/29/usaid-solicitation-sustainable-wash-systems/, accessed November 2019.

Water Services Act 1997 (No.108 of 1997) South Africa: Government of South Africa.

WWAP (United Nations World Water Assessment Programme)/UN-Water (2018) *The United Nations world water development report 2018: nature-based solutions for water*. Paris: UNESCO.

15 "Just wrong", "disgusting", "grotesque"

How to deal with public rejection of new potentially life-saving technologies

Dan Weijers

Introduction: "Just wrong", "disgusting", "grotesque"

Dear researchers and policy makers, imagine that you have just finalized plans for the rollout of a new game-changing technology – one that could significantly improve security and save lives. At the first press conference for your project, some reporters start asking questions you had not anticipated. The next morning you awake to a deluge of media alerts about your project. Excited, you hurriedly begin to read them…

> "just wrong"
> "ridiculous… grotesque"
> "absurd… disgusting"
> "morally reprehensible"
> "despicable… callous… retarded."[1]

Oh dear, a tidal wave of vitriol, a veritable flood of moral repugnance! This was certainly not what you had hoped for. Why would the public respond like this to your new life-saving technology? Should you ignore this moral outrage or engage with it? Can you turn public perception back around, or is it too late? Unfortunately for the team behind Policy Analysis Market (PAM), they did not have the liberty of considering these questions before the program was terminated at the highest levels of the US government. PAM was a proposed prediction market funded by the US Defense Advanced Research Projects Agency (DARPA) in 2003 that had the potential to prevent terrorist attacks. After an unexpected public outcry, PAM was canceled without even sending a "please explain" to the team behind it (Hanson, 2007).

This chapter begins by explaining how PAM was supposed to work and how public expressions of moral repugnance led to the project being shut down. It then introduces and explains the psychology of moral repugnance and discusses a taxonomy of kinds of moral repugnance that might result from the rollout of a new or disruptive technology for security purposes. A major contribution of this chapter is the discussion of what these different kinds of repugnance should mean for new disruptive technology proposals. Using PAM as the central example, the chapter works through a taxonomy of kinds of repugnance to identify the responses that would be appropriate for teams to use when researching and implementing new technologies. The chapter closes with some words of advice about managing public perception of disruptive new security technologies.

Prediction markets and the policy analysis market

The terrorist attack on New York's World Trade Center in 2001 reportedly cost the US $3.3 trillion (Carter & Cox, 2011) and, far worse, the horrifying loss of 2,977 lives (Amadeo, 2019). In the wake of this devastating attack, DARPA, an agency tasked with generating breakthrough technologies and capabilities to aid US national security efforts, began investigating new methods to better understand international security threats, including terrorism (Hanson, 2007). A small part of this investigation was the plan to trial a prediction market, called PAM, based on the work of Robin Hanson (Hanson, 2007).

As explained by Weijers (2013a), modern prediction markets (PMs) are electronic marketplaces that facilitate the purchase and sale of shares in predictions about real-world outcomes. Participants browse a list of specific predictions and buy or sell shares in predictions they think are priced too high or too low. The price of a share in a PM varies between $0.00 and the payout price for predictions that turn out to be true. For example, in a PM that pays out $1.00 for each share in an accurate prediction, prices will vary between about $0.01 and $0.99. The most recent sale price of a share in a PM is an indication of how likely the prediction is to be true according to the collective wisdom of the participants in the market. For example, PredictIt is currently running a PM on "Who will win the 2020 U.S. presidential election?", with several specific predictions available to trade on, such as "Donald Trump will win the 2020 US presidential election". As at September 26, 2019, the collective wisdom about Trump's reelection chances is that he has a 42% chance.[2] Any participant in the market who believes Trump has a much lower chance of a second term as president would sell (or short) shares in the prediction, hoping for a payout if Trump is not elected president in 2020.[3] If Trump is elected president in 2020, then participants holding shares in the prediction will be paid out $1.00 per share, netting them $1.00 minus the average price they paid for their shares for each share they hold.

Prediction markets have proven themselves to be very well-calibrated prediction machines (Surowiecki, 2004). In other words, predictions with high share prices usually end up being true and those with low prices usually end up being false. For example, the predictions facilitated by the Iowa Electronic Markets regularly outperform polls in predicting the outcomes of national elections (Berg, Nelson, & Rietz, 2008). Encouraged by the success of PMs in many domains, several researchers have argued that carefully constructed PMs could be used to fight terrorism (Hanson, 2006a; Looney, 2004; Surowiecki, 2004; Weijers & Richardson, 2014a; 2014b; Weijers, 2016; Yeh, 2006). It has also been argued that an anti-terrorism PM, such as PAM, could be run for about the same cost as increasing the number of active spies by one (Weijers, 2016). These considerations likely encourage DARPA and Hanson to work toward a trial of PAM. The main idea behind PAM was to use technology to create a new avenue to gather information that may not have been collected by existing intelligence and security measures (Hanson, 2007). A PM that anyone from around the world could participate in might harvest the ambient wisdom of globally dispersed crowds and insights from individuals with concrete information that they would not share through existing channels (Hanson, 2006a; Weijers & Richardson, 2014b).[4]

In 2003, PAM was presented to a small audience to explain the basics of how the trial and the intended PM would work (Hanson, 2007). As was widely documented at the time, a slide showing example predictions included the assassination of Yasser Arafat and a North Korean missile attack (Hanson, 2006b). These predictions were cited by Senators Ron Wyden and Byron Dorgan (2003a) in their rebuke of PAM the day after the presentation.[5] The following media cycle was a flood of moral outrage, laced with a spattering of reasons why PAM was morally problematic and unlikely to work (Weijers & Richardson, 2014b). PAM was canceled two days later.

Moral repugnance – more than a feeling?

When someone criticizes a disruptive, possibly life-saving, security technology very strongly and very emotionally, they are probably expressing moral repugnance about it. *Moral* repugnance is potentially much more threatening to a new technology than *regular* repugnance. Garden-variety, or regular, repugnance is best thought of as intense revulsion – finding something extremely disgusting. Moral repugnance, as discussed by Weijers and Keyser (2016), has the added complexity of the disgust being explicitly or implicitly linked to one or more *moral* reasons why the target of the disgust is worthy of further contempt. Part of the complexity of moral repugnance is the often nebulous or ill-defined interplay between moral emotion and moral reason that usually occurs within people as a morally-charged intuition – a feeling that the proposed technology is abhorrent in the extreme.

As any good modern politician knows, reasons are often not as powerful as emotions – knowledge that is backed up by various experiments, including those by Taber and Lodge (2006), which demonstrated political reasoning is biased by moral emotions. Unfortunately, winning over a crowd, or a populace, seems much more easily achieved by eliciting positive emotions toward yourself than by expressing a sound line of reasoning (Lodge & Taber, 2013). Leon Kass (the former chair of President G.W. Bush's Council on Bioethics) might support this feature of politics. He has argued that emotions and intuitions, even when they are not backed by explicit reasons, are often a better guide to morality and policy than reason-based arguments (Kass, 1998). Discussing moral repugnance,[6] Kass (1998, p. 687) claims that it can be the "emotional expression of deep wisdom", and that lack of a good moral reason to explain the intuition does not impede its authority. Kass assumes there is some good moral reason for widespread moral repugnance, even if none come to mind (for any of those people!). If Kass is right about the deep wisdom of widespread moral repugnance even when it lacks explicit moral reasoning, then this is how moral repugnance might be founded upon *implicit* moral reasons.

It is reasonable to assume that most philosophers, and any people inspired by the goals of the enlightenment, would find it easier to engage in debates about the moral acceptability of new security technologies if the alleged reasons for the technologies' immorality were made explicit. Consider moral debates at home or at work. Reasons offered for opposing views provide the opportunity to gain a new perspective, to learn, and possibly refine one's own view on the issue. But, amenability to debate is not the main difficulty for those like Kass who find it acceptable to win a moral argument with feelings alone. A huge amount of research in social and moral psychology and behavioral economics has demonstrated the many ways in which our intuitions, including moral intuitions, are subject to pernicious biases.[7]

Consider the following example of moral bias.[8] Security forces are planning a rescue mission in three remote locations. Traditional methods can be employed (Plan A) or a new and relatively untested technology, such as fully autonomous drones, can be used (Plan B). Here are the best predictions of the outcomes for each plan:

1) Plan A results in 200 people being saved
2) Plan B results in a one-third chance of 600 people being saved and a two-thirds chance of saving no one

Which plan is morally preferable? Most people think Plan A is preferable because it seems morally better to have the guarantee of saving lives or to avoid the possibility of saving no one.[9]

Imagine now another remote rescue situation. Again, you are presented with two options, the second of which employs a new technology. The best predictions of the outcomes for each plan are:

1) Plan C results in 400 deaths
2) Plan D results in a one-third chance of no deaths and a two-thirds chance of 600 deaths

Which plan is morally preferable? Most people prefer Plan D in this second remote rescue scenario because it seems morally repugnant to choose a policy that results in 400 deaths when there was an option that included a chance of no one dying.[10]

The reason this is an example of how moral judgments can be biased is revealed by noticing that people tend to prefer the riskier new technology plan in the second scenario, but not in the first, despite both scenarios and sets of plans being identical in substance. The only difference between the first and second scenarios is the framing of the plans. It appears that framing a plan as saving lives or permitting deaths affects our moral intuitions about the plans. People tend to *feel* that one plan is better than another, but rational inspection of the details reveals that if those feelings are based on deep wisdom, then deep wisdom seems like a poor guide for national security decision-making in this context.

Since moral intuitions can be biased, and therefore unhelpful, policy makers should use widespread moral repugnance as a *red flag* – a warning that a thorough moral investigation may be warranted. The purpose of the investigation is to search for reasons which indicate that pushing ahead with the technology would be immoral and to carefully tease those out from negative intuitions and feelings of disgust that are not based on important moral considerations. For example, the ice cream cone is a technology that allows people to walk around while eating ice cream without generating wasteful packaging. Kass revealed in his book about nourishing the soul through eating that he finds this technology morally repugnant:

> licking an ice cream cone… offends those who know eating in public is offensive. ... This doglike feeding… ought to be kept from public view, where, even if *we* feel no shame, others are not compelled to witness our shameful behavior.
>
> (Kass, 1999, p.148–149; emphasis in original)

Despite the strong moral language used by Kass to describe eating an ice cream, a thorough moral investigation of his repugnance may well reveal no compelling moral reasons for his repugnance.

The public is outraged! What should I do?

So, you have discovered that the public is up in arms about your new disruptive technology. What should you do? Depending on whether you are a researcher or a policy maker, you will have several important interests to balance, especially ethics, professional expectations, political implications, geopolitical imperatives (such as keeping up with expected developments by strategic rivals),[11] and, to a lesser extent, prudence. The best way to respond to moral repugnance depends very much on whether or how the repugnance is, or could be, justified. The next section sets out a taxonomy of kinds of moral repugnance that makes reference to relevant contextual aspects from the point of view of researchers and policy makers proposing disruptive new technologies.

Before turning to the taxonomy, the general procedure for action should be noted. Figure 15.1, below, outlines the appropriate steps. Exactly how to carry out each step is highly context-dependent and, so, will only be discussed briefly here. This procedure for action is, as noted, a general one, so application of it should be attempted with an open mind and a decent serving of common sense. This procedure should be followed whenever moral repugnance is expected or detected. Especially for disruptive new technologies, being prepared to deal with potential moral repugnance before it occurs could easily make the difference between your idea never getting off the shelf and your new technology saving lives. So, it is best to run this procedure within your team and then with a focus group well before planning any public events or disseminating any information.

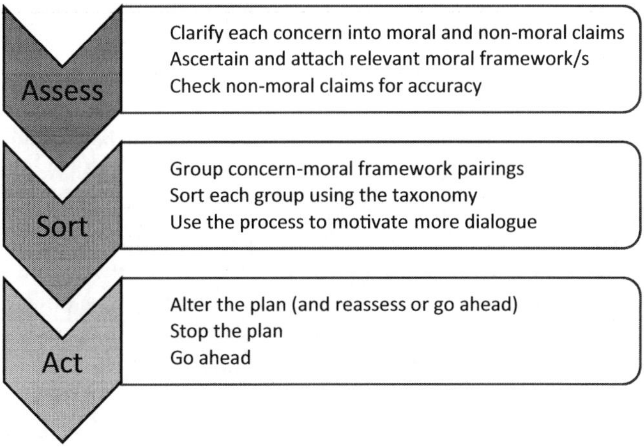

Figure 15.1 A general procedure for action when moral repugnance is expected or detected

Gather and assess the potential moral repugnance

The first step is to assess any apparent moral repugnance. Combining in-house brainstorming and primary and secondary research, gather all of the instances of potential moral repugnance. Once all of the actual and likely utterances of moral repugnance are gathered, they need to be clarified. Clarifying the concern requires identifying what the reason is for the repugnance and whether that reason is moral. It is also useful to clarify how strong and widespread the repugnance based on this reason is. Trending and similar search engine and social media metrics may help with this.

When seeking to clarify potential moral repugnance, it is important to check for records and read all of what a complainant has said. Focusing on just the attention-grabbing headlines or snippets of what someone has said can cause an unhelpful lack of clarity. The context around the snippets is where the potential moral reasons are likely to be stated, or at least implied. The expressions of moral repugnance made by Senators Ron Wyden and Byron Dorgan (2003a, no page) during their news conference on the "terror financing scheme", PAM, are an illuminating example of this. Amidst calling PAM "ridiculous… grotesque" (Wyden) and "absurd… disgusting" (Dorgan), the senators also presented several moral claims and arguments. These comments from Dorgan, for example, strongly imply reasons for his moral repugnance:

> This betting parlor on the Internet will include wagers, for example…: Will Mr. Arafat be assassinated? Will there be missile attacks from North Korea? … And those predictive bets will then give intelligence presumably to the Department of Defense. I think this is unbelievably stupid. … It is a tragic

> waste of taxpayers' money. … It is offensive. And it is, in my judgment, nearly useless.
>
> (2003a, no page)

Dorgan clearly believes that PAM will not be effective in helping intelligence forces learn anything they do not already know. Dorgan also claims that betting on death is offensive. But, clarifying the potential moral concern is not always as easy as reading a transcript. In other instances, influential people may need to be sought out and offered the opportunity to clarify their comments themselves.

In the process of clarifying each moral concern, care should be taken to distinguish between moral and non-moral claims. Moral claims link some action or state of the world to a moral value, thereby imbuing that action or state with a special force that is thought to be weightier than nearby non-moral claims. For example, "betting on death is morally wrong" is a moral claim that links the action of betting on death to a negative moral value, one that is more powerful than a mere violation of etiquette or convention. Non-moral claims do not link moral values with the world in this way. For example, the statement from earlier that the events at New York's World Trade Centre on September 11, 2001, caused 2,977 people to die does not link a moral value to any action or state of the world.[12]

Consider Dorgan's concern about PAM's ineffectiveness, which can be captured by the following premises:

1) PAM will not produce any new anti-terrorist information
2) PAM costs a lot of taxpayers' money
3) Spending a lot of taxpayers' money on something that doesn't work is wasteful of taxpayers' money
4) Being wasteful of taxpayers' money is morally wrong
5) Therefore PAM is morally wrong

Premises 1 and 2 are non-moral empirical claims – claims about how the world is or will be. If either of these claims are false, then PAM would not fit the definition of "wasteful of taxpayers' money" in premise 3. Premise 4 is the moral claim in the argument – it links being wasteful to a moral value, in this case a negative one.

The separation of moral and non-moral claims is important because the truth of non-moral claims is usually established in a different way to the truth of moral claims. Indeed, philosophers are still debating whether it is possible to establish the truth of moral claims, or even sensibly apply a truth value (e.g. true or false) to them (e.g. Nolan, Restall, & West, 2005). Given this important difference, moral and non-moral claims are treated differently in the taxonomy, which requires their prior separation.

Also, while in the process of clarifying each moral concern, it is important to ascertain the relevant complainants' moral frameworks and ensure the concerns are coupled with the relevant frameworks during the process. As discussed by Weijers and Keyser (2016, p. 101), a moral framework is a "systematic set of moral beliefs that can be reasoned through". The framework needs to be accessible via introspection, such that the complainant in question could ask themselves, "do I believe in any moral principles that are relevant to this situation?" and articulate those principles. Noting the moral framework linked to each potential complainant is important because the moral frameworks feed into the sorting process and can cause different responses to be specified to the repugnance.

The final task in the "assess" step is to investigate whether the non-moral claims are true, plausible, or false. Many non-moral empirical claims can be checked directly against current scientific findings or, if they are claims about the future, deemed plausible or otherwise based on a similar check. For example, several experts on prediction markets and security have argued that PAM would plausibly have gathered some new and useful information not otherwise available to security and intelligence forces (see Hanson, 2006b; Looney, 2004; Yeh, 2006).

Non-moral conceptual claims should also be checked. For example, a disruptive new technology may be labeled expensive or wasteful, like PAM was. Complete understanding of these concepts usually relies on some background knowledge about the domain they are being applied to. It is important to check whether the concept is being employed in a sense that is appropriate for the relevant domain of inquiry. For example, "expensive" in the domain of federal security policy means something quite different in dollar terms to the same concept used in the domain of grocery shopping. Thorough investigations of the non-moral conceptual claims should check for the inappropriate use of concepts, be they context-blind or otherwise idiosyncratic. For example, the budget for the trial or full rollout of PAM could be compared to similar projects to support or undermine claims about how "expensive" the program is. Weijers (2016) did exactly this, finding that PAM would have cost about US$1 million to set up and run, which is approximately equal to the cost of deploying one more spy in the field.

Remember to go through this process with members of your team very early on in the development process. If the team finds your new technology morally repugnant, then it is unlikely to stand a chance once the public discovers it and should not be pursued. In April 2018, over 3,000 Google employees successfully revolted against Google's involvement with Project Maven (Conger, 2018). A Pentagon initiative, Project Maven involves harnessing machine learning to help drones distinguish between humans and objects (Work, 2017). This case demonstrates the potential for security-related technologies to cause moral outrage and the importance of (at least) getting your own team onboard before developing the new technology. It is entirely possible that the outrage over Project Maven, and recognition that the Pentagon will seek to deploy other AI-enhanced military technologies in the near future, has contributed to the US Department of Defense's newly formed Joint Artificial Intelligence Center seeking to hire an ethicist to guide development and deployment of AI technologies. After all, the Center's director, Jack Shanahan, led Project Maven (Lopez, 2019).

Sort the potential moral repugnance using the taxonomy

As indicated in Figure 15.1, the first task in the "sort" step is to group the potential expressions of moral repugnance such that within groups both the concern and the relevant moral framework are the same. This sorting will likely result in several groups that can each be considered a singular concern–framework pairing from here on. Resource-based decisions may need to be made at this point; you may wish to proceed with only the groups of concerns that seem popular or contain compelling moral reasons.

The resulting groups are then subjected to the taxonomy of moral repugnance discussed in the next section. Figure 15.2, below, presents a flow chart to make using the taxonomy easier.

Ideally, this process should also be worked through in anticipation of potential public outrage as well as in response to any actual public outrage. Engaging with

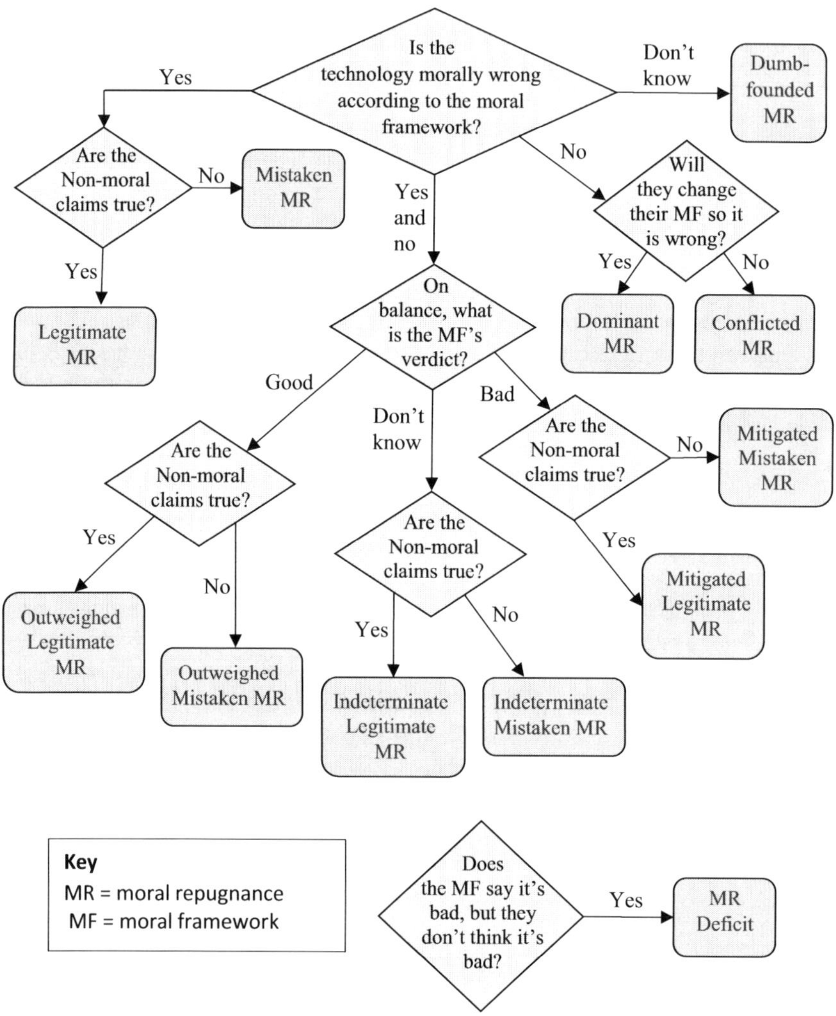

Figure 15.2 A flowchart for the taxonomy of moral repugnance

the process ahead of time makes it easier to find the time to enter a dialogue with people that may hold the identified concerns. As will be discussed in the next section, the taxonomy recommends ignoring some concerns. By explaining this to potential complainants, it may motivate them to provide more reasons and fewer unjustified expressions of disgust. This is helpful for your team because reasons can be considered, and changes to the plan more nuanced than "shelve it forever" or "just go ahead" can be considered.

Reassess the plan and put it into action

Correct use of the taxonomy will result in a general recommendation for action for each of the concern–framework pairings. Some recommendations will be to ignore the concern, but others will suggest changing the plan or even scrapping the whole project. Scrapping a project halfway or further through development can be heartbreaking; but if it is recommended by the taxonomy, then it is the right thing to do. In many cases, the process of using the taxonomy of moral repugnance on a disruptive technology will produce mixed recommendations. Here, as elsewhere, common sense should be applied.

If use of the taxonomy was in response to actual expressions of concern, then action should be as swift as possible. The plan should be changed in accordance with the recommendations, and the changes should be communicated to the relevant stakeholders. If use of the taxonomy was in anticipation of potential expressions of concern, then the recommendations should also inform purely cosmetic and communicative decisions about how the new technology is presented to the public.

A taxonomy of moral repugnance

The taxonomy of moral repugnance presented in this section is based on the detailed conceptual work of Weijers and Keyser (2016). A major contribution of this chapter is the discussion of what these different kinds of repugnance should mean for new disruptive technology proposals. The first subsection discusses the simpler kinds of potential moral repugnance to introduce readers to the main issues. The second subsection discusses more complex, but also more realistic, kinds of potential moral repugnance. The final subsection briefly discusses some less common kinds of moral repugnance.

Some basic varieties of moral repugnance and how to deal with them

This subsection discusses three basic varieties of moral repugnance: Legitimate Moral Repugnance, Mistaken Moral Repugnance, and Dumbfounded Moral Repugnance.

Legitimate Moral Repugnance (Legitimate MR) is the most concerning of all the types of repugnance for the team behind the new technology. Legitimate MR is repugnance backed by a logically valid argument that contains only true or plausible non-moral claims and moral claims that cohere with the complainant's moral framework (Weijers & Keyser, 2016). In other words, Legitimate MR does not result from any empirical mistakes or conceptual peculiarities. Rather, Legitimate MR is the result of a well-formed moral concern.

For example, Wyden and Dorgan might argue that:

1) PAM would encourage and enable betting on death (non-moral claim)
2) Encouraging and enabling betting on death is always immoral (moral claim)
3) Therefore, PAM is immoral

Defenders of PAM might quibble a little about Premise 1, urging for the use of "trading" or "investing" instead of "betting". However, prediction markets tend to operate more effectively when some traders have little or no new information (Wolfers & Zitzewitz, 2004) and trade in a way that is difficult to distinguish from betting. So, Premise 1 seems highly plausible. The logical form of the argument is valid, so the conclusion can be confirmed if Premise 2 is true.

For the purposes of the taxonomy, the initial check of any moral premise is done earlier in the process, when the concern is clarified with the relevant complainant. The next check on any moral premise is against the moral framework of the complainant. This check tests for whether the proposed moral claim is in tension with any other moral commitments that the complainant might have. If the moral claim coheres with the complainant's other moral commitments, then we have an example of Legitimate MR. Presumably, Wyden and Dorgan endorse the moral claim that encouraging and enabling betting on death is always immoral, and that claim coheres with their other moral beliefs.

This description of Legitimate MR allows for a range of possible moral concerns. Even if only one person expresses Legitimate MR for a particular reason, a deeper investigation is warranted. The key questions to answer in such an investigation are: What proportion of people are likely to agree with this criticism if they heard it? And, does the team behind the new technology agree with it?

In the case above, it seems likely that many people would consider betting on death to be morally bad (but not "almost everyone" as Dorgan (2003a, no page) implies).[13] Furthermore, it's not at all clear what proportion of people believe betting on death is something that should *never* be permitted, meaning that it cannot be done, even if it produced some greater good, such as saving lives from terrorist attacks. The moral claim in the argument attributed to Wyden and Dorgan above is absolute, meaning it permits no exceptions. But, as Surowiecki (2004) mentions in his discussion of PAM, national security is a fairly brutal business. As such, good people regularly have to resolve moral dilemmas by trading off various bad actions and consequences (e.g. invasions of privacy, cost to taxpayers) in order to achieve a more important good (e.g. saving lives by keeping innocent people safe). So, it is unlikely the team behind PAM would agree with the absolute moral claim that betting on death is *always* morally wrong. Weijers and Richardson (2014a) investigate this issue in depth, arguing that it is unreasonable to believe that betting on death is always morally wrong, especially when it could save lives. So it seems that, although possibly legitimate and widespread, this concern of

Wyden and Dorgan should be thought of as an issue to manage rather than a reason to shut the program down. However, the more reasonable version of this moral claim, that betting on death is morally bad *but not absolutely bad*, is complex and will be revisited in the next subsection.

After investigations are complete, widespread Legitimate MR that the team behind the new technology also find compelling, requires a thorough reworking of the product to avoid the moral concern. If the product cannot be reworked in a way that avoids the moral concern, then the project should be scrapped.

In cases where Legitimate MR is possibly widespread, but apparently based on the widespread belief in a moral claim that the team thinks is wrong, then the team has to decide what balance of changing the product and the communication strategy will best achieve their goal of saving lives. Changes that alleviate the legitimate moral concern without detracting from the technology's main benefit should probably be made, but this is not always possible.

For example, the team behind PAM could not have removed the betting element at the heart of their technology.[14] But they might have pointed out to reporters and politicians, at that fateful presentation, that although betting on death is not nice, it is not so morally bad that it should never be permitted; for example, when that betting can save lives. This may have changed key communicators' beliefs about the absolute nature of the moral claim in Premise 2, and possibly pushed them toward Outweighed Moral Repugnance (discussed below).

Finally, some investigations will reveal that, although one person finds the new technology legitimately morally repugnant, it is based on a highly idiosyncratic moral framework and a moral claim that would be rejected as immoral or amoral by nearly all other moral frameworks. For example, believing that prediction markets are immoral because making predictions is morally wrong may well fit into someone's moral framework somewhere, but it would be rejected by the vast majority. The team behind the new technology should not treat cases like these as reasons to stop or change their project.

Mistaken Moral Repugnance (Mistaken MR), is morally charged repugnance that results from the combination of a moral claim and a false non-moral claim (that, if true, would trigger the moral claim) (Weijers & Keyser, 2016). In other words, someone who has Mistaken MR, would see the technology as morally permissible if they understood all of the relevant non-moral facts. This kind of repugnance is likely to be common, especially given the emerging prevalence of technologically incompetent talking heads (Dawkins, 1998) and purposeful misinformation via news and social media (Lazer et al, 2018).

As mentioned earlier, a major concern with PAM is whether it would have been successful at generating security-relevant information that would not have otherwise been gathered by existing methods (e.g. Richey, 2005; Stiglitz, 2003; Wyden & Dorgan, 2003a). However, many subject experts have argued that anti-terrorist prediction markets would likely be effective (Hanson, 2006a; Looney, 2004; Surowiecki, 2004; Weijers & Richardson, 2014a; 2014b; Weijers, 2016; Yeh, 2006). Given the extent of disagreement over this empirical issue, careful investigation of such non-moral premises is important. Depending on the severity of the moral claim attached to the non-moral claim under investigation, feasibility studies can be a good option for resolving such disagreements.

In other cases, it will be easy to establish whether the non-moral claim is too implausible to distract the team behind the new technology. For example, a Flat-Earther's moral repugnance at a round-the-world yachting race based on the non-moral premise "since the Earth is flat, a round-the-world yachting race would result in the deaths of all the competitors" could be dismissed without so much as a Google search.

Assuming the non-moral premise can confidently be dismissed as false or implausible, then the project should proceed. However, depending on the depth and scope of the misinformation among stakeholders, a communications effort may have to be launched alongside the technical development of the product.

Dumbfounded Moral Repugnance (Dumbfounded MR) occurs when someone very strongly believes that the new technology is morally wrong, but they cannot find within their moral framework a moral principle that justifies their belief (Weijers & Keyser, 2016). Based on the findings of Haidt, Bjorklund, and Murphy (2000) it seems that a person experiencing Dumbfounded MR might say that they don't know *why* it's morally wrong, but they are sure that it *is*.

If Dumbfounded MR is widespread, then the team should consider investigating potential causes for the dumbfounding that explain but not justify it. For example, someone with an unconscious racist bias may have inadvertently acquired it through an upbringing devoid of real contact with people of that race, but replete with negative media coverage of them. This conditioning account explains how the racism came to be, but it does not morally excuse racist behavior in an age rich with information about racial equality.

The likelihood of the cause of the intuition being irrelevant to any reasonable moral framework, such as in the racism case, seems high for any intractable Dumbfounded MR. So, contrary to Kass's advice, if despite best efforts, no plausible reason for the Dumbfounded MR can be found, the team can ignore such complainants. Further, I would use Kass's moral repugnance at eating ice cream in public as an example of the possibility of atavistic notions inadvertently polluting our current moral intuitions. This is especially true of discourse about new technologies, since our intuitions are less trustworthy when they are about things that are unlike anything we have previously experienced (Lieberman, 2000; Myers, 2004; Weijers, 2013b; Woodward & Allman, 2007). Wisdom about new technologies ultimately comes from reasons, not feelings.[15]

Some complex varieties of moral repugnance and how to deal with them

In many real-life instances of following the procedure outlined here, individuals that have considered the proposed technology, and their own moral framework, carefully will often answer "yes and no" to the question from Figure 15.2, "Is the technology morally wrong according to [your] moral framework?". Their answer to the follow-up question – about how the moral concerns balance out – will decide which of the following three categories their moral repugnance falls into:

Outweighed Moral Repugnance, Indeterminate Moral Repugnance, or Mitigated Moral Repugnance.

Outweighed Moral Repugnance (Outweighed MR) occurs when a complainant, after consulting their own moral framework, views the technology as having both up- and downsides, morally speaking, and sees the moral upsides as outweighing the downsides (Weijers & Keyser, 2016). In other words, the new technology is seen as overall morally positive, despite the perception of one or more moral problems. The first step is to investigate any non-moral claims, including those that form part of the argument *in favor* of the new technology, to see whether they are mistaken. Any mistaken aspects of the moral repugnance should be dealt with as recommended in the subsection above.

The second step is to treat all of the perceived downsides, that do not include mistakes, as individual instances of Legitimate MR. Depending on the moral assessment of the team and the expected level of stakeholder agreement, these instances of Legitimate MR may result in changes to, and even the cessation of, the project (despite them being outweighed in one or more complainants). In many cases, this process will result in something like the example of betting on death used in the subsection above; the moral concern checks out as legitimate, and is shared by others, but is outweighed by a more pressing moral concern. By encouraging and enabling betting on death, PAM was being disrespectful to human life; which is morally bad but not so bad that it overrides all other considerations, such as PAM's potential to save innocent lives from terrorist attacks. Of course, once a legitimate moral concern has been identified, minor changes to the new technology that can reduce that concern without jeopardizing the main benefit should be investigated.

Indeterminate Moral Repugnance (Indeterminate MR) occurs when someone can't decide whether the moral upsides they see in the technology outweigh the moral downsides they see in it (Weijers & Keyser, 2016). Indeterminate MR should be dealt with in much the same way as Outweighed MR. Although, it is likely that the team behind the new technology will have to consider changes to their product more seriously for Indeterminate MR than for Outweighed MR.

Mitigated Moral Repugnance (Mitigated MR) occurs when someone believes that the moral upsides they see in the technology make it more acceptable to them, but that the moral downsides are so significant that they cannot, on balance, support the new technology (Weijers & Keyser, 2016). Again, this kind of repugnance should be dealt with by first checking the non-moral moral claims for mistakes and then investigating any Legitimate MR as outlined in the subsection above. This investigation may result in the team having to dramatically rework the project – or even abandon it.

After investigating the legitimate concerns for popularity and resonance with the team behind the new technology, careful thought needs to be given to whether public perception could be shifted from Mitigated MR to Outweighed MR by framing the moral issues in a clearer or more relatable way. Trading off different moral values is a difficult task, often with no explicit rules. As simple a thing as publishing one or two photographs can considerably tilt public perception on something as major as a war or a humanitarian crisis.[16]

Some unusual varieties of moral repugnance and how to deal with them

In this subsection, three more categories of moral repugnance, and what to do about them, are briefly discussed.

Dominant Moral Repugnance (Dominant MR) occurs when someone's moral repugnance about a technology is so strong that they will change their existing moral framework to accommodate it (Weijers & Keyser, 2016). In response to Dominant MR, the team behind the new technology should take the issue very seriously. It is possible that the disruptive technology is so new that members of the public initially overreact out of fear of change or the unknown. The team behind the technology should focus this investigation on the argument behind the Dominant MR, including the moral and non-moral claims. Mistaken non-moral claims and peculiar moral claims may not change the team's plans, but any reasonable and widespread Dominant MR probably means the project needs to be radically changed or stopped.

Conflicted Moral Repugnance (Conflicted MR) occurs when someone finds a new technology morally repugnant, but when consulting their own moral framework, can only find reasons that the technology is morally good (Weijers & Keyser, 2016). Conflicted MR should be treated in a very similar way to Dumbfounded MR, except more effort could be put into communicating the moral benefits of the new technology in order to encourage these people to accept their own moral reasons for why the new technology is morally good.

Moral Repugnance Deficit (MR Deficit) occurs when someone does not feel like a technology is morally bad, but when consulting their own moral framework, finds overwhelming reasons to consider the technology morally bad.[17] If MR Deficit is discovered within the team, the team should take seriously the moral concerns captured in the person's moral framework, if not their feelings, because members of the team are very well-versed with how the technology will work. When dealing with MR Deficit, the concerns should be investigated as though they were Legitimate MR. Going ahead as planned with a major project that is unethical is a morally and, most likely in the long run, prudentially bad idea.

Planning and framing public perceptions

So many social and political issues are debated nearly exclusively within echo chambers, such as groups on social media (Garrett, 2009). This complicating factor should be taken into account when developing disruptive security technologies. The framing of the initial message about a new morally praiseworthy technology can go a long way toward ensuring it is accepted by the public.

For example, if PAM had been called Anti-Terrorism Markets (even though the original purpose was wider than that), then reporters at the original presentation may not have been so quick to think that the technology was disrespectful. A more convincing discussion of how the markets might attract novel information might also have prevented moral repugnance about the technology being a waste of money.

Ethics screening for disruptive security technologies should not be dominated by cherry-picking advisors expected to rubber stamp whatever is proposed. By using the process outlined here, teams behind disruptive life-saving technologies might more effectively work out whether their idea is morally positive and how best to communicate that to others. In other words, the taxonomy on moral repugnance could help organizations like DARPA deal with the information they receive when "engaging a variety of experts and stakeholders with varying points of view – both to hear what they and their professional communities of practice have to say and to help convey to those communities DARPA's insights about what technology can and cannot do" (DARPA).

Notes

1 The "just wrong" comment was made by Senator Thomas Daschle in Congress (Congressional Record Vol. 149, No. 114, July 29, 2003: S10082–S10083). The "ridiculous… grotesque" comments were made by Senator Ron Wyden (Wyden & Dorgan 2003a). The "absurd… disgusting" comment was made by Senator Byron Dorgan the next day (Wyden & Dorgan 2003b). The "morally reprehensible" comment was made by the editors in "Pentagon drowns in its own 'dead pool'", *The Virginian Pilot*, July 31, 2003, B10. The "despicable… callous… retarded" remark was made by blogger Greg Saunders (See Meirowitz and Tucker (2004) for more detail on the reactions to PAM).

2 See: https://www.predictit.org/markets/detail/3698/Who-will-win-the-2020-US-presidential-election.

3 For more information on PMs, see: Cowgill, Wolfers, and Zitzewitz, (2009), Surowiecki, (2004), Wolfers and Zitzewitz (2004), or Weijers (2013b; 2018).

4 For more information on how a PM could be set up to effectively help fight terrorism, see: Hanson (2006a), Weijers (2016), and Weijers and Richardson (2014a).

5 Senators Ron Wyden and Byron Dorgan (2003a) were the prominent and very vocal leaders of the pushback against PAM, which they suggested would be "offensive to almost everyone" (no page). See Meirowitz and Tucker (2004) for more detail on the reactions to PAM.

6 Kass calls it "repugnance", but makes it clear he views the term as morally loaded (Kass, 1998).

7 For psychology, see especially the work of Jonathan Haidt (2001; 2007). For behavioral economics, see especially the work of Kahneman and Tversky, summarized in Kahneman (2011). For a summary of moral biases and their potential impact on policy and law, see Sunstein (2005).

8 This case was inspired by similar cases, such as Kahneman and Tversky's famous flu case (1984). See Sunstein (2004) for other examples.

9 This assumption about what most people would prefer is based on the complete structural similarity between my scenario and Kahneman and Tversky's famous flu case (1984).

10 Again, this assumption about what most people would prefer is based on the complete structural similarity between my scenario and Kahneman and Tversky's famous flu case (1984).

11 A mounting concern in the US is that potential authoritarian adversaries do not face the same domestic or normative constraints when it comes to developing and deploying, for civilian or military purposes, new technologies.

12 To be clear, this claim likely arouses negative feelings and possibly moral outrage within us, but the claim *itself* does not even mention a moral value.

13 Contrary to Dorgan's implication, the popularity of death pools seems to show that many people do not find betting on death at all offensive (Weijers & Richardson, 2014a). "Death pools" are competitions in which participants, for fun or financial reward, pick a list of celebrities that they think will die in the next year (Weijers & Richardson, 2014a). Unfortunately, several of these death pools already allow or even encourage play or real betting on when current and past presidents will die (e.g. ptrradio.com/deathpool, stiffs.com, etc).

14 Note that removing the financial incentives of betting for play money or "bragging points" instead would be unlikely to alleviate the moral concerns about the frivolous and disrespectful nature of betting on death.

15 To be clear, Dumbfounded MR should certainly be investigated during the process to see whether good moral reasons can justify it (even if the complainant could not see those reasons themselves). *Intractable* Dumbfounded MR, however, does not constitute a reason to stop or change the plan for the new technology.

16 For example, see Astor's (2018, no page) discussion of photographs that "Changed the Course of the Vietnam War".

17 This kind of moral repugnance does not appear in Weijers and Keyser (2016).

References

Amadeo, K (2019) 'How the 9/11 attacks affect the economy today?: what is their lasting damage?', *The Balance*. January 26. https://www.thebalance.com/how-the-9-11-attacks-still-affect-the-economy-today-3305536, accessed February 20, 2019.

Astor, M (2018) 'A photograph that changed the course of the Vietnam war', *New York Times*. February 1. https://www.nytimes.com/2018/02/01/world/asia/vietnam-execution-photo.html, accessed February 20, 2019.

Berg, JE, Nelson, FD and Rietz, TA (2008) 'Prediction market accuracy in the long run', *International Journal of Forecasting*, 24(2), pp. 285–300.

Carter, S and Cox, A (2011) 'One 9/11 tally: $3.3 trillion', *New York Times*. September 8. https://archive.nytimes.com/www.nytimes.com/interactive/2011/09/08/us/sept-11-reckoning/cost-graphic.html?_r=1, accessed February 20, 2019.

Conger, K (2018) 'Google plans not to renew its contract for project maven, a controversial Pentagon drone AI imaging program', *Gizmodo*. https://www.gizmodo.com.au/2018/06/google-plans-not-to-renew-its-contract-for-project-maven-a-controversial-pentagon-drone-ai-imaging-program/, accessed September 27.

Cowgill, B, Wolfers, J, and Zitzewitz, E (2009) 'Using prediction markets to track information flows: evidence from Google', in Das, S, Ostrovsky, M, Pennock, D, Szymanski, BK (eds) *Auctions, market mechanisms and their applications*. Lecture Notes of the Institute for Computer Sciences, Social Informatics and Telecommunications Engineering, Volume 14. Berlin: Springer, pp. 3–33.

DARPA (no date) *Ethics & societal implications*. https://www.darpa.mil/program/our-research/ethics, accessed September 27, 2019.

Dawkins, R (1998) 'What's wrong with cloning?', in Nussbaum, MC and Sunstein, CR (eds) *Clones and clones: facts and fantasies human cloning*. New York: W.W. Norton & Company.

Garrett, RK (2009) 'Echo chambers online? Politically motivated selective exposure among internet news users', *Journal of Computer-Mediated Communication*, 14(2), pp. 265–285.

Haidt, J (2001) 'The emotional dog and its rational tail: a social intuitionist approach to moral judgment', *Psychological Review*, 108(4), pp. 814–834.

Haidt, J (2007) 'The new synthesis in moral psychology', *Science*, 316(5827), pp. 998–1002.

Haidt, J, Bjorklund, F and Murphy, S (2000) *Moral dumbfounding: when intuition finds no reason*. (Unpublished manuscript, University of Virginia). http://faculty.virginia.edu/haidtlab/articles/manuscripts/haidt.bjorklund.working-paper.when%20intuition%20finds%20no%20reason.pub603.doc, accessed February 20.

Hanson, R (2006a) 'Designing real terrorism futures', *Public Choice*, 128(1), pp. 257–274.

Hanson, R (2006b) 'Decision markets for policy advice', in Gerber, AS and Patashnik, EM (eds) *Promoting the general welfare: new perspectives on government performance*. Washington, DC: Brookings Institution Press.

Hanson, R (2007) 'The policy analysis market (a thwarted experiment in the use of prediction markets for public policy)', *Innovations: Technology, Governance, Globalization*, 2(3), pp. 73–88.

Kahneman, D (2011) *Thinking, fast and slow*. New York, NY: Farrah, Straus and Giroux.

Kahneman, D and Tversky, A (1984) 'Choices, values, and frames', *American Psychologist*, 39, pp. 341–350.

Kass, L (1998) 'The wisdom of repugnance: why we should ban the cloning of humans', *Valparaiso University Law Review*, 32(2), pp. 679–705.

Kass, L (1999) *The hungry soul: eating and the perfecting of our nature*. Chicago, IL: University of Chicago Press.

Lazer, DM, Baum, MA, Benkler, Y, Berinsky, AJ, Greenhill, KM, Menczer, F and Schudson, M (2018) 'The science of fake news', *Science* 359(6380), pp. 1094–1096.

Lieberman, M (2000) 'Intuition: a social cognitive neuroscience approach', *Psychological Bulletin*, 126, pp. 109–137.

Lodge, M and Taber, CS (2013) *The rationalizing voter*. Cambridge: Cambridge University Press.

Looney, RE (2004) 'DARPA's policy analysis market for intelligence: outside the box or off the wall?', *International Journal of Intelligence and Counterintelligence*, 17(3), pp. 405–419.

Lopez, TC (2019) *DOD seeks ethicist to guide artificial intelligence deployment*. US Department of Defense. https://www.defense.gov/explore/story/Article/1950724/dod-seeks-ethicist-to-guide-deployment-of-artificial-intelligence/source/GovDelivery/, accessed October 1.

Meirowitz, A and Tucker, JA (2004) 'Learning from terrorism markets', *Perspectives on Politics*, 2(2), pp. 331–336.

Myers, DG (2004) *Intuition: its powers and perils*. New Haven, CT: Yale University Press.

Nolan, D, Restall, G and West, C (2005) 'Moral fictionalism versus the rest', *Australasian Journal of Philosophy*, 83(3), pp. 307–330.

Richey, M (2005) 'Thoughts on the theory and practice of speculative markets qua event predictors', *Essays in Philosophy*, 6(1), Article 26. http://ssrn.com/abstract=1670777, accessed March 26, 2013.

Sunstein, CR (2004) 'Lives, life-years, and willingness to pay', *Columbia Law Review*, 104, pp. 205–252.

Sunstein, CR (2005) 'Moral heuristics', *Behavioral and Brain Sciences*, 28, pp. 531–573.

Stiglitz, J (2003) 'Terrorism: there's no futures in it', *Los Angeles Times*. July. http://mason.gmu.edu/~rhanson/PAM/PRESS2/LATstiglitz-7-31-03.htm, accessed October 1, 2019.

Surowiecki, J (2004) *The wisdom of crowds*. New York: Doubleday.

Taber, CS and Lodge, M (2006) 'Motivated skepticism in the evaluation of political beliefs', *American Journal of Political Science*, 50(3), 755–769.

Weijers, D (2013a) 'Prediction markets', *Observatory for responsible research and innovation in ICT*. http://www.danweijers.com/pdf/Technology-Prediction_Markets(Dan_Weijers)pdf, accessed October 1, 2019.

Weijers, D (2013b) 'Intuitive biases in judgements about thought experiments: the experience machine revisited', *Philosophical Writings*, 41(1), pp. 17–31.

Weijers, D (2016) 'Prediction markets as an alternative to one more spy', in Galliott, J and Reed, W (eds) *Ethics and the future of spying: technology, intelligence collection and national security*. Routledge.

Weijers, D (2018). 'Prediction Markets' in Kolb, RW (ed.) *The SAGE encyclopedia of business ethics and society* (Vols. 1–7). Thousand Oaks, CA: SAGE Publications, Inc., pp. 2716–2718.

Weijers, D and Keyser, V (2016) 'The varieties and dynamics of moral repugnance: prediction markets and betting on matters of life and death', *The Humanities and Technology Review*, 3, pp. 91–129.

Weijers, D and Richardson, J (2014a) 'Is the repugnance about betting on terrorist attacks misguided?', *Ethics and Information Technology*, 16(3), pp. 251–262.

Weijers, D and Richardson, J (2014b) 'A moral analysis of effective prediction markets on terrorism', *International Journal of Technoethics*, 5(1), pp. 28–43.

Wolfers, J and Zitzewitz, E (2004) 'Prediction markets', *Journal of Economic Perspectives*, 18(2), pp. 107–126.

Woodward, J and Allman, J (2007) 'Moral intuition: its neural substrates and normative significance', *Journal of Physiology-Paris*, 101(4–6), pp. 179–202.

Work, R (2017) *Establishment of an algorithmic warfare cross-functional team (project maven)*. Government Executive. https://www.govexec.com/media/gbc/docs/pdfs_edit/establishment_of_the_awcft_project_maven.pdf, accessed September 27, 2019.

Wyden, R and Dorgan, B (2003a) *Senators Ron Wyden and Byron Dorgan hold news conference on a terror financing scheme*. http://hanson.gmu.edu/PAM/govt/senator-wyden-dorgan-pressconf-7-28-03.txt, accessed February 9, 2019.

Wyden, R and Dorgan, B (2003b) *U.S. Senator Ron Wyden holds news conference following closed policy luncheon*. http://hanson.gmu.edu/PAM/govt/Senator-Wyden-Dorgan-pressconf-7-29-03.txt, accessed February 9, 2019.

Yeh, PF (2006) 'Using prediction markets to enhance US intelligence capabilities', *Studies in Intelligence*, 50(4), pp. 137–149.

Conclusion

Society, security, and technology: Mapping a fluid relationship

Simona R. Soare

Introduction

Technological progress has historically been at the center of social, economic, and military change, bringing with it both huge opportunity and commensurate challenges. The world is yet again experiencing rapid and far-reaching change as a result of the fast-paced development, adoption and diffusion of emerging technologies, including artificial intelligence (AI), quantum computing (QC), robotics and automation, 3D printing, deepfakes, and blockchain. As this book has shown, like other technologies before them, emerging technologies hold great promise to enable human well-being but also give rise to great uncertainty around their proliferation and (mis)use.

This volume has explored and mapped out the expected impact of these technologies on national and international security from a multidisciplinary perspective. In the process, it has also highlighted the way these technologies challenge enduring intra- and interstate relationships. Particularly, the volume has explored the impact of emerging technologies on systemic change, on national and international security. It has investigated the opportunities and vulnerabilities they create, and the revolutionary impact they are expected to have on society, state and international relations, in security and defense and beyond. The volume has also considered the technical, legal, and ethical implications of the adoption of emerging technologies. This book's novel contribution to the debate about the impact of emerging technologies in the International Relations (IR) literature can be distilled into three main areas: The arguably revolutionary impact of emerging technologies; the persistent importance of human factors and political contexts in the development, adoption and use of emerging technologies; and the blurring of lines and boundaries between the traditional levels of analysis – sub-state, state, and inter-state.

Revolutionary, really?

Individual contributions in this volume expect emerging technologies to transform, if not revolutionize, international politics and warfare. The speed, scope, and systemic impact of emerging technologies promise to bring about a Fourth Industrial Revolution. However, there is no consensus on how broad this revolutionary impact will be or when it will be most acutely felt, socially, militarily, and politically. The political impact of emerging technologies is far from being fully grasped, despite these technologies being commercially available already. This creates both uncertainty and the potential for technological and operational surprise, which in IR theory is considered to be destabilizing to the international order.

In his chapter, James Johnson highlights the role American perceptions play in interpreting the challenge posed by Chinese technological progress in the quest for "AI supremacy". These perceptions are shaped by inferred Chinese intentionality and expected technological capacity rather than the actual Chinese use of "AI weapons". AI technologies in natural language processing, computer vision, and other areas are already mature and commercially widespread, and technological breakthroughs are regularly occurring. The world appears to be preparing for the scaling down of Moore's law where computer power *more than* doubles every two years (NSCAI, 2020, p. 9, 45). The premium in countries across the world is on rapidly adopting AI. The US emphasizes the rapid adoption of AI in military applications, and society at large, to maximize first-mover advantage (DoD, 2019). Yet the American experience in scaling the use of emerging technologies in military applications should inform Washington's perceptions about other international actors, who will face similar challenges. In this context, further research is needed on whether it makes a difference to actors' perceptions if AI and other emerging technologies are progressively adopted and integrated in military applications, rather than rapidly and suddenly. As Joe Burton argues in the first chapter of this volume, incremental change often disguises the transformational impact of technological progress. It can also affect strategic surprise, lessen uncertainty, and reduce the perception of threats to international security and stability.

Further interdisciplinary research is needed to help us understand what emerging technologies, such as AI and QC, offer to the world of policy making beyond their speed and scale, and immediate quantifiability. If China or the US start using AI and QC on a wide scale in support of sectoral or strategic decision-making, how will we be able to trace and understand the differences compared to today's policy making in these states? Measuring whether a specific technology is revolutionary is a challenging task. Computers, apps, and algorithms are already an important mediator of human experiences. Over 80% of Netflix and Spotify account holders follow recommendations provided by these platforms' inbuilt algorithms (Plummer, 2017; Beuscart et al, 2019), although they are often unaware of their existence. All big tech companies are actively investing in social studies to better understand how their products are received and used by individuals and how they can better tailor them to create or respond to user needs. In such cases, the technological tools that enable qualitative improvements in human experiences are largely imperceptible to individuals. The same may also apply in other sectors, including in national security. For example, wargaming data shows that in potential adversarial interactions between great powers, the opponent would not be sure if military action was being taken by an algorithm, an autonomous robot, or a human. Geographical proximity of forces equipped with AI-enabled and fully autonomous capabilities further complicates deterrence and escalation dynamics (Wong et al, 2020).

The impact of the *use* of emerging technologies, as opposed to their technological features, is another important distinction. As Sean Ainsworth highlights, the Russian use of automated bots and trolls in massive disinformation campaigns is actively shaping power relations within societies as well as in international relations. The same applies to deepfakes and synthetic media. As Curtis Barnes and Tom Barraclough argue, these technologies appear to be breaking ceilings on legitimate political and journalistic tools and are capable of overwriting physical reality into a virtual one, but they have not been used to great effect yet. The same cannot be said of the more physical manifestations of the use of emerging technologies in managing natural resources, like water, as highlighted by Nathan Cooper, which are more visible and measurable. Andrew Colarik and Richard Wilson's contribution on blockchain technology also highlights how internet-based inequality can have observable economic effects and quantifiable impact on power dynamics.

Is the revolutionary impact of emerging technologies to do with whether the specific technology changes the offense–defense balance? Ryan Ko's chapter, as well as Aiden Warren and Alek Hillas's contribution, both suggest emerging technologies are potentially offense-dominant technologies (i.e. fully automated cyber tools, lethal infantry robots, drones, munitions, and hypersonic vehicles). However, AI is an enabler for exponentially more efficient automated cyber defense, as much as it is a facilitator of the increased frequency, scope, and efficiency of cyber-attack vectors. One important lesson is that we should not assume these capabilities are inherently offense-dominant or defense-dominant merely by reading their technical specifications. Not all emerging technologies have the same impact on security, policy, and society, and not all of them have the same impact in the physical and virtual realms. For example, Peter Cook points out that 3D printed firearms can be incredibly disruptive but, despite their availability on the dark web and the relatively low cost of producing them, there is still no evidence they are being used to any significant degree by non-state actors to pursue their agendas. By contrast, the number of active automated bots and trolls, as well as the number of actors using them are rapidly proliferating.

Sean Ainsworth's analysis of the Russian information operations raises questions about how we measure their impact – by the number of Facebook shares of a misinformation piece, by the number of online supporters of a given extremist party or populist political leader, or by their impact on political polarization and radicalization? It is worth reflecting on how online (self-) radicalization can result in mass shootings and terrorist attacks whereas the same online echo chamber appears to be less effective in translating online conspiracy theories and disinformation into support for extremist political parties in the ballot. Our current level of knowledge on these issues is not advanced enough to help us understand these complex and sophisticated relationships.

Equally, more research is required into how and why individuals have different perceptions about the power relations between themselves and private-sector corporations on the one hand, and themselves and governmental authorities on the other. For example, individuals consent to give their private data to multinational and big tech corporations in order to use their services and platforms, even when they do not know and cannot control how their data is used, but they become immediately skeptical of sharing the same private data with governments. Few individuals protest the widespread commercial use of facial and voice recognition technology on their smartphones – and the active collection of biometric data it relies on – but they become inherently suspicious of the technology when it is wielded by governments. As William Hoverd argues, power dynamics in the virtual realm – such as attribution of cyber-attacks – may even alienate public support by increasing public suspicion about governmental actions and intentions.

The transparency of cyber, AI, and QC tools may obscure their influence on policy making with repercussions for accountability and political responsibility. Beyond debates about the specific transformational impact of individual emerging technologies, one must equally ponder the possibility that it will be their *cumulative* effect that will be truly revolutionary.

(Still) human-dominant technology

The book also contains important insights into the relationship between humans and machines in international security. In his chapter, Ryan Ko raises questions about the impact of a progressive tendency toward automating cyber-attacks and the implications this carries for cyber security. "Automating the hacker", Ko argues, will enable international and non-state actors to distribute or fend off up to four cyber-attacks per second, well beyond the limits of human responsiveness. His analysis points to an area where the impact of AI is likely to be both faster and more widespread by comparison to the physical domain, where the fully automated use of force is still regarded with skepticism, as Warren and Hillas explain in their contribution on Lethal Autonomous Weapon Systems (LAWS). Lethal, fully autonomous weapons are still politically, legally, ethically, and militarily problematic, despite the fact that technological solutions have enabled full automation of military platforms for decades. There are over 150 military platforms in use today with varying degrees of autonomy that could potentially be upgraded to full autonomy, but humans are still in the loop in all of them. The drone market is particularly relevant, for there are military platforms already available with fully autonomous modes preloaded that have never been used (Kania, 2020). The use of remotely piloted drones has undermined US policy in the Middle East and drawn international criticism, as Francis Okpaleke and Joe Burton highlight in their chapter. That international actors are reluctant to relinquish political and military control over technological tools and enablers is a powerful argument that their revolutionary impact can be shaped, avoided and/or postponed through regulatory processes. This is why the national and international "legislative gap" on practical applications of emerging technologies is so consequential. It also requires further consideration of how effective regulation can develop in parallel and in coordination with the rapid technological progress rather than perpetually trying to catch up with it. That is not to say fully autonomous weapons will not revolutionize warfare and politics but, rather, that when and to what degree they do so is also a function of our political choices.

This puts a premium on international actors' policy performance in adopting AI and other emerging technologies rather than on the qualitative improvements in policy making. Simona R. Soare's analysis of the way democratic and authoritarian regimes shape the development and use of emerging technologies underlines the importance of perennial political interests in guiding the process, to very different ends. Politics is an inherently human activity. Technology adoption is a political (and commercial) process which is dependent on the existence of a broader enabling ecosystem. The use of emerging technologies by state and non-state actors should always be understood in the context of political goals and systems.

Similarly, as a result of human and political choice, emerging technologies can have an asymmetrically transformational impact, as Reuben Steff argues in his chapter on small states and AI. This is important, Steff argues, not just for how small international actors can leverage more power out of the use of emerging technologies but also for how these technologies shape and redraw international partnerships and alliances and create new strategic dependencies between small states and their great power patrons. The proliferation of emerging technologies, from autonomous drones to computer vision algorithms for intelligence processing, from smart city and 5G network infrastructure to smart policing tools increasingly reshapes domestic and international power, potentially creating "technospheres" of political influence (Imbrie et al, 2020). China's rapidly expanding Belt and Road (BRI) network and Russia's "near abroad", much like the American system of alliances, act as transmission belts for the acceleration of advanced technology transfers, but more research is needed on whether technology transfers also enable and accelerate ideological transfers.

The role of human agency in the adoption and use of emerging technologies is increasingly challenging to discern but remains as important as ever. With the advent of AI and autonomy, agency is increasingly divided between human and machine, even when we assume the constancy of a "human in the loop". This is progressively blurring the roles of human and machine. Missile defense systems have been fully automated for decades. China, the US, the EU, and NATO all use automated intelligence analysis and decision-making support. The US Air Force is experimenting with algorithms that construct alternative actionable scenarios to support decision-makers (DOD, 2019). In multi-domain operations or in mosaic warfare, an algorithm or a collection of algorithms working collaboratively will choose and deploy the best force configuration for a given mission (Jensen & Paschkewitz, 2019). China is working on deploying AI-enabled target acquisition technology on its hypersonic glide vehicles, essentially making them "fire-and-forget" capabilities (Kania, 2020). As highlighted by the UN CCW GGE in 2019 and by Warren and Hillas in their chapter, the principle of state responsibility and the applicability of international law to emerging technologies stand. Human–machine symbiosis is required in relation to emerging digital technologies and autonomous robotics, but American, European, Russian, and Chinese leaders have a very different understanding of what "meaningful human control" of fully autonomous weapons means, which is a consequence of the political and legal constraints of their systems. While there is agreement among political and military elites that strategic decision-making, including when it comes to the use of nuclear weapons, will not be delegated to machines anytime soon, it is uncertain whether different states will find autonomous decision-making acceptable at lower decision-making levels.

There are many variables in the development and use of emerging technologies. Countless articles and reports point to the importance of investment gaps among international actors for the development of AI or underline the role played by institutional capacity or strategic cultures in the adoption of emerging technologies. Some suggest export controls are immensely consequential. Nevertheless, the truly consequential variable of the development of emerging technologies has been and remains human – human knowledge, human ingenuity, human skills. Numerous reports and studies underline not just the strategic implications of AI on military power but also the human skills gap in developing and widely using AI as well as the challenges of attracting and retaining human AI talent. It is arguable at what point human ingenuity will be surpassed by quantum computers and digital engineering algorithms in quality rather than speed, but as with general AI, we are not there yet. Until we get to that point, humans are still at the center of the development, use, and consequences of emerging technologies. Measuring these technologies' effectiveness is essentially an exercise in measuring their impact on human life. Determining who is winning the technological race for "AI dominance" or "quantum supremacy" would be meaningless without the political interests and structures associated with these concepts.

Blending levels of analysis

The issue with claiming that AI and other emerging technologies are transformational is that it obscures a more nuanced analysis of how these technologies revolutionize different sectors of our lives and at what levels of analysis they are most influential. This volume has set out an analysis of the three Walzian levels of analysis – sub-state, state, and interstate – to offer a broad picture of the impact of emerging technologies and give the reader a detailed view of how they are already transforming different aspects of international politics. Emerging technologies are certainly shaping each one of Waltz's traditional levels. Reuben Steff and Khusrow Akkas Abbasi argue that AI will impact the balance of power between the US and China by influencing their economic and technological capacity and their military power. James Johnson also argues AI and other emerging technologies will shape the balance of power between Beijing and Washington. However, further consideration should be given to how these emerging technologies affect the speed of international change. Will the adoption of AI make systemic transitions and the accompanying shifts in the global balance of power faster and, therefore, more uncertain? The answer to this question depends on whether the advantage created by emerging technologies – individually or collectively – is enduring or transitory, resulting in different patterns of international change than the traditional century-long hegemonic cycles of the past. While great powers invest significantly more in emerging technologies, Steff argues AI can act as a force multiplier for smaller international actors, helping them punch above their weight economically and militarily in international politics. Singapore and Israel, for example, are small states that are leading in the rapid adoption of AI and the reasons for this deserve more scrutiny in studies about the impact of AI on society and security. These dynamics may also affect the pace and trajectory of transition at the systemic level. While Steff and Abbasi are concerned with the impact of AI on a bipolar configuration of the balance of power between the US and China, Johnson suggests there are elements of multipolarity at play that should also be considered. Simona R. Soare argues emerging technologies are creating opportunities and constraints for democracies and authoritarian regimes who are both trying to shield their societies from undue hostile influence. Again, this may introduce breaks and hurdles to the pace of change. Dan Weijers argues narrative is crucial in shaping individual acceptance of emerging technologies, particularly as it relates to perceptions of morals and ethics. At the human level, therefore, the rate of change will be influenced by the pace of our individual, institutional, and societal acceptance of emerging technologies. Whether an individual is *empowered* or *overpowered* by emerging technologies is as much a result of how public discourse shapes our perceptions of these new technologies as it is of technological, structural and political factors.

This brings us to the role of the private sector and civil society, including academia. Big tech exponentially outspends governments. The EU, the second largest economy in the world (before the COVID-19 pandemic) spends less on AI in one year than Microsoft does on AI-enabling digital infrastructure alone. Engaging the private sector in the endeavor of adopting emerging technologies into the public sector is proving challenging – indeed, this is a more challenging undertaking for democracies than it is for authoritarian regimes who foster clientelist relations with their industry. This has led to discussions about the expanding limits of what states consider to be critical infrastructure. Governmental actions to restrict Huawei's role in building 5G in Europe and the US have been met with Chinese threats of retaliation. Chinese technology theft and espionage led to US arrests of American and foreign nationals accused of allegedly selling sensitive technology and information to China. Similarly, Russian nationals were indicted in 2019 by US authorities for interfering with the 2016 election process, now legally considered by Washington to be a part of its national critical infrastructure. While Western authorities are resisting criticism of their scrutiny over private-sector foreign investment, the result of their actions is to increasingly securitize private-sector science and technology. Chinese Foreign Direct Investment (FDI) in the high-tech sector is closely scrutinized by US and European government agencies and, more often than not, is blocked for reasons of national security. In all but name, these high-tech companies are being assimilated into governmental understandings of national critical infrastructure. Equally, governmental exploitation and weaponization of digital networks and technological transfer networks will likely result in the dawn of "technospheres" of political influence in which dependency on a technology provider, and especially on technology-as-service providers such as Amazon, Alibaba, Google and Huawei, can influence a state's information environment, intra-state and institutional relations, and international relations.

What the chapters in this volume collectively reveal is that treating the impact of emerging technologies in isolation, at singular levels of analysis, is a thing of the past. There is an unprecedented synergy between the effects emerging technologies produce at different levels of analysis. Indeed, the patterns of technological diffusion, especially in the case of general-purpose technologies, like AI, mean impact can hardly ever be contained within the borders or the corporate offices of a single international actor. Local and national governments' performance in adopting emerging technologies have international consequences. Using emerging technologies to manage water resources can lead to international conflicts and escalation dynamics, as Nathan Cooper explains. Emerging technologies now intermediate our democratic politics. In doing so, they help redefine the relationship between state and citizen as well as challenging the limits of traditional relations between states and the citizens of another polity. Russian disinformation operations in the West are not just about interstate relations, the balance of power, and national security. They are equally about the technology-enabled relationship between Moscow and citizens of Western states, which bypasses their sovereign, national institutions. Okpaleke and Burton argue the US use of drones in the Middle East has undermined American political interests in the region because of the arguably illegal effects of applying an emerging technology in the context of a local conflict. Strategic events can thus have disproportionate local effects and vice-versa. Likewise, international actors can respond to strategic moves with tactical ones, such as the Western response to Russian disinformation and hybrid threats.

Conclusion

The collective contributions in this book suggest the adoption of AI, quantum, automation, or advanced manufacturing techniques like 3D printing will introduce great uncertainty in international and national processes for the foreseeable future. We appreciate the irony of saying this at the end of a book that we see as one of the most comprehensive analysis of emerging technologies and their impact on security and defence. Mass disruption, technological surprise, and distributed physical, virtual, and cognitive effects resulting from the application of emerging technologies are here to stay. How international actors act and react to these technologies to exploit or manage technological progress and the associated uncertainty will be a crucial variable of national and international security.

This book has set out to answer pressing questions about the impact of emerging technologies on national and international security as well as on the security of societies and non-state actors. It has been a complex, multidisciplinary effort, but it is far from a complete one. Further research is required at the intersection of sociology, psychology, international relations, security studies, political science, crime science, intelligence studies, and computer science to understand the impact emerging technologies (will) have in shaping the world moving forward, as well as the role and scope of human agency in shaping the technologies themselves. This volume has contributed to the debate by highlighting that emerging technologies are challenging long-held assumptions about the use of coercive power, about agency and political accountability, about the role of institutions in domestic and international domains, about the interplay of domestic politics and international security, and even about how reality is constructed at the confluence of physical, virtual, and biological realms. Mapping the fluid relationship between society, security, and technology looks at this point in history to be an effort to trace lines in rapidly running and increasingly turbulent waters.

References

Beuscart, JS, Coavoux, S and Maillard, S (2019) 'Les algorithmes de recommandation musicale et l'autonomie de l'auditeur: Analyse des écoutes d'un panel d'utilisateurs de streaming', *Réseaux*, 1(213), pp. 17–47. https://www.cairn.info/revue-reseaux-2019-1-page-17.htm, accessed 14 July 2019.

Defense Innovation Unit (2019) *Portofolio*. https://www.diu.mil/solutions/portfolio, accessed 14 July 2019.

Department of Defense (2019) *Summary of the 2018 department of defense artificial intelligence strategy: harnessing AI to advance our security and prosperity*. February 19. https://media.defense.gov/2019/Feb/12/2002088963/-1/-1/1/SUMMARY-OF-DOD-AI-STRATEGY.PDF

Imbrie, A, Fedasiuk, R, Aiken, C, Chhabra, T and Chahal, H (2020) *Agile alliances: how the US and its allies can deliver a emocratic way of AI, center for security and emerging technology*. February. https://cset.georgetown.edu/wp-content/uploads/CSET-Agile-Alliances.pdf

Jensen, B and Paschkewitz, J (2019) 'Mosaic warfare: small and scalable are beautiful', *War on the Rocks*. December 23. https://warontherocks.com/2019/12/mosaic-warfare-small-and-scalable-are-beautiful/, accessed 14 July 2019.

Kania, EB (2020) 'AI weapons', in *China's military innovation*. Brookings Institution. April. https://www.brookings.edu/research/ai-weapons-in-chinas-military-innovation/, accessed 14 July 2019.

National Security Commission on Artificial Intelligence (2020) *First quarter recommendations*. March. https://drive.google.com/file/d/1wkPh8Gb5drBrKBg6OhGu5oNaTEERbKss/view, accessed 14 July 2019.

Plummer, L (2017) 'This is how Netflix's top-secret recommendation system works', *Wired*. August 22. https://www.wired.co.uk/article/netflix-data-personalisation-watching, accessed 14 July 2019.

US Army (2018) *The U.S. army in multi-domain operations 2028*. TRADOC Pamphlet 525-3-1. December 6. https://www.tradoc.army.mil/Portals/14/Documents/MDO/TP525-3-1_30Nov2018.pdf, accessed 14 July 2019.

Wong, YH et al. (2020) *Deterrence in the age of thinking machines*. RAND Corporation. https://www.rand.org/content/dam/rand/pubs/research_reports/RR2700/RR2797/RAND_RR2797.pdf, accessed 14 July 2019.

Index

Page numbers in **bold** denote tables, those in *italic* denote figures.